T0315146

# Genomic Signal Processing

Princeton Series in Applied Mathematics

Editors

Ingrid Daubechies (Princeton University); Weinan E (Princeton University); Jan Karel Lenstra (Eindhoven University); Endre Süli (University of Oxford)

The Princeton Series in Applied Mathematics publishes high-quality advanced texts and monographs in all areas of applied mathematics. Books include those of a theoretical and general nature as well as those dealing with the mathematics of specific applications areas and real-world situations.

Titles in the Series

*Chaotic Transitions in Deterministic and Stochastic Dynamical Systems: Applications of Melnikov Processes in Engineering, Physics, and Neuroscience* by Emil Simiu

*Selfsimilar Processes* by Paul Embrechts and Makoto Maejima

*Self-Regularity: A New Paradigm for Primal-Dual Interior-Point Algorithms* by Jiming Peng, Cornelis Roos, and Tamás Terlaky

*Analytic Theory of Global Bifurcation: An Introduction* by Boris Buffoni and John Toland

*Entropy* by Andreas Greven, Gerhard Keller, and Gerald Warnecke

*Auxiliary Signal Design for Failure Detection* by Stephen L. Campbell and Ramine Nikoukhah

*Thermodynamics: A Dynamical Systems Approach* by Wassim M. Haddad, VijaySekhar Chellaboina, and Sergey G. Nersesov

*Optimization: Insights and Applications* by Jan Brinkhuis and Vladimir Tikhomirov

*Max Plus at Work* by Bernd Heidergott, Geert Jan Olsder, and Jacob van der Woude

*Genomic Signal Processing* by Ilya Shmulevich and Edward R. Dougherty

# Genomic Signal Processing

Ilya Shmulevich and Edward R. Dougherty

PRINCETON UNIVERSITY PRESS

PRINCETON AND OXFORD

Published by Princeton University Press, 41 William Street,
Princeton, New Jersey 08540

In the United Kingdom: Princeton University Press,
3 Market Place, Woodstock, Oxfordshire OX20 1SY

Library of Congress Control Number: 2006936719

ISBN-13: 978-0-691-11762-1
ISBN-10: 0-691-11762-4

British Library Cataloging-in-Publication Data is available

Printed on acid-free paper. ∞

pup.princeton.edu

Printed in the United States of America

1  3  5  7  9  10  8  6  4  2

10  9  8  7  6  5  4  3  2  1

# Contents

# Preface

Recent methods facilitate large-scale surveys of gene expression in which transcript levels can be determined for thousands of genes simultaneously. In particular, expression microarrays result from a complex biochemical-optical system incorporating robotic spotting and computer image formation and analysis. Since transcription control is accomplished by a method that interprets a variety of inputs, we require analytical tools for expression profile data that can detect the types of multivariate influences on decision making produced by complex genetic networks. Put more generally, signals generated by the genome must be processed to characterize their regulatory effects and their relationship to changes at both the genotypic and phenotypic levels. Two salient goals of functional genomics are to screen for the key genes and gene combinations that explain specific cellular phenotypes (e.g., disease) on a mechanistic level and to use genomic signals to classify disease on a molecular level.

Owing to the major role played in genomics by transcriptional signaling and the related pathway modeling, it is only natural that the theory of signal processing should be utilized in both structural and functional understanding. Genomic signal processing (GSP) is the engineering discipline that studies the processing of genomic signals. Analogously, one might think of digital signal processing (DSP), which studies the processing of digital signals. Both GSP and DSP employ the mathematical methods associated with signal processing in general; however, neither is defined by these methods but rather by the physical signals to which they are applied. The aim of GSP is to integrate the theory and methods of signal processing with the global understanding of functional genomics, with special emphasis on genomic regulation. Hence, GSP encompasses various methodologies concerning expression profiles: detection, prediction, classification, control, and statistical and dynamical modeling of gene networks. Overall, GSP is a fundamental discipline that brings to genomics the structural model–based analysis and synthesis that form the basis of mathematically rigorous engineering.

Application is generally directed toward tissue classification and the discovery of signaling pathways. Accomplishment of these aims requires a host of signal-processing approaches. These include signal representation relevant to transcription, such as wavelet decomposition and more general decompositions of stochastic time series, and system modeling using nonlinear dynamical systems. The kind of correlation-based analysis historically used for understanding pairwise relations between genes or cellular effects cannot capture the complex network of nonlinear information processing based upon multivariate inputs from inside and outside the genome. Regulatory models require the kind of nonlinear dynamics studied in signal processing and control, and in particular the use of stochastic data flow networks common to distributed computer systems with stochastic inputs. This is not to say that existing model systems suffice. Genomics requires its own model systems, not simply straightforward adaptations of currently formulated models. New systems must capture the specific biological mechanisms of operation and distributed regulation at work within the genome. It is necessary to develop appropriate mathematical theory, including optimization for the kinds of external controls required for therapeutic intervention and approximation theory, to arrive at nonlinear dynamical models that are sufficiently complex to adequately represent genomic regulation for diagnosis and therapy and not too complex for the amounts of data experimentally feasible or for the computational limits of existing computer hardware.

From a wide view point, GSP is driven by genomics as a scientific discipline. Whereas a global perspective may not tell us which mathematical structures apply to specific problems, it does focus our attention on the proper role of mathematics within genomic science. Indeed, this role is not peculiar but rather is a reflection of the general role of mathematics within science—and this latter role is dictated by our understanding of scientific epistemology.

Models compose the expressed content of science. A model is a logical construct in which the variables and relations between the variables represent our appreciation of a selected portion of physical reality at a certain level of specification. It is a skeleton that reflects the salient features of a physical situation of interest to the scientist. It is a conceptualization of a part of nature, a logical apparatus that bears a connection to nature through our ability to utilize it as a determining factor in the construction of experiments and the prediction of outcomes resulting from those experiments. The test of a model is its accuracy in the prediction of sensory events, such as the movement of a needle on a meter. A model is a contingent hypothesis that perpetually stands open to rejection should its predictions fail. The epistemology and method of science are united in the model concept. A model gains its legitimacy from data. At issue is the concurrence of observed and predicted measurements. The particular mathematical apparatus of the model is secondary. Given two models, the better one is the one that provides better prediction. Two different mathematical systems that lead to equally accurate predictions are of equal validity. Arguments over whether a continuous or a discrete model is better in a particular setting can be settled only by which provides better prediction, not by which is more intuitively appealing. In *QED, The Strange Theory of Light and Matter*, Richard Feynman writes, "It is whether or not the theory gives predictions that agree with experiment. It is not a question of whether a theory is philosophically delightful, or easy to understand, or perfectly reasonable from the point of view of common sense. They theory of quantum electrodynamics describes Nature as absurd from the point of view of common sense. And it agrees fully with experiment. So I hope you can accept Nature as She is—absurd." If Nature is absurd, then model validity cannot possibly be decided by reasoning. Reasoning may produce the model, or perhaps it appears in an apparition, but the model's validity rests solely with its ability to predict experimental outcomes.

An intuitive, nonmathematical appreciation might help a scientist in exploratory thinking, but it is unimportant relative to the actual object of science, which is the mathematical model. Any attempt to force a nonmathematical understanding creates the risk of having a diminished (or erroneous) description of physical reality. James Jeans puts the matter unambiguously in *The Mysterious Universe*: "The final truth about phenomena resides in the mathematical description of it; so long as there is no imperfection in this, our knowledge is complete. We go beyond the mathematical formula at our own risk; we may find a [nonmathematical] model or picture which helps us to understand it, but we have no right to expect this, and our failure to find such a model or picture need not indicate that either our reasoning or our knowledge is at fault." It is implicit in Jeans' statement that one should beware of applying mathematical methods without first understanding the mathematics at a deep level. The scientific meaning of an analysis lies in the meaning of the mathematics behind the analysis. In the absence of an understanding of the mathematics, there is ipso facto a lack of understanding regarding the results. The danger here is not only a lack of understanding but also a misunderstanding that leads to faulty applications. This problem is especially acute when data are processed by some complex iterative algorithm whose outcome depends on many factors, some of which are left unspecified or are unknown, as in the case of applying a classification rule to a small sample without regard to its expected behavior in the context of small samples, or when a statistical procedure is applied without thought given to the characteristics of the governing distribution, as in the case of applying a large-variance error estimator.

The predictive validity of a model requires that the symbols be tied to observations by some semantic rules that relate to the general principles of the mathematical model or to conclusions drawn from the principles. These *operational definitions* are an intrinsic part of the theory, for without them there would be no connection between the principles and the observations. The general principles must have consequences that can be checked via their relation to sensory observations. The mathematical structure may relate abstract symbols, but there must be a well-defined procedure for relating consequences of the structure to quantifiable observations.

Since the validity of the model involves observation and measurement, scientific knowledge is limited by the limits of measurement. While we might not be confronted with an intrinsic limitation, such as that embodied in the Heisenberg uncertainty principle, at any given time we are limited by the ability to observe in accordance with existing technology, both in the kind of data that can be obtained and in its accuracy. Nothing scientific can be said about a system for which no measurements are possible at the scale of the theory. Erwin Shrödinger puts the matter this way in *Science Theory and Man*: "It really is the ultimate purpose of all schemes and models to serve as scaffolding for any observations that are at all conceivable. The prohibition against clothing them with details, that are by no possible means observable, is a matter for [no regret]. . . . If Heisenberg's assertion be correct, and if it appears at first sight to make gaps in our picture of the world which cannot be filled, then the obvious thing to do is to eliminate the regions which refuse to be filled with thought; in other words, to form a view of the world which does not contain those regions at all." It makes no sense to apply a mathematical method that either depends on or utilizes unobservable measurements. This does not imply that one should not build mathematical systems that require data not yet available; indeed, such systems formalize one's thinking and are critical to guiding the development of necessary technology. It is just that, until suitable data become available to test the theory, it remains unvalidated, or perhaps partially validated, depending on what data are currently available.

While scientific knowledge ultimately resides in a mathematical system of relationships, with validity based on observation, this does not imply that exploratory experimentation is unimportant. The observation of phenomena and the assessment of quantitative experimental data are the grist for the creative mill that produces the variables and the systems of relationships among them. One's hands need to be dirty, down in the dirt, digging for the roots with directed, not random, digging,

The aspects of Nature reflected by a scientific theory are ultimately at the discretion of the scientist—limited by the availability of suitable technology. The worth of the theory is related to choices made at the outset. The scientist posits hypotheses in the form of a structure consisting of elements and relations between these elements. Some relations will be judged as valid according to observation, and others will not. The main point here is that the set of potential relations is not random; rather, it is formulated by the human intellect. Schrödinger: "A selection has been made on which the present structure of science is built. That selection must have been influenced by circumstances that are other than purely scientific." Science is anthropomorphic. Its direction and success depend upon humanly made choices.

Science is neither abstract mathematics nor the perusal of data acquired without careful consideration of the manner in which the data are to be used to validate relationships. Science requires scientists. It is in the person of the scientist that a hypothetical network of relations is postulated, and it is only in the context of such a network of relations that one can propose an efficient experimental design to identify valid relations through statistical means.

Genomic science requires engineers, mathematicians, and statisticians who appreciate the full depth of the models and algorithms they develop and apply. This applies both to their known properties and to the uncertainties concerning them. Only then can the scientific content of statements involving these models be apprehended. The purpose of this book is to discuss central issues of GSP from the mathematical side. In doing so it has four major aims: (1) provide rigorous mathematical definitions and propositions regarding important issues, (2) discuss the extent to which models provide accurate and precise statements regarding observations, (3) explain the real-world problems of GSP in a manner compatible with the nature of the mathematical methods employed, and (4) give direction to the role of GSP in providing diagnostic and therapeutic solutions.

## DEDICATION

We dedicate this book to Jeffrey M. Trent, whose vision and support for a mathematically rigorous engineering-systems approach to genomics has been instrumental in bringing us, and many other electrical engineers, into genomics. From the beginning, he has maintained a clear directive: "Think ten to twenty years in the future, but always remember we are here to cure patients." This is a directive from which we have tried never to waiver.

## ACKNOWLEDGMENTS

We would like to take this opportunity to acknowledge the following institutions for generously supporting our research in genomic signal processing: the National Human Genome Research Institute, the National Cancer Institute, the National Institute for General Medical Sciences, the National Science Foundation, the Translational Genomics Research Institute, the University of Texas M. D. Anderson Cancer Center, and the Institute for Systems Biology. We would also like to acknowledge the more than 200 persons who have collaborated with us in our genomics research. Although there are certainly too many to name, there are two people who deserve special recognition. We express our sincerest appreciation to Michael L. Bittner and Wei Zhang, two outstanding partners who have spent untold hours with us in a joint effort to explain genomics in the kind of operational form necessary for the systems approach inherent to engineering.

# Chapter One

## Biological Foundations

No single agreed-upon definition seems to exist for the term *bioinformatics*, which has been used to mean a variety of things ranging in scope and focus. To cite but a few examples from textbooks, Lodish et al. (2000) state that "bioinformatics is the rapidly developing area of computer science devoted to collecting, organizing, and analyzing DNA and protein sequences." A more general and encompassing definition, given by Brown (2002), is that bioinformatics is "the use of computer methods in studies of genomes." More general still: "bioinformatics is the science of refining biological information into biological knowledge using computers" (Draghici, 2003). Kohane et al. (2003) observe that the "breadth of this commonly used definition of bioinformatics risks relegating it to the dustbin of labels too general to be useful" and advocate being more specific about the particular bioinformatics techniques employed.

While it is true that the field of bioinformatics has traditionally dealt primarily with biological data encoded in digital symbol sequences, such as nucleotide and amino acid sequences, in this book we will be mainly concerned with extracting information from gene expression measurements and genomic signals. By the latter we mean any measurable events, principally the production of messenger ribonucleic acid (RNA) and protein, that are carried out by the genome. The analysis, processing, and use of genomic signals for gaining biological knowledge and translating this knowedge into systems-based applications is called *genomic signal processing*.

In this chapter, our aim is to place this material into a proper biological context by providing the necessary background for some of the key concepts that we shall use. We cannot hope to comprehensively cover the topics of modern genetics, genomics, cell biology, and others, so we will confine ourselves to brief overviews of some of these topics. We particularly recommend the book by Alberts et al. (2002) for a more comprehensive coverage of these topics.

## 1.1 GENETICS

Broadly speaking, genetics is the study of genes. The latter can be studied from different perspectives and on a molecular, cellular, population, or evolutionary level. A gene is composed of deoxyribonucleic acid (DNA), which is a double helix consisting of two intertwined and complementary nucleotide chains. The entire set of DNA is the *genome* of the organism. The DNA molecules in the genome are assembled into *chromosomes*, and genes are the functional regions of DNA.

Each gene encodes information about the structure and functionality of some protein produced in the cell. Proteins in turn are the machinery of the cell and the major determinants of its properties. Proteins can carry out a number of tasks, such as catalyzing reactions, transporting oxygen, regulating the production of other proteins, and many others. The way proteins are encoded by genes involves two major steps: transcription and translation. *Transcription* refers to the process of copying the information encoded in the DNA into a molecule called messenger RNA (mRNA). Many copies of the same RNA can be produced from only a single copy of DNA, which ultimately allows the cell to make large amounts of proteins. This occurs by means of the process referred to as *translation*, which converts mRNA into chains of linked amino acids called *polypeptides*. Polypeptides can combine with other polypeptides or act on their own to form the actual proteins. The flow of information from DNA to RNA to protein is known as the *central dogma* of molecular biology. Although it is mostly correct, there are a number of modifications that need to be made. These include the processes of reverse transcription, RNA editing, and RNA replication.

Briefly, *reverse transcription* refers to the conversion of a single-stranded RNA molecule to a double-stranded DNA molecule with the help of an enzyme aptly called *reverse transcriptase*. For example, HIV virus consists of an RNA genome that is converted to DNA and inserted into the genome of the host. *RNA editing* refers to the alteration of RNA after it has been transcribed from DNA. Therefore, the ultimate protein product that results from the edited RNA molecule does not correspond to what was originally encoded in the DNA. Finally, *RNA replication* is a process whereby RNA can be copied into RNA without the use of DNA. Several viruses, such as hepatitis C virus, employ this mechanism. We will now discuss some preliminary concepts in more detail.

### 1.1.1 Nucleic Acid Structure

Almost every cell in an organism contains the same DNA content. Every time a cell divides, this material is faithfully replicated. The information stored in the DNA is used to code for the expressed proteins by means of transcription and translation. The DNA molecule is a polymer that is strung together from monomers called deoxyribonucleotides, or simply *nucleotides*, each of which consists of three chemical components: a sugar (deoxyribose), a phosphate group, and a nitrogenous base. There are four possible bases: adenine, guanine, cytosine, and thymine, often abbreviated as A, G, C, and T, respectively. Adenine and guanine are *purines* and have bicyclic structures (two fused rings), whereas cytosine and thymine are *pyrimidines*, and have monocyclic structures. The sugar has five carbon atoms that are typically numbered from $1'$ to $5'$. The phosphate group is attached to the $5'$-carbon atom, whereas the base is attached to the $1'$ carbon. The $3'$ carbon also has a hydroxyl group (OH) attached to it.

Figure 1.1 illustrates the structure of a nucleotide with a thymine base. Although this figure shows one phosphate group, up to three phosphates can be attached. For example, adenosine $5'$-triphosphate (ATP), which has three phosphates, is the molecule responsible for supplying energy for many biochemical cellular processes.

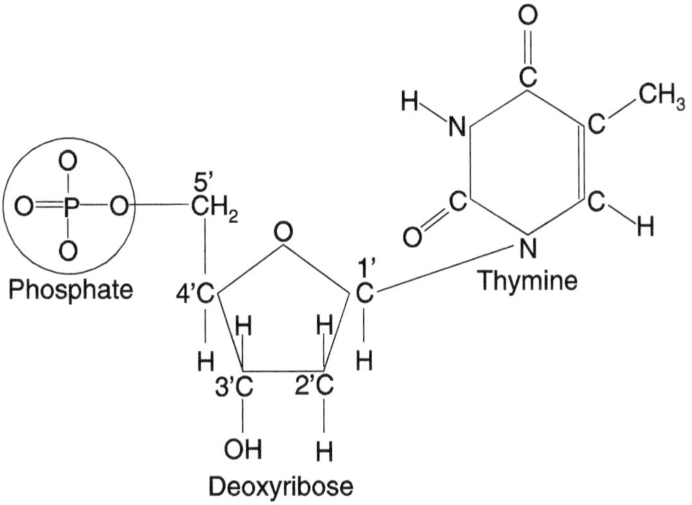

Figure 1.1. The chemical structure of a nucleotide with a thymine base.

Ribonucleic acid is a polymer that is quite close in structure to DNA. One of the differences is that in RNA the sugar is ribose rather than deoxyribose. While the latter has a hydrogen at the 2′ position (figure 1.1), ribose has a hydroxyl group at this position. Another difference is that the thymine base is replaced by the structurally similar uracil (U) base in a ribonucleotide.

The deoxyribonucleotides in DNA and the ribonucleotides in RNA are joined by the covalent linkage of a phosphate group where one bond is between the phosphate and the 5′ carbon of deoxyribose and the other bond is between the phosphate and the 3′ carbon of deoxyribose. This type of linkage is called a *phosphodiester bond*. The arrangement just described gives the molecule a 5′ → 3′ polarity or directionality. Because of this, it is a convention to write the sequences of nucleotides starting with the 5′ end at the left, for example, 5′-ATCGGCTC-3′. Figure 1.2 is a simplified diagram of the phosphodiester bonds and the covalent structure of a DNA strand.

DNA commonly occurs in nature as two strands of nucleotides twisted around in a double helix, with the repeating phosphate–deoxyribose sugar polymer serving as the backbone. This backbone is on the outside of the helix, and the bases are located in the center. The opposite strands are joined by hydrogen bonding between the bases, forming *base pairs*. The two backbones are in opposite or antiparallel orientations. Thus, one strand is oriented 5′ → 3′ and the other is 3′ → 5′. Each base can interact with only one other type of base. Specifically, A always pairs with T (an A · T base pair), and G always pairs with C (a G · C base pair). The bases in the base pairs are said to be *complementary*. The A · T base pair has two hydrogen bonds, whereas the G · C base pair has three hydrogen bonds. These bonds are responsible for holding the two opposite strands together. Thus, if a DNA molecule contains many G · C base pairs, it is more stable than one containing many A · T base pairs. This also implies that DNA that is high in G · C content requires a higher temperatue

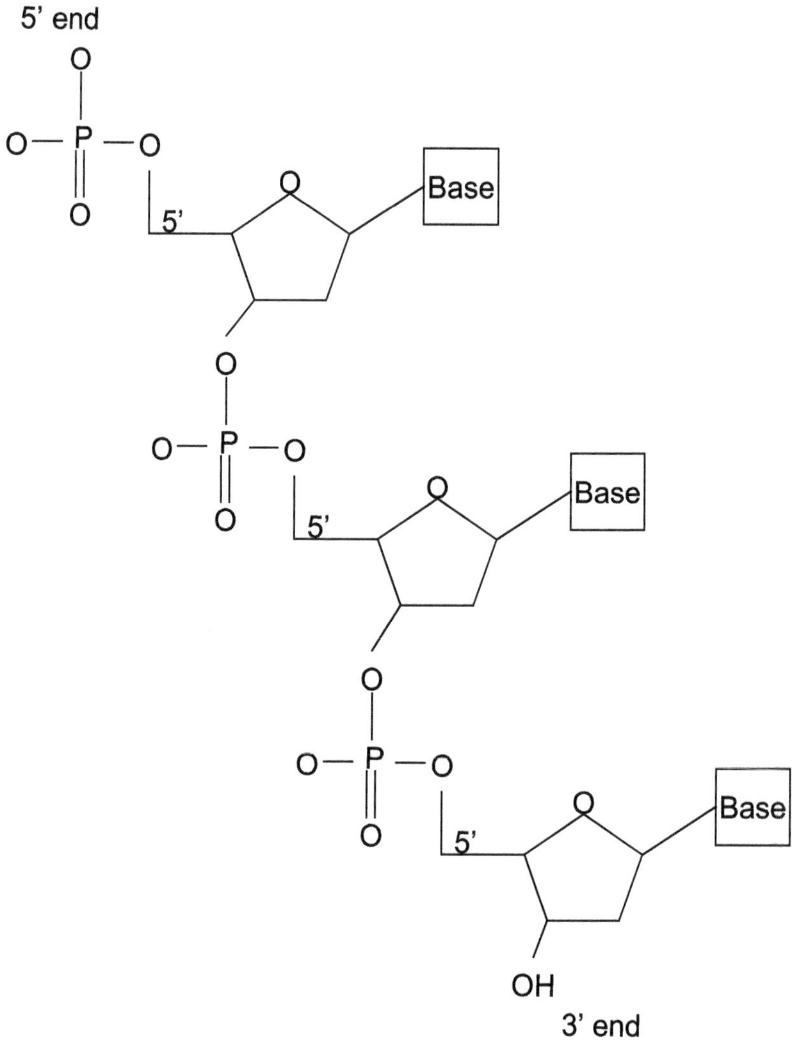

5' end

Figure 1.2. A diagram of one DNA strand showing the sugar-phosphate backbone with the phosphodiester bonds.

to separate, or *denature*, the two strands. Although the individual hydrogen bonds are rather weak, because the overall number of these bonds is quite high, the two strands are held together quite well.

Although in this book we will focus on gene expression, which involves transcription, it is important to say a few words about how the DNA molecule duplicates. Because the two strands in the DNA double helix are complementary, they carry the same information. During replication, the strands separate and each one acts as a *template* for directing the synthesis of a new complementary strand. The two new double-stranded molecules are passed on to daughter cells during

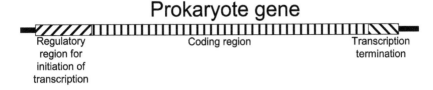

Figure 1.3. The gene structure in prokaryotes and eukaryotes.

cell division. The DNA replication phase in the cell cycle is called the *S* (synthesis) *phase*. After the strands separate, the single bases (on each side) become exposed. Thus, they are free to form base pairs with other free (complementary) nucleotides. The enzyme that is responsible for building new strands is called *DNA polymerase*.

### 1.1.2 Genes

Genes represent the functional regions of DNA in that they can be transcribed to produce RNA. A gene contains a *regulatory region* at its upstream (5') end to which various proteins can bind and cause the gene to initiate transcription in the adjacent RNA-encoding region. This essentially allows the gene to receive and then respond to other signals from within or outside the genome. At the other (3') end of the gene, there is another region that signals termination of transcription.

In eukaryotes (cells that have a nucleus), many genes contain *introns*, which are segments of DNA that have no information for, or do not code for, any gene products (proteins). Introns are transcribed along with the coding regions, which are called *exons*, but are then cut out from the transcript. The exons are then spliced together to form the functional messenger RNA that leaves the nucleus to direct protein synthesis in the cytoplasm. Prokaryotes (cells without a nucleus) do not have an exon/intron structure, and their coding region is contiguous. These concepts are illustrated in figure 1.3.

The parts of DNA that do not correspond to genes are of mostly unknown function. The amount of this type of *intergenic* DNA present depends on the organism. For example, mammals can contain enormous regions of intergenic DNA.

### 1.1.3 RNA

Before we go on to discuss the process of transcription, which is the synthesis of RNA, let us say a few words about RNA and its roles in the cell. As discussed above, most RNAs are used as an intermediary in producing proteins via the process of translation. However, some RNAs are also useful in their own right in that they can carry out a number of functions. As mentioned earlier, the RNA that is used to make proteins is called messenger RNA. The other RNAs that perform various functions are never translated; however, these RNAs are still encoded by some genes.

One such type of RNA is transfer RNA (tRNA), which transports amino acids to mRNA during translation. tRNAs are quite general in that they can transport amino acids to the mRNA corresponding to any gene. Another type of RNA is ribosomal RNA (rRNA), which along with different proteins comprises *ribosomes*. Ribosomes coordinate assembly of the amino acid chain in a protein. rRNAs are also general-purpose molecules and can be used to translate the mRNA of any gene. There are also a number of other types of RNA involved in splicing (snRNAs), protein trafficking (scRNAs), and other functions. We now turn to the topic of transcription.

### 1.1.4 Transcription

Transcription, which is the synthesis of RNA on a DNA template, is the first step in the process of gene expression, leading to the synthesis of a protein. Similarly to DNA replication, transcription relies on complementary base pairing. Transcription is catalyzed by an *RNA polymerase* and RNA synthesis always occurs from the $5'$ to the $3'$ end of an RNA molecule. First, the two DNA strands separate, with one of the strands acting as a template for synthesizing RNA. Which of the two strands is used as a template depends on the gene. After separation of the DNA strands, available ribonucleotides are attached to their complementary bases on the DNA template. Recall that in RNA uracil is used in place of thymine in complementary base pairing. The RNA strand is thus a direct copy (with U instead of T) of one of the DNA strands and is referred to as the *sense* strand. The other DNA strand, the one that is used as a template, is called the *antisense* strand. This is illustrated in figure 1.4.

Transcription is initiated when RNA polymerase binds to the double-stranded

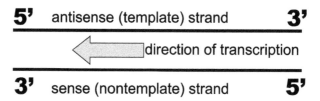

Figure 1.4. The direction of transcription. The antisense strand, oriented $3' \rightarrow 5'$, is used as a template to which ribonucleotides base-pair for synthesizing RNA, which always grows in the $5' \rightarrow 3'$ direction.

DNA. The actual site at which RNA polymerase binds is called a *promoter*, which is a sequence of DNA at the start of a gene. Since RNA is synthesized in the $5' \rightarrow 3'$ direction (figure 1.5), genes are also viewed in the same orientation by convention. Therefore, the promoter is always located upstream (5'side) of the coding region. Certain sequence elements of promoters are conserved between genes. RNA polymerase binds to these common parts of the sequences in order to initiate transcription of the gene. In most prokaryotes, the same RNA polymerase is able to transcribe all types of RNAs, whereas in eukaryotes, several different types of RNA polymerases are used depending on what kind of RNA is produced (mRNA, rRNA, tRNA).

In order to allow the DNA antisense strand to be used for base pairing, the DNA helix must first be locally unwound. This unwinding process starts at the promoter site to which RNA polymerase binds. The location at which the RNA strand begins to be synthesized is called the *initiation site* and is defined as position +1 in the gene. As shown in figure 1.5, RNA polymerase adds ribonucleotides to the $3'$ end of the RNA. After this, the helix is re-formed once again.

Since the transcript must eventually terminate, how does the RNA polymerase know when to stop synthesizing RNA? This is accomplished by the recognition of certain specific DNA sequences, called *terminators*, that signal the termination of transcription, causing the RNA polymerase to be released from the template and ending the RNA synthesis. Although there are several mechanisms for termination, a common direct mechanism in prokaryotes is a terminator sequence arranged in such a way that it contains self-complementary regions that can form *stem-loop* or *hairpin* structures in the RNA product. Such a structure, shown in figure 1.6, can cause the polymerase to pause, thereby terminating transcription. It is interesting to note that the hairpin structure is often GC-rich, making the self-complementary base pairing stronger because of the higher stability of $G \cdot C$ base pairs relative to $A \cdot U$ base pairs. Moreover, there are usually several U bases at the end of the hairpin structure, which, because of the relatively weaker $A \cdot U$ base pairs, facilitates dissociation of the RNA.

In eukaryotes, the transcriptional machinery is somewhat more complicated than in prokaryotes since the primary RNA transcript (pre-mRNA) must first

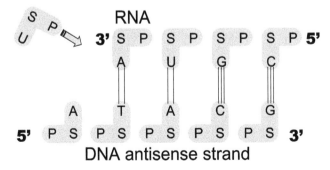

Figure 1.5. The addition of a uracil to the $3'$ end of the synthesized RNA. P stands for phosphate and S stands for sugar.

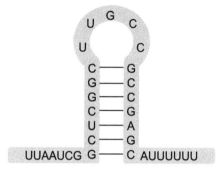

Figure 1.6. RNA hairpin structure used as a termination site for RNA polymerase in prokaryotes.

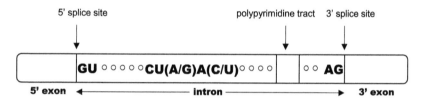

Figure 1.7. The splice sites, the branchpoint sequence, and the polypyrimidine tract.

be processed before being transported out of the nucleus (recall that prokaryotes have no nucleus). This processing first involves *capping*—the addition of a 7-methylguanosine molecule to the 5′ end of the transcript, linked by a triphosphate bond. This typically occurs before the RNA chain is 30 nucleotides long. This cap structure serves to stabilize the transcript but is also important for splicing and translation. At the 3′ end, a specific sequence (5′-AAUAAA-3′) is recognized by an enzyme which then cuts off the RNA at approximately 20 bases further down, and a *poly(A) tail* is added at the 3′ end. The poly(A) tail consists of a run of up to 250 adenine nucleotides and is believed to help in the translation of mRNA in the cytoplasm. This process is called 3′ *cleavage* and *polyadenylation*.

The final step in converting pre-mRNA into mature mRNA involves *splicing*, or removal of the introns and joining of the exons. In order for splicing to occur, certain nucleotide sequences must also be present. The 5′ end of an intron almost always contains a 5′-GU-3′ sequence, and the 3′ end contains a 5′-AG-3′ sequence. The AG sequence is preceded by a *polypyrimidine tract*—a pyrimidine-rich sequence. Further upstream there is a sequence called a *branchpoint sequence*, which is 5′-CU(A/G)A(C/U)-3′ in vertebrates. The splice sites as well as the conserved sequences related to intron splicing are shown in figure 1.7.

The actual splicing reaction consists of two steps. In the first step, the G at the 5′ splice site is attacked by the 2′-hydroxyl group in the adenine (fourth base) in the branchpoint sequence, which creates a circular molecule called a *lariat*, thereby freeing the exon upstream of the intron. In the second step, the 3′ splice site is cleaved, which releases the lariat and joins the two exons. The released intron is then degraded. We should briefly mention at this point that in eukaryotes it is possible for a particular pre-mRNA to produce several types of mature mRNA by means of *alternative splicing*. In other words, certain exons can be removed by splicing and therefore are not retained in the mature mRNA product. These alternative mRNA forms ultimately give rise to different proteins. A good example of this is the fibronectin gene, which by means of alternative splicing can produce two different protein isoforms. One isoform of this protein is produced by fibroblasts (a connective tissue cell), whereas the other is secreted by hepatocytes (an epithelial liver cell). Two exons that are responsible for fibronectin adhering to cell surfaces are found in mRNA produced in fibroblasts, while in hepatocytes, they are spliced out. Consequently, the fibronectin produced by hepatocytes cannot adhere to cell surfaces and can easily be transported in the serum.

### 1.1.5 Proteins

A protein is a chain of linked *amino acids*. A total of 20 amino acids can occur, each having unique properties. In a protein, they are linked by covalent bonds called *peptide bonds*, which are formed by the removal of water molecules between the amino acids. Although the picture of amino acids linked in a chain suggests a simple linear arrangement, proteins are structurally quite complex. The linear sequence of amino acids constitutes the protein's *primary structure*. However, various forces acting between atoms at different locations within the protein cause it to take on a specific shape. The *secondary structure* of a protein refers to the regular arrangement of the amino acids within localized regions of the protein. The most common types of secondary structure are the $\alpha$ helix (a spiral structure) and the $\beta$ sheet (a planar structure).

The term *tertiary structure* describes the three-dimensional arrangement of all amino acids. In most proteins, various combinations of $\alpha$ helices and $\beta$ sheets can fold into compact globular structures called *domains*, which are the basic units of tertiary structure. It is often the case that amino acids that are far apart in the linear sequence are in close proximity in the tertiary structure. Different domains in large proteins are often associated with specific functions. For example, one domain might be responsible for catalytic activity, while another may modulate DNA-binding ability. Often, several identical or different types of folded (tertiary) structures bind together to form a *quaternary structure*.

What is important to remember is that the function of a protein is determined by its three-dimensional structure, which in turn is determined by the sequence of amino acids. Also important is the fact that most proteins are chemically modified after they are produced. This may have the effect of altering their life span, activity, or cellular location. Some modifications, such as those involving linkage of various chemical groups to the protein, are often reversible, whereas others, such as removal of entire peptide segments, are not.

Table 1.1  The 20 Amino Acids and Their Corresponding Codons

| Amino Acid | Codons |
|---|---|
| Alanine | GCA, GCC, GCG, GCU |
| Arginine | AGA, AGG, CGA, CGC, CGG, CGU |
| Asparagine | AAC, AAU |
| Aspartic acid | GAC, GAU |
| Cysteine | UGC, UGU |
| Glutamic acid | GAA, GAG |
| Glutamine | CAA, CAG |
| Glycine | GGA, GGC, GGG, GGU |
| Histidine | CAC, CAU |
| Isoleucine | AUA, AUC, AUU |
| Leucine | UUA, UUG, CUA, CUC, CUG, CUU |
| Lysine | AAA, AAG |
| Methionine | AUG |
| Phenylalanine | UUC, UUU |
| Proline | CCA, CCC, CCG, CCU |
| Serine | AGC, AGU, UCA, UCC, UCG, UCU |
| Threonine | ACA, ACC, ACG, ACU |
| Tryptophan | UGG |
| Tyrosine | UAC, UAU |
| Valine | GUA, GUC, GUG, GUU |
| STOP | UAA, UAG, UGA |

The last row shows the three STOP codons. As can be seen, 18 of the amino acids are specified by more than one codon. Codons that specify the same amino acid are said to be synonymous.

### 1.1.6 Translation

The sequence of amino acids in a protein is determined from the sequence of nucleotides in the gene that encodes the protein. Recall that the sequence of nucleotides in DNA is transcribed into an mRNA sequence. After this, ribosomes, which are large macromolecular assemblies, move along the mRNA in $5' \rightarrow 3'$ direction and read the mRNA sequence in nonoverlapping chunks of three nucleotides. Each nucleotide triplet, called a *triplet codon*, codes for a particular amino acid. As there are 4 different possible nucleotides, there are $4^3 = 64$ possible codons. Since there are only 20 possible amino acids and 64 possible codons, most amino acids can be specified by more than one triplet. The *genetic code* specifies the correspondence between the codons and the 20 amino acids, as shown in table 1.1.

Each codon is recognized by an *anticodon*—a complementary triplet located on the end of a transfer RNA molecule—having a cloverleaf shape. The anticodon binds with a codon by the specific RNA-RNA base complementarity. Note that since codons are read in the $5' \rightarrow 3'$ direction, anticodons are positioned on tRNAs in the $3' \rightarrow 5'$ direction. This is illustrated in figure 1.8.

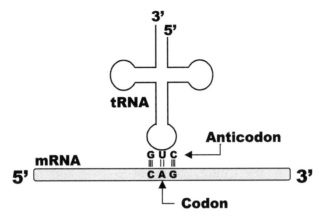

Figure 1.8.  The anticodon at one end of a tRNA molecule binds to its complementary codon
in mRNA.

Each tRNA molecule carries a specific amino acid attached to its free $3'$ end.
Moreover, each tRNA is designed to carry only one of the 20 possible amino acids.
However, some amino acids can be carried by more than one type of tRNA molecule. Also, certain tRNA molecules require accurate base-pair matching at only
the first two codon positions and can tolerate a mismatch, also called a *wobble*, at
the third position. tRNAs are essentially "adaptor" molecules in the sense that they
can convert a particular sequence of three nucleotides into a specific amino acid, as
specified in table 1.1.

The tRNAs and mRNA meet at the ribosome, which contains specific sites at
which it binds to mRNA, tRNAs, and other factors needed during protein synthesis.
It is important that the *reading frame* progresses accurately, three nucleotides at a
time, without skipping any nucleotides, for if this were to occur, the triplets would
code for the wrong amino acids and in turn would synthesize the wrong protein,
which might have highly adverse phenotypic effects. The ribosome ensures this
precise movement.

The ribosome has an aminoacyl-tRNA-binding site (*A site*), which holds the incoming tRNA molecule that carries an amino acid, and a peptidyl-tRNA-binding
site (*P site*), which holds the tRNA molecule linked to the growing polypeptide
chain. These two sites are located quite close together so that the two tRNA molecules bind to two adjacent codons on the mRNA. After an aminoacyl-tRNA molecule binds to the A site by forming base pairs with the codon, the amino acid linked
to the tRNA molecule positioned at the P site becomes uncoupled from its host
tRNA molecule and forms a peptide bond to the new amino acid attached to the
tRNA that just arrived at the A site. Then, the ribosome moves exactly three nucleotides along the mRNA, the tRNA molecule at the A site is moved to the P site,
the old tRNA that used to be at the P site (without its amino acid) is ejected from the
ribosome and reenters the tRNA pool in the cytoplasm. This leaves the A site free
for the newly arriving aminoacyl-tRNA molecule. This three-step process, referred
to as the *elongation* phase of protein synthesis, is illustrated in figure 1.9.

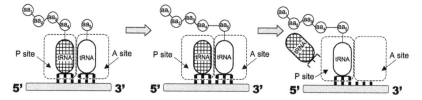

Figure 1.9. The three-step process of the elongation phase in protein synthesis.

The last row in table 1.1 does not contain an amino acid; rather, it shows the three possible codons (UAA, UAG, and UGA), called *stop codons*, that are used to terminate translation. If the ribosome is positioned in such a way that the stop codon is at the A site, certain proteins called *release factors* recognize and bind directly to this stop codon. By a series of steps, this terminates transcription and releases the polypeptide chain. The ribosome, having finished its job, also dissociates and is ready to reassemble on a new mRNA molecule to begin protein synthesis again.

We have now discussed how the ribosome accurately moves one codon at a time along the mRNA and how it terminates translation when it reaches a stop codon. But we have not yet discussed how the ribosome initiates the process of protein synthesis or, more specifically, how it knows where to begin. Indeed, depending on where the process starts, the subsequent nonoverlapping nucleotide triplets code for completely different polypeptide chains. Thus, there are three possible reading frames. The initiation process of protein synthesis consists of several rather complex steps catalyzed by proteins called *initiation factors* and involves the ribosome, the mRNA, initiator tRNA (which carries methionine), and guanosine 5'-triphosphate (GTP). In eukaryotes, the initiator tRNA binds to the triplet AUG, which codes for methionine, as table 1.1 shows. In *Escherichia coli*, for example, AUG or GUG can serve as the initiation codon. Since there could in principle be many AUGs, how is the correct initiation codon selected? In prokaryotes, there is a conserved purine-rich sequence 8–13 nucleotides in length, called the *Shine-Dalgarno sequence*, which is located upstream of the initiation codon. This sequence is believed to position the ribosome correctly by base-pairing with one of the subunits of the ribosome.

### 1.1.7 Transcriptional Regulation

All cells contain the same DNA content. What, then, differentiates a liver cell from a white blood cell? The properties of a cell, including its architecture and ability to participate in various activities while interacting with its environment, are determined by gene activities and, in particular, the expressed proteins. This implies that there must be a control mechanism, internal, external, or both, that regulates expression of the proteins characterizing the cell type or the functional state in which a particular cell is found. The process of gene regulation can be extremely complex, especially in higher eukaryotes.

As an example of why it is important for a cell to produce certain proteins in response to various environmental conditions, let us consider a bacterium. Sugar metabolism involves a number of enzymes. There are various sugars, such as lactose, glucose, and galactose, that can be used as an energy source. However, depending on the type of sugar, a different set of enzymes is needed to enable the sugar to enter the cell and then break down the sugar. Of course, one possibility would be to have all these enzymes available so that when a particular sugar is presented, the necessary enzyme would be readily available for it. Such a strategy, however, would be highly wasteful and inefficient, requiring too much energy for producing all the enzymes, many of which might never be needed. Thus, the strategy a bacterium uses is to synthesize only the needed enzymes by activating the genes that encode these enzymes and inactivating or repressing the genes that encode unnecessary enzymes. These genes can therefore be activated and inactivated in response to various environmental conditions.

We have already discussed one of the mechanisms necessary for transcription to occur: RNA polymerase must bind to the promoter of a gene. However, various other DNA-binding proteins can determine whether or not transcription can take place. In prokaryotes, for example, certain sites in the vicinity of the promoter can be bound by regulatory proteins called *activators* or *repressors*. An activator protein, as the name suggests, must bind to its target site in order for transcription to occur. A repressor protein, on the other hand, must not bind to its target site in order to enable transcription—if it binds, transcription is blocked. One of the mechanisms by which activators and repressors are able to alter the transcription of a gene involves physical interaction with RNA polymerase (bound to the nearby promoter); a bound repressor can physically interfere with RNA polymerase binding, whereas an activator can assist the RNA polymerase in attaching to the DNA.

If the binding of a regulatory protein can determine whether a gene transcript is produced, what determines this protein's ability to bind to its target site? In many cases, this ability is modulated by the interplay between the protein's DNA-binding domain (a region on the protein that can directly bind to specific DNA sequences) and the protein's *allosteric site* (a site on a protein where small molecules, called *allosteric effectors*, can bind and cause conformational changes). When an allosteric effector binds to the allosteric site on a regulatory protein, the structure of the DNA-binding domain can change, thereby enabling or disabling the protein's ability to bind its target site and ultimately determining whether transcription can occur. For example, in an activator, the presence of an allosteric effector might enable the protein's ability to bind DNA. On the other hand, a repressor might be able to block transcription (by binding to its target site) only in the absense of the effector molecule. For instance, in lactose metabolism in *E. coli*, the *lac* repressor loses its affinity for its target site and falls off the DNA when lactose (its allosteric effector) binds to it. This in turn allows the RNA polymerase situated nearby to do its job and transcribe the nearby genes encoding certain enzymes needed for the metabolism of lactose (Figure 1.10).

The *lac* repressor itself is encoded by a regulatory gene (*lac*I). Thus, this is a good example of how the product of one gene is able to control the transcription

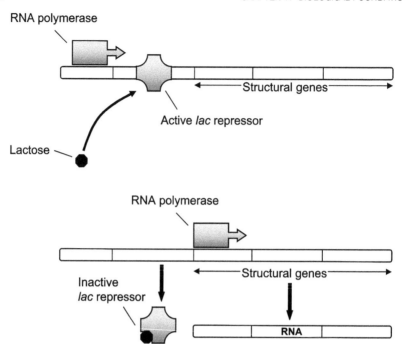

Figure 1.10. Lactose serving as an allosteric effector of the *lac* repressor protein. The repressor binds to the DNA, blocking RNA polymerase and inhibiting transcription of the nearby structural genes encoding enzymes. When lactose is introduced, it binds to the *lac* repressor and causes a change in its conformation, causing it to fall off the DNA and thereby freeing the RNA polymerase to quickly initiate transcription.

of other genes. However, such control mechanisms can be much more complicated even in prokaryotes, involving a number of other genes and factors in a multivariate fashion—a topic that we will deal with later in the book when we study models of genetic networks. As an example of an additional control mechanism, let us again consider the lactose system in *E. coli*. It is known that the cell is able to capture more energy from the breakdown of glucose than of lactose. Thus, if both lactose and glucose are present, the cell will favor glucose. Consequently, the cell will not require the production of *lac* enzymes to metabolize lactose until it runs out of glucose. This implies that there is an additional level of control due to the concentration of glucose.

Indeed, when glucose concentration is high, a substance called cyclic adenosine monophosphate (cAMP) is present at low concentrations. However, when glucose begins to run out, the level of cAMP increases, serving as an alert signal for the cell. cAMP then forms a complex with the cAMP receptor protein (also called catabolite activator protein or CAP), and this complex binds to its target site just upstream of the RNA polymerase site. Without cAMP, CAP cannot bind to its target site. Thus, cAMP is an allosteric effector, and CAP is an activator protein. After the cAMP-CAP complex binds, it induces a bend in the DNA, and this is believed to enhance the affinity of RNA polymerase for the *lac* promoter. Putting all this together, we see that low levels of glucose induce the transcription of genes encoding the enzymes necessary for the metabolism of lactose. We can conclude that the presence of lactose can activate these genes, but this is not a sufficient condition; glucose must also not be present.

The regulation of transcription in eukaryotes is fundamentally similar to that in prokaryotes. The regulatory proteins, often called *trans-acting elements*, bind to specific sequences in DNA, which are often called *cis-acting elements*. Unlike the situation in prokaryotes, however, cis-acting elements can be located at great distances from the transcription start sites. There are also several classes of cis-acting elements.

The RNA polymerase–binding region constitutes the core promoter. This region usually contains a *TATA box*—a highly conserved sequence approximately 30 base pairs (bp) upstream of the transcription initiation site. In some eukaryotes, instead of a TATA box, there is a promoter element called an initiator. Nearby are the promoter-proximal cis-acting sequences; as in prokaryotes, the proteins that bind to these sites can physically interact with RNA polymerase and help it bind to the promoter. Promoter-proximal elements are typically within 100–200 bp of the transcription initiation site. These elements also contain conserved sequences, such as CCAAT. Finally, there can be cis-acting elements located very far away from the promoter. These fall into two classes: enhancers and silencers. *Enhancers* increase transcription rates, whereas *silencers* inhibit transcription. Enhancers and silencers can act as far as thousands of base pairs away from the start site. They can also be located upstream or downstream of the promoter. It is also interesting that many enhancers are cell-type-specific, as are some promoter-proximal elements. This means that they are able to exert their effect only in particular differentiated cell types. The mechanism by which these distant cis-acting elements can execute control over transcription is believed to involve DNA loops, whereby the trans-acting regulatory protein bound at a distant enhancer site is brought into close proximity with other proteins in the promoter-proximal cis-acting elements. This can also explain why enhancers and silencers can be either upstream or downstream of the promoter. Figure 1.11 illustrates this mechanism.

Finally, we should mention that regulation by feedback is common and often necessary for proper cell functioning. For example, positive feedback refers to the situation where a regulatory protein activates its own transcription. More complicated feedback mechanisms are possible, such as when protein A regulates the transcription of gene B whose gene product regulates the transcription of gene A. Negative feedback is also common. Simple positive self-feedback can allow a gene to be switched on continuously. For example, the *myoD* gene, which begins to be expressed when a cell differentiates into a muscle cell, is kept continually expressed by means of its own protein binding to its own cis-acting elements. Thus, the cell continues to synthesize this muscle-specific protein and remains differentiated, passing on the protein to its daughter cells.

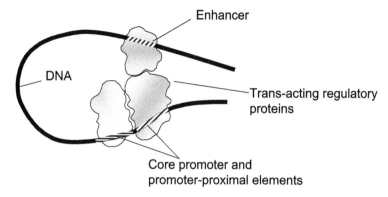

Figure 1.11. How a DNA loop can bring a trans-acting regulatory protein bound at a distant
           enhancer site into close proximity with other regulatory proteins bound at the
           core promoter and promoter-proximal elements.

## 1.2 GENOMICS

The field of genomics is concerned with investigating the properties and behavior
of large numbers of genes, typically in a high-throughput manner. The overarch-
ing goal of this enterprise is to understand how the genome functions as a whole.
That is, how do genes and proteins work together, regulating the functions of cells
and the development of organisms, and what goes wrong when they do not work
together as they should in diseases such as cancer?

Genomic studies can be conducted on different levels roughly corresponding to
DNA, RNA, and protein. On the DNA level, most genomic studies have addressed
the problem of gene mapping and sequencing for entire genomes, such as in the
Human Genome Project. A popular method, called comparative genomic hybridization (CGH), can be used to produce a map of DNA sequence copy numbers as
a function of chromosomal location throughout the entire genome. Thus, it can be
used to perform genomewide scanning of differences in DNA sequence copy
numbers with many applications in cancer research. High-throughput CGH
methods using array technology are now available.

Several recently developed technologies now allow us to study the *transcriptome*—the aggregation of mRNAs present in a cell at the time of measurement—in
a high-throughput manner. This involves identifying the mRNAs and quantifying
their abundances or, in other words, measuring gene expressions on the transcript
level. First, a few words about cDNA libraries.

A direct way to determine what transcripts are produced would be to collect the
mRNA produced in cells and create from it, a *complementary DNA* (cDNA) library
that can be subsequently sequenced. Recall that reverse transcriptase is the enzyme
that uses RNA as a template to produce cDNA. The reason for constructing such
a cDNA library is that most of the DNA in the genome (introns, promoters, intergenic regions) is not needed for determining gene expressions. Thus, the cDNAs
contain no intron sequences and cannot be used to express the protein encoded by

the corresponding mRNA.

After reverse transcription of the mRNA, the synthesized cDNA is introduced into bacteria. This is accomplished by using *cloning vectors*: a DNA fragment is joined to, say, a chromosome of a bacterial virus (phage), and the whole recombinant DNA molecule is introduced into bacteria for replication (recall that a virus also uses the host cell for its own replication). In addition to phage vectors, plasmid vectors—small, circular molecules of double-stranded DNA derived from naturally occurring bacterial cells—are often used. In either case, the bacterial host cells are then *transfected* with the vectors, which means that the recombinant DNA molecules are inserted into the cells, which then keep dividing and produce more and more of the foreign DNA. Cloning vectors typically incorporate a gene that allows those cells that contain the vector to be selected from those that do not contain it. This selectivity gene usually provides resistance to some toxin, such as an antibiotic. Thus, only the cells with inserted vectors ultimately survive. Each bacterial colony in a culture dish houses a unique vector that contains the cDNA counterpart of a particular mRNA. Today, large sequenced cDNA clone libraries are commercially available.

Although in principle every clone in a cDNA library can be sequenced to study the transcriptome of a given cell system, this approach requires a large amount of work and is not practical. One high-throughput approach for studying the transcriptome is called serial analysis of gene expression or SAGE. This method is based on the fact that a short (e.g., 12-bp) nucleotide sequence, called a *SAGE tag*, can uniquely identify a gene as long as the sequence is always from a certain position in the transcript. Indeed, 12 bp can, in theory, distinguish up to $4^{12} = 16\,777\,216$ different transcripts. The abundance of each SAGE tag is then quantified and used as a measurement of the level of expression of the corresponding gene. Many SAGE tags can be concatenated (covalently linked together within a single clone) head to tail to produce a single *concatamer*, which can then be sequenced. This step reduces the number of sequencing reactions that need to be carried out; information on 30 or more genes can be obtained with one sequencing reaction.

## 1.2.1 Microarray Technology

Another extremely popular high-throughput technique for studying transcriptomes is the DNA *microarray* (DNA *chip*). For completely sequenced genomes, DNA microarrays allow the screening of whole-genome expression, meaning that every gene in the organism can be simultaneously monitored. This is already possible for the human genome. Although there are several kinds of microarrays reflecting the method of their construction, such as in situ synthesized oligonucleotide arrays and spotted cDNA arrays, on a basic level a microarray is a gridlike arrangement of thousands of unique DNA molecules, called *probes*, attached to some support surface such as glass or a nylon membrane. We will not attempt to cover in detail different microarray technologies and all aspects of microarray production. Instead, we refer the reader to several books, such as Schena (2003), Blalock (2003), Hardiman (2003), and Zhang et al. (2004). In this section, we will outline the generic principles of microarrays and use the popular spotted cDNA or long oligo arrays as our example.

The DNA fragments are arrayed (spotted) on a simple glass microscope slide by a robot. The robot typically dips its pins into solutions, typically placed in 96-well plates, that contain the DNA. Then the tiny amounts of solution that adhere to the pins are transferred to a support surface, most commonly glass. Each such transfer by a pin results in a tiny printed spot containing DNA molecules, and the printed array consists of a gridlike arrangement of such spots. The pins can be either solid, similar to a sharp needle, or quill-type (also called split-pin), which have narrow slits as in a quill pen. In the former case, the robot must redip the pins into the solution before touching down on the support surface, whereas in the latter case, the robot can perform multiple printings without redipping. As a result, solid-pin robots are slower, but quill-pin robots are more prone to pin damage and clogging, resulting in printing dropouts. Yet another robotic printing approach is the inkjet method, which is the same method used in standard printers, whereby a precise amount of sample is expelled from a miniature nozzle equipped with a piezoelectric fitting by applying an electric current. Robots are now capable of printing upward of 50,000 spots on a standard microscope glass slide. The slides are often precoated with polylysine or polyamine in order to immobilize the DNA molecules on the surface. The DNA is denatured so that each spot contains deposits of single-stranded DNA.

In spotted microarrays, each spot contains a different DNA sequence (probe) corresponding to the spliced mRNA (exons) of a specific gene (in the case of eukaryotic genes). Thus, different spots can serve as detectors of the presence of mRNAs that were transcribed from different genes in the sample of interest. This is accomplished as follows. The RNA is first extracted from the cells whose transcriptome we wish to study. Then, the RNA is converted to cDNA and amplified by the reverse transcriptase–polymerase chain reaction (rtPCR). During this process, fluorescent molecules (tags) are chemically attached to the DNA. If a specific mRNA molecule was produced by the cells in the sample, then the corresponding fluorescently labeled cDNA molecule will eventually stick to its complementary single-stranded probe via base pairing. The rest of the cDNA molecules that are unable to find their complementary "partners" on the array will be washed away during a washing step. Since the fluorescent tags will still be attached to the molecules that base-pair with the probes, the corresponding spots will fluoresce when measured by an instrument that provides fluorescence excitation energy and detects the level of emitted light, yielding a digital image of the microarray in which the level of measured fluorescence is converted to pixel intensities. After an image processing step intended to segment the spots and summarize the overall pixel intensity in each spot, a data file containing measurements of gene expressions (and a host of other measurements, such as background intensity, spot area, signal-to-noise ratio, and so on) is produced. These data are then subjected to statistical analysis and modeling.

In the case of cDNA microarrays, two distinct fluorescent dyes are commonly used to compare relative gene expressions between two samples. In this scenario, one dye is incorporated into the cDNA of one sample (say, tumor) and the cDNA from the other sample (say, normal) is tagged with the other dye. Commercial scanners use multiple laser sources to excite each of the dyes at their specific

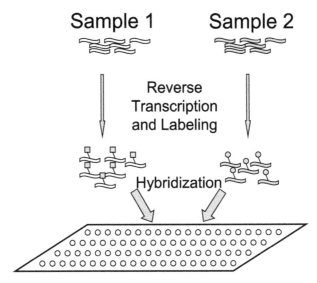

Figure 1.12. Two-color hybridization. The RNA extracted from each sample is reverse-transcribed and labeled to make cDNA so that each sample is labeled with a different fluorescent dye. The labeled cDNAs are then hybridized to the probes on the microarray so that the fluorescent signal intensities corresponding to the abundances of mRNA transcripts produced in each of the samples can be quantified by a laser scanner and image processing.

wavelengths so that their fluorescent emissions can be detected. This "two-color" microarray approach allows one to directly compare relative transcript abundance between two samples and provides a type of self-normalization in that possible local variations in the array affecting only some of the spots affect both samples equally. The two-color approach is illustrated in figure 1.12.

Many of the current approaches in computational and systems biology that we will discuss in this book make use of microarray data. Although this technology is undergoing rapid change, an understanding of the fundamental principles behind microarray experiments is essential for investigators wishing to apply computational, mathematical, and statistical methods to microarray data. Each of the multitude of steps, including biological sample handling, slide preparation and printing, labeling, hybridization, scanning, and image analysis, presents its own challenges and can greatly influence the results of the experiment. Without an appreciation of the ways each of these steps can affect the data produced in a microarray experiment, computational and modeling efforts run the risk of yielding unsound conclusions. We strongly recommend that the interested reader, especially one who intends to work with microarray data, consult the available literature on microarray experiments, quality control, and basic data analysis (e.g., Schena, 2003; Baldi and Hatfield, 2002; Zhang et al., 2004).

## 1.3 PROTEOMICS

As we have discussed, genomics is concerned with studying large numbers of genes and, in particular, how they function collectively. However, genes execute their functions via the transcriptome and the *proteome*, the latter pertaining to the collection of functioning proteins synthesized in the cell. Thus, a comprehensive understanding of how the genome controls the development and functioning of living cells and how they fail in disease requires the study of transcriptomics and proteomics. While these disciplines specifically refer to the study of RNA and proteins, respectively, we take the broader view that genomics must encompass all these levels precisely because the genetic information stored in DNA is manifested on the RNA and protein levels. Indeed, transcriptional profiling using a high-throughput technology such as microarrays is already commonly considered part of genomics. Thus, many of the methods and models that we describe in this book are applicable not only to transcriptional but also to protein expression measurements.

Transcriptional measurements can show us which genes are expressed in a cell at a given time or under a given condition, but they cannot give us accurate information about the synthesized proteins despite the fact that RNA is translated into proteins. One reason for this is that different mRNAs can be translated at different times and a transcriptional measurement at a given time may not reflect the protein content. Furthermore, while some proteins are being synthesized, others are being degraded in a dynamical process. Thus, direct measurements of protein activity are also necessary.

Two common techniques are two-dimensional gel electrophoresis and mass spectrometry. The first method is used for separating the individual proteins in the proteome. It is based on standard polyacrylamide gel electrophoresis, which is used to separate proteins according to their molecular weights but also involves a second step where the gel is rotated 90° and the proteins are then separated according to their charges. This procedure results in a two-dimensional arrangement of spots, each one corresponding to a different protein. This method can be used to detect differences between two proteomes by comparing the presence or intensity of spots in corresponding locations. Another application of this procedure in proteomics is the isolation of proteins for further characterization by mass spectroscopy.

The protein in a given spot can be further analyzed and identified by mass spectrometry, which is a technique used to separate ions according to their charge-to-mass ratios. A common technique used in proteomics is matrix-assisted laser desorption ionization time of flight (MALDI-TOF). In this technique, protein ions are accelerated through an electric field. The smallest ones arrive at the detector first as they travel through the flight tube. This time of flight in the electric field is used as a measure of the charge-to-mass ratio. Mass spectrometry has greatly enhanced the usefulness of two-dimensional gels.

Another important goal in proteomics is the identification of interactions between two or more proteins. This information can be highly useful in characterizing the function of a new or uncharacterized protein. For example, if the unknown protein interacts with another protein known to be located on the cell surface, this may indicate that the unknown protein is involved in cell-cell signaling. There are several popular methods, such as phage display and the yeast two-hybrid system, that can be used in a high-throughput fashion to detect protein-protein interactions. These techniques can generate protein interaction maps, which show all the interactions between the proteins. Let us give a brief outline of the yeast two-hybrid method.

This technique is based on the ability of two physically bound proteins to activate transcription of a reporter gene. The reporter gene, integrated into the yeast genome, is expressed only when a suitable transcription factor is present. Key to this method is the fact that transcription factors require two separate domains, a DNA-binding domain and an activation domain. Although these two domains are usually part of the same molecule, an artificial transcription factor can be constructed from two proteins, each carrying one of the two domains. First, the gene encoding the target protein is cloned in a vector that contains the cDNA sequence for the binding domain. Then, a library of uncharacterized cDNA sequences, each one cloned in a vector containing the cDNA sequence for the activation domain, is created. Thus, the target protein is fused to the binding domain, and each of the uncharacterized proteins is fused to the activation domain. These are all mixed together and cotransfected into a yeast strain containing an integrated reporter gene. If a protein-protein interaction occurs between the target protein and one of the uncharacterized proteins, the corresponding fused binding domain and activation domain are brought together, forming a transcription factor that can activate the reporter gene. The cells where the reporter is expressed can then be selected in order to identify the specific protein-protein interaction that took place.

# Bibliography

Alberts B, Johnson A, Lewis J, Raff M, Roberts K, Walter P. (2002) *Molecular Biology of the Cell*, 4th ed., Garland Science, New York.

Baldi P, Hatfield GW. (2002) *DNA Microarrays and Gene Expression*, Cambridge University Press, Cambridge, UK.

Blalock EM, ed. (2003) *A Beginner's Guide to Microarrays*, Kluwer Academic Publishers, Norwell, MA.

Brown TA. (2002) *Genomes*, 2nd ed., John Wiley & Sons, New York.

Draghici S. (2003) *Data Analysis Tools for DNA Microarrays*, Chapman & Hall/CRC, Boca Raton, FL.

Hardiman G, ed. (2003) *Microarrays Methods and Applications: Nuts & Bolts*, DNA Press, Skippack, PA.

Kohane IS, Kho A, Butte AJ. (2003) *Microarrays for an Integrative Genomics*, MIT Press, Cambridge, MA.

Lodish H, Berk A, Zipursky SL, Matsudaira P, Baltimore D, Darnell JE. (2000) *Molecular Cell Biology*, 4th ed., W. H. Freeman & Co., New York.

Schena M. (2003) *Microarray Analysis*, John Wiley & Sons, Hoboken, NJ.

Zhang W, Shmulevich I, Astola J. (2004) *Microarray Quality Control*, John Wiley & Sons, Hoboken, NJ.

# Chapter Two

## Deterministic Models of Gene Networks

A deterministic model of a genetic regulatory network can involve a number of different mechanisms that capture the collective behavior of the elements constituting the network. The models can differ in numerous ways:

- What physical elements are represented in the model (e.g., genes, proteins, other factors)
- Whether the model is capable of representing a dynamical process (i.e., Is there a notion of time?)
- At what resolution or scale is the behavior of the network elements captured (e.g., Are genes discretized, such as being either on or off, or do they take on continuous values?)
- How the network elements interact (e.g., interactions can be either present or absent or they can have a quantitative nature).

What deterministic models have in common, however, is that there is no inherent notion of randomness or stochasticity in the model once it is specified, say, by inferring it from data. The existing stochastic models of gene networks are often generalizations of deterministic models where some aspects of the model, such as the rules of interaction or the values of the network elements, acquire stochastic components.

## 2.1 GRAPH MODELS

One of the simplest approaches in representing and modeling genetic regulatory interactions involves constructing various kinds of graphs. A rich mathematical theory exists around graphs (graph theory), which has countless applications in many diverse fields. Biologists have been representing knowledge about regulatory interactions using graphs for quite some time. Informally speaking, this amounts to representing various elements of a regulatory system, such as genes, proteins, or other molecules, as nodes or vertices in a graph and connecting these elements by possibly weighted arrows, where the weights can be used to represent the particular kinds of relationships between the corresponding elements connected by the arrows. For example, the weight of an arrow can be either $+1$ or $-1$, corresponding to activation or inhibition, respectively. Undirected connections can also be used, simply representing interactions between the elements without capturing directionality, such as in protein-protein interaction networks (Ito et al., 2001; Uetz et al., 2000; Przulj et al., 2004; Yook et al., 2004; also see section 1.3).

There are a number of publicly available databases that contain information about such interactions, including KEGG: Kyoto Encyclopedia of Genes and Genomes (Kanehisa and Goto, 2000), GeNet (Samsonova et al., 1998), GeneNet (Ananko et al., 2002), and many others. A number of tools for visualizing and navigating such networks are available (Han et al., 2004; Serov et al., 1998), with many databases offering integrated visualization tools online. Graph models can be highly useful for predicting various functional relationships in biological networks from various structural properties of the graph. They can also be useful in predicting the effects of specific perturbations on individual elements in the network or on the network as a whole. Let us now state several definitions and discuss applications to genetic and protein interaction networks.

A *directed graph* (often called a *digraph*) $G$ is a pair $(V, E)$, where $V$ is a set of *vertices* and $E \subseteq V \times V$ is a set of *edges*, where each edge is an ordered pair of vertices. The notation $(a, b)$ denotes an edge in a directed graph $G$ that *leaves* vertex $a$ (tail) and *enters* vertex $b$ (head). We also say that $b$ is *adjacent* to $a$. This relationship can be represented in an *adjacency matrix* $\mathbf{A} = (a_{ij})$ corresponding to a graph $G = (V, E)$, where

$$a_{ij} = \begin{cases} 1 & \text{if } (i, j) \in E, \\ 0 & \text{otherwise,} \end{cases}$$

and where the vertices are numbered as consecutive integers $1, 2, \ldots, |V|$. If a directed graph represents genetic regulatory relationships, then the edge $(a, b)$ can be used to represent the situation where element $a$ (e.g., a protein) controls or affects element $b$ (e.g., a gene). Self-regulation corresponds to a *loop* in the graph, which is an edge $(a, a)$ between a vertex and itself or, correspondingly, a 1 on the diagonal of the adjacency matrix $\mathbf{A}$. Spectral methods using adjacency matrices have been used to predict functions of uncharacterized proteins (Bu et al., 2003).

In many instances, directionality may be unimportant, such as in the protein-protein interactions mentioned above where an interaction between two proteins either occurs or does not occur. In an *undirected graph* (often called just a *graph*), the set of edges $E$ consists of unordered pairs of vertices. In this case, the adjacency relation is symmetric (if $a$ is connected to $b$, then $b$ is connected to $a$) and antireflexive (there are no loops). The symmetry property implies that the adjacency matrix is symmetric: $\mathbf{A} = \mathbf{A}^T$. By convention, in an undirected graph, the edge $(a, b)$ is the same as $(b, a)$, although strictly speaking an edge in this case is a set $\{a, b\}$, where $a, b \in V$ and $a \neq b$.

Let $G = (V, E)$ be a digraph and let $v \in V$ be one of its vertices. The *out-degree* of vertex $v$ is the number $d^+(v)$ of edges of $G$ leaving vertex $v$. Similarly, the *in-degree* of vertex $v$ is the number $d^-(v)$ of edges entering vertex $v$. It is straightforward to determine the number $d^+(v)$ or $d^-(v)$ from the adjacency matrix $\mathbf{A}$ by simply summing the row or column, respectively, corresponding to $v$. For example, the out-degree of $v$ may correspond to the number of genes regulated by protein $v$, while the *in-degree* of gene $u$ may represent the number of proteins regulating $u$. Obviously, in an undirected graph, $d^+(v) = d^-(v) = d(v)$, which is simply called the *degree* of $v$. In a protein-protein interaction network, this represents the number

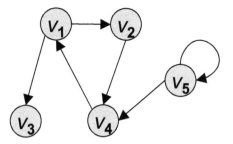

Figure 2.1. An example of a directed graph.

of proteins interacting with $v$. If $d(v) = 0$, then the vertex $v$ is said to be *isolated*, meaning that this protein does not interact with any other protein.

**Example 2.1** *Consider the graph depicted in figure 2.1. The vertex set is* $V = \{v_1, v_2, v_3, v_4, v_5\}$, *and the edge set is*

$$E = \{(v_1, v_2), (v_1, v_3), (v_2, v_4), (v_4, v_1), (v_5, v_4), (v_5, v_5)\}.$$

*The out-degrees of the vertices are* $(2, 1, 0, 1, 2)$, *and the in-degrees are* $(1, 1, 1, 2, 1)$. *The adjacency matrix corresponding to this graph is*

$$\mathbf{A} = \begin{bmatrix} 0 & 1 & 1 & 0 & 0 \\ 0 & 0 & 0 & 1 & 0 \\ 0 & 0 & 0 & 0 & 0 \\ 1 & 0 & 0 & 0 & 0 \\ 0 & 0 & 0 & 1 & 1 \end{bmatrix}.$$

A *path* of length $k$ from vertex $a$ to vertex $b$ is a sequence $(v_0, v_1, \ldots, v_k)$ of vertices such that $a = v_0$, $b = v_k$, and $(v_{i-1}, v_i) \in E$ for $i = 1, \ldots, k$. We say that $b$ is *reachable* from $a$ if there exists a path from $a$ to $b$. A *simple* path is one that does not contain repeated vertices. A *cycle* is a path containing at least one edge and in which $v_0 = v_k$. A graph that contains no cycles is said to be *acyclic*. In a genetic network, a path between two genes can represent part of a regulatory pathway or series of interactions. For example, a cascade of several protein kinase reactions in a signal transduction pathway, such as the mitogen-activated protein kinase (MAPK) cascade, can be represented by a path. A cycle can indicate a feedback relationship. It is easy to use the adjacency matrix $\mathbf{A}$ to test whether a path exists between any two vertices. In the matrix $\mathbf{A}^k$, which is the matrix product of $k$ copies of $\mathbf{A}$, the element $a_{ij}^{(k)}$ is equal to the number of paths of length $k$ from vertex $v_i$ to vertex $v_j$.

In an undirected graph, we say that vertices $a$ and $b$ are *connected* if there is a path between them. It is easy to see that the "connected" relation defines equivalence classes of the vertices. These equivalence classes are called the *connected components* of the graph. This is an important concept in protein interaction networks since a connected component represents a subnetwork in which a group of proteins interact with each other but not with the rest of the network. In real protein networks, it is often the case that a *giant component* exists in addition to many other small components (Wagner, 2001). For a digraph, there is a notion of *strongly connected components*, which are the equivalence classes defined by the relation "are mutually reachable." That is, $a$ and $b$ are in the same equivalence class if there exists a path from $a$ to $b$ and from $b$ to $a$. If a metabolic network is represented by a digraph, where the substrate of a reaction is connected with the product by a directed edge, then a strongly connected component is a subnetwork in which the metabolites can be converted to each other (Ma and Zeng, 2003). For example, in the metabolic network in *E. coli*, 29 strongly connected components containing no less than three metabolites have been identified, with the largest of these containing 274 vertices (Ma and Zeng, 2003). There are efficient linear-time algorithms, in terms of the numbers of vertices and edges, for computing strongly connected components of a directed graph using the depth-first search algorithm (Bang-Jensen and Gutin, 2001).

As mentioned earlier, each edge can have an associated weight given by a weight function $w : E \to S$, where $S$ is the set of possible weights. If $S = \{+1, -1\}$, the weight simply represents activation/inhibition or other dichotomous relationships, whereas if $S = \mathbb{R}$ (the set of real numbers), then the strength of interaction can also be captured (Noda et al., 1998; Moriyama et al., 1999). Dynamics can be introduced into the network by assigning a state $g(v)$ to each vertex (gene) $v$ and updating the states for all the genes using certain deterministic or stochastic rules. For example, if $g : V \to \{0, 1\}$, then each gene can be considered to be either inactive (0) or active (1). This paves the way toward various dynamical models of genetic networks, such as Boolean networks, which we will discuss later.

Weighted undirected graphs can also be used to represent strengths of relationships between genes, such as the correlations between their expressions (mRNA levels). Consider, for example, transcriptional measurements of $n$ genes over $p$ experimental conditions using DNA microarrays. The latter can be different time points (i.e., time-course data), different drug treatments, or genetic alterations such as deletions. We can compute the correlation coefficient or some other measure of concordance between all pairs of genes. This produces a *complete graph* (a graph that has an edge between every pair of vertices), the edges of which are weighted by the corresponding correlations between the genes connected by these edges. Now suppose that we choose a threshold $\tau$ such that if the absolute value of the weight (correlation coefficient) exceeds $\tau$, we keep the edge, otherwise we remove it. Having performed this operation, it is possible that two genes that are not highly correlated, and hence do not have an edge connecting them, may still be connected via some path. Such a path in the graph may uncover the possible relationships between the constituent genes, such as being involved in the same biological pathway. If there is more than one path, the shortest path (sum of the weights of its edges) may provide the most parsimonious explanation of the dependence between the two genes in light of the expression data. The well-known Dijkstra's algorithm can be used to find the shortest path in a weighted graph (Bang-Jensen and Gutin, 2001).

Such an approach was taken by Zhou et al. (2002) to predict the function of unknown genes from known genes lying on the same shortest path, using *Saccharomyces cerevisiae* (yeast) gene expression data. The obvious advantage of such an approach is that while clustering (chapter 6) based only on pairwise correlations or distances cannot reveal functional relationships between genes whose expression profiles are not highly correlated, the graph-based path approach may reveal more subtle "transitive coexpression." In other words, two genes in the same pathway may not be highly correlated with each other but may be strongly correlated with some other gene, thus establishing a path of length 2.

The graph formalism has also proven itself to be highly useful in studying general topological properties of various biological networks such as metabolic networks, genetic networks, and protein-protein interaction networks. A characterization of global graph properties can yield useful insight into the underlying organizational structure of the biological network. Let us now consider several such properties.

One interesting concept is that of a *small-world* network. This idea was popularized by Milgram (1967), who noted that in a social network, where persons correspond to vertices and edges correspond to acquaintances between persons, most pairs of individuals are typically separated by "six degrees of separation". This idea was later explored in a play by John Guare and made into the film *Six Degrees of Separation* in 1993. A similar property holds for networks of movie actors, where two actors are connected by an edge if they have appeared together in a movie. A popular movie trivia game called Six Degrees of Kevin Bacon tries to connect a movie actor to the actor Kevin Bacon in as few connections as possible. The answers can be found at the University of Virginia's "Oracle of Bacon" (http://oracleofbacon.org/). Amazingly, most actors have a "Bacon number" of 2 or 3. Returning to cellular networks, the small-world property implies that most nodes in the network are separated by only a few intermediates. For example, Jeong et al. (2000) found that in metabolic networks compiled from 43 organisms, where the vertices are substrates and the edges are chemical reactions, the average path length was 3.3. This gives strong evidence and support for the well-known pleiotropic and linked nature of biological processes (Fraser and Marcotte, 2004).

Another global measure of a graph's structure is the *clustering coefficient* (Watts and Strogatz, 1998). Recall that $d(v)$ denotes the *degree* of vertex $v$, which is equal to the number of other vertices connected to $v$. Some of these vertices may be connected to each other. Let $E(v)$ be the number of edges connecting these vertices. The maximum number of edges that can possibly exist between them is

$$\binom{d(v)}{2} = \frac{d(v) \cdot [d(v) - 1]}{2}.$$

The clustering coefficient of vertex $v$ is defined as

$$C(v) = \frac{2E(v)}{d(v) \cdot [d(v) - 1]},$$

which is the ratio of the actual number of edges connecting the neighbors of $v$ and the total possible number of such edges. Finally, the clustering coefficient $C$

for an entire graph is the average of all the clustering coefficients across all the vertices. Intuitively, $C$ measures the "cliquishness" of a network; for example, a high value of $C$ implies that a person's acquaintances are also highly likely to know each other. Interestingly, most real networks exhibit a higher clustering coefficient than a comparable random network with the same number of vertices and edges (Albert and Barabási, 2002).

Another very important measure is the *degree distribution* of a graph, which is simply the distribution of the degree $d(v)$ of a randomly selected vertex $v$. In a directed graph, we have in- and out-degree distributions corresponding to $d^-(v)$ and $d^+(v)$, respectively. Let us consider the famous Erdős-Rényi random graph model, where in a graph containing $N$ vertices each pair of vertices is connected by an edge with probability $p$. It is easy to see that such a graph contains approximately

$$p\binom{N}{2} = \frac{pN(N-1)}{2}$$

edges. Similarly, the clustering coefficient for such a graph equals $p$. For a randomly chosen vertex $v$, the probability that its degree $d(v)$ is equal to $k$ is given by the binomial distribution

$$P(d(v) = k) = \binom{N-1}{k} p^k (1-p)^{N-1-k},$$

since out of a total of $(N-1)$ remaining vertices, there are $\binom{N-1}{k}$ ways to choose $k$ neighbors and for each such choice the probability that $v$ is connected to them is $p^k$ and not connected to any of the others is $(1-p)^{N-1-k}$. When $N$ is large and $p$ is relatively small, such that $Np$ is of moderate size, the binomial distribution is well approximated by a Poisson distribution with parameter $\lambda = Np$. Thus, the degree of a random vertex $v$ is given by

$$P(d(v) = k) \approx e^{-Np} \frac{(Np)^k}{k!}.$$

It is a fact that most real networks, such as the World Wide Web (WWW), the Internet, metabolic networks, and many others have degree distributions that significantly differ from the Poisson distribution and are thus not well approximated by a random graph model (Albert and Barabási, 2002). The degree distribution of many real networks follows a power law given by

$$P(d(v) = k) \sim k^{-\gamma},$$

and networks with this property are often called *scale-free* (Barabási and Albert, 1999). When plotted on a log-log scale, the power law distribution (probability mass function) appears as a straight line with a slope equal to $-\gamma$. In networks represented by directed graphs, it is possible for the in- and out-degree distributions both to be power laws but with different parameters $\gamma_{in}$ and $\gamma_{out}$. For example,

*E. coli* metabolic networks were found to be scale-free with $\gamma_{in} \approx \gamma_{out} \approx 2.2$, and the scale-free topology appears to be ubiquitous in metabolic networks of all organisms in all domains of life (Jeong et al., 2000).

The topology of a scale-free network is dominated by a few highly connected vertices (hubs) that link the rest of the less connected vertices. A good example is the familiar network of airline routes shown in the magazines in every seat pocket on an airplane. Every airline has several cities designated hubs that are connected by direct flights to many other cities. Thus, if cities are represented by vertices and direct flights by directed edges, then most vertices have small in- and out-degrees, whereas hubs have very large in- and out-degrees. Consequently, flights between two nonhub cities often require the passenger to transfer via a hub city. This is a general property of scale-free networks, and in genetic or metabolic networks, certain substrates or genes play the role of such highly connected hubs. We should note that in Erdős-Rényi random graphs, the probability of a very large vertex degree is very low.

It is interesting to consider how scale-free topology can emerge in real networks (Barabási and Albert, 1999). One of the most important and natural mechanisms of network growth is via the addition of new vertices that link to vertices already present in the network. For example, new airports obviously become linked, via new flights, to existing airports. The other intuitive mechanism of network growth is called *preferential attachment*. That is, new flights are most likely to be directed or routed through existing highly connected hubs. Similarly, frequently cited papers are more likely to be cited again, and a highly connected WWW page is likely to be linked by a new page. Thus, there is a higher probabilty of linking to a vertex with a high degree.

It is intuitively obvious that the removal of such highly connected hubs can have a drastic effect on the entire network: the shutdown of a hub city airport because of bad weather conditions can affect a large number of flights and badly disrupt the operations of the entire airline, whereas the shutdown of an airport that has only a relatively small number of incoming or outgoing flights generally does not affect the entire airline network. In biological systems, such hubs may correspond to important biological processes such as splicing machinery and DNA polymerase (Fraser and Marcotte, 2004). At the same time, it is worth noting that scale-free networks are in fact extremely robust against random errors. However, they are also quite vulnerable to targeted attacks against highly connected vertices (Albert et al., 2000). Jeong et al. (2000) found that in metabolic networks of several organisms, random removal of up to 8 percent of the substrates did not lead to an increase in the average path length. However, the average path length increased by 500 percent when 8 percent of the most connected nodes were removed. A similar finding was obtained by Jeong et al. (2001) for yeast protein networks.

The graph models of networks also form the basis of many other models that endow the static structure of a graph with dynamical processes. We will now discuss some of these model classes. An important point to keep in mind is that the topological properties of the underlying graph, such as the degree distribution, often play major roles in determining the dynamical behavior of the dynamical model based on the graph structure.

## 2.2  BOOLEAN NETWORKS

Consider a directed graph $G(V, E)$, where the vertices $V$ represent genes and the directed edges $E$ represent the actions of genes, or rather their products, on other genes. For example, directed edges from genes A and B into gene C indicate that A and B act jointly on C. The specific mechanism of action is not represented in the graph structure itself, so an additional representation is necessary. One of the simplest representation frameworks assumes that genes are binary-valued entities, indicating that they can be in one of two possible states of activity (on or off) at any given point in time and that they act on each other by means of Boolean functions. Thus, in the above example, gene C is determined by the output of a Boolean function whose inputs are A and B. The underlying directed graph merely represents the input-output relationships. Let us now present this idea more formally.

A *Boolean network*, originally introduced by Kauffman (1969a,b, 1974, 1993) is defined by a set of nodes (genes) $\{x_1, \ldots, x_n\}$ and a list of Boolean functions $\{f_1, f_2, \ldots, f_n\}$. Each gene $x_i \in \{0, 1\}$ $(i = 1, \ldots, n)$ is a binary variable whose value at time $t + 1$ is completely determined by the values of genes $x_{j_1}, x_{j_2}, \ldots, x_{j_{k_i}}$ at time $t$ by means of a Boolean function $f_i : \{0, 1\}^{k_i} \to \{0, 1\}$. That is, there are $k_i$ regulatory genes assigned to gene $x_i$ that determine the "wiring" of that gene. Thus, we can write

$$x_i(t + 1) = f_i(x_{j_1}(t), x_{j_2}(t), \ldots, x_{j_{k_i}}(t)). \tag{2.1}$$

In a *random Boolean network* (RBN), the functions $f_i$, sometimes called couplings (Aldana et al., 2002), are selected randomly, as are the genes used as their inputs. Such an approach has been used to study the ensemble properties of such dynamical systems. Each $x_i$ represents the state (expression) of gene $i$, where $x_i = 1$ represents the fact that gene $i$ is expressed and $x_i = 0$ means that it is not expressed. A given gene transforms its inputs (regulatory factors that bind to it) into an output, which is the state or expression of the gene itself at the next time point. All genes are assumed to update synchronously in accordance with the functions assigned to them, and this process is then repeated. The artificial synchrony simplifies computation while preserving the generic properties of global network dynamics (Huang, 1999; Kauffman, 1993). It is clear that the dynamics of the network are completely determined by equation (2.1). Let us give an example.

Consider a Boolean network consisting of 5 genes $\{x_1, \ldots, x_5\}$ with the corresponding Boolean functions given by the truth tables in table 2.2. The *maximum connectivity* $K = \max_i k_i$ is equal to 3 in this case. Note that $x_4(t + 1) = f_4(x_4(t))$ is a function of only one variable and is an example of autoregulation.

The dynamics of this Boolean network are shown in figure 2.2. Since there are 5 genes, there are $2^5 = 32$ possible states that the network can be in. Each state is represented by a circle, and the arrows between states show the transitions of the network according to the functions in table 2.2. It is easy to see that because of the inherent deterministic directionality in Boolean networks, as well as only a finite number of possible states, certain states will be revisited infinitely often if, depending on the initial starting state, the network happens to transition into them.

Table 2.1 Truth Tables of the Functions in a Boolean Network with Five Genes.

| | $f_1$ | $f_2$ | $f_3$ | $f_4$ | $f_5$ |
|---|---|---|---|---|---|
| | 0 | 0 | 0 | 0 | 0 |
| | 1 | 1 | 1 | 1 | 0 |
| | 1 | 1 | 1 | — | 0 |
| | 1 | 0 | 0 | — | 0 |
| | 0 | 0 | 1 | — | 0 |
| | 1 | 1 | 1 | — | 0 |
| | 1 | 1 | 0 | — | 0 |
| | 1 | 1 | 1 | — | 1 |
| $j_1$ | 5 | 3 | 3 | 4 | 5 |
| $j_2$ | 2 | 5 | 1 | — | 4 |
| $j_3$ | 4 | 4 | 5 | — | 1 |

The indices $j_1$, $j_2$, and $j_3$ indicate the input connections for each of the functions.

Such states are called *attractors*, and the states that lead into them comprise their *basins of attraction*. For example, in figure 2.2, the state (00000) is an attractor and the seven other (transient) states that eventually lead into it comprise its basin of attraction.

The attractors represent the *fixed points* of the dynamical system that capture its long-term behavior. The attractors are always cyclical and can consist of more than one state. Starting from any state on an attractor, the number of transitions needed to return to itself is called the *cycle length*. For example, the attractor (00000) has cycle length 1, while the states (11010) and (11110) comprise an attractor of length 2.

Real genetic regulatory networks are highly stable in the presence of perturbations of the genes. Within the Boolean network formalism, this means that when a minimal number of genes transiently change value (say, by means of some external stimulus), the system transitions into states that reside in the same basin of attraction from which the network eventually "flows" back to the same attractor. Generally speaking, large basins of attraction correspond to higher stability. Such stability of networks in living organisms allows cells to maintain their functional state within the tissue environment (Huang, 1999).

Although in developmental biology, epigenetic, heritable changes in cell determination have been well established, it is now becoming evident that the same types of mechanisms may also be responsible in carcinogenesis and that gene expression patterns can be inherited without the need for mutational changes in DNA (MacLeod, 1996). In the Boolean network framework, this can be explained by *hysteresis*, which is a change in the system's state caused by a stimulus that is not changed back when the stimulus is withdrawn (Huang, 1999). Thus, if the change in some particular gene does in fact cause a transition to a different attractor, the network often remains in the new attractor even if the gene is switched off. Thus, the attractors of a Boolean network also represent a type of memory of the dynamical system (Huang, 1999).

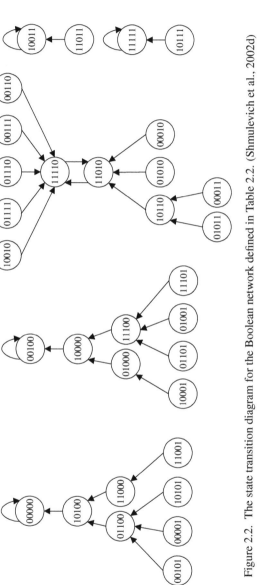

Figure 2.2. The state transition diagram for the Boolean network defined in Table 2.2. (Shmulevich et al., 2002d)

### 2.2.1 Cell Differentiation and Cellular Functional States

Boolean networks reflect the nature of complex adaptive systems in that they are "systems composed of interacting agents described in terms of rules" (Holland, 1995). A central concept in dynamical systems is *structural stability*, which is the persistent behavior of a system under perturbation. Structural stability formally captures the idea of behavior that is not destroyed by small changes in the system. This is most certainly a property of real genetic networks since a cell must be able to maintain homeostasis in metabolism and in its developmental program in the face of external perturbations and stimuli. Boolean networks naturally capture this phenomenon, as the system usually flows back into the attractors when some of the genes are perturbed. Real gene regulatory networks exhibit spontaneous emergence of ordered collective behavior of gene activity. Moreover, recent findings provide experimental evidence for the existence of attractors in real regulatory networks (Huang and Ingber, 2000). At the same time, Wolf and Eeckman (1998) have shown that dynamical system behavior and stability of equilibria can be largely determined from regulatory element organization. This suggests that there must be certain generic features of regulatory networks that are responsible for their inherent robustness and stability. In addition, since there are many different cell types in multicellular organisms, despite the fact that each cell contains exactly the same DNA content, the cellular "fate" is determined by which genes are expressed.

This was the insight pursued by Kauffman in his pioneering studies on genetic regulatory networks (Kauffman, 1993). The idea was to generate random Boolean networks with certain properties and then systematically study the effects of these properties on the global dynamical behavior of the networks. For example, random Boolean networks were studied with varying average connectivity, bias, which is the probability that a function outputs the value 1, and different classes of Boolean functions, such as canalizing functions. A Boolean function $f : \{0, 1\}^n \to \{0, 1\}$ is called *canalizing* in its $i$th input if there exist $y$ and $z$ such that for all $(x_1, x_2, \ldots, x_n) \in \{0, 1\}^n$ with $x_i = y$, we have $f(x_1, x_2, \ldots, x_n) = z$. We will return to these questions shortly.

Kauffman's intuition was that the attractors in the Boolean networks should correspond to cellular types. This interpretation is quite reasonable if cell types are characterized by stable, recurrent patterns of gene expression.

It is widely believed that many complex and adaptable systems such as the genome operate in a zone between order and disorder, or on the "edge of chaos." In the *ordered* regime, attractors are quite short (i.e., their length is small), there are only a few of them, and they are stable, meaning that transient perturbations are likely to return the system to the same attractor. This last property—homeostasis— is mainly due to the fact that the presence of only a few short attractors implies large basins of attraction. In addition, the small cycle lengths imply the existence of large *frozen* components, which are sets of genes that do not change value as the network progresses through time. There are only isolated islands of genes that change values, but they cannot "communicate" or transfer information to each other because of the large frozen components. Thus, the network is highly resistant to perturbations (changes in the value of a gene) as well as to wiring mutations (changes in the

Boolean functions). It is unlikely that living systems operate deeply in the ordered regime because evolution demands that there be some sensitivity to perturbations and mutations.

On the other hand, in the *chaotic* regime, the cycle length of attractors grows exponentially as a function of the number of genes, and perturbation of a gene propagates to many other genes in an avalanche-like manner. Unlike in the ordered regime, there are few small islands of frozen genes with a large proportion of genes exhibiting variation. Thus, networks in the chaotic regime are very sensitive to initial conditions and perturbations, implying that organisms cannot be in the chaotic regime either.

Systems operating along the boundary between order and chaos are said to be in the *complex* or *critical* regime. As Kauffman (1995) puts it, "A living system must first strike an internal compromise between malleability and stability. To survive in a variable environment, it must be stable to be sure, but not so stable that it remains forever static." In the critical phase, the unfrozen components (genes that are changing over time) break up into isolated islands separated by frozen components. Thus, there are many genes that are unfrozen, and many that are frozen. As the network goes around the attractor cycle, representing the cell cycle, the genes that are unfrozen are presumably the "cell cycle genes" responsible for cell cycle regulation. For example, recent studies with microarray technology have revealed catalogs of genes whose transcript levels vary periodically within the cell cycle (Spellman et al., 1998; Whitfield et al., 2002).

Another interpretation of the attractors in Boolean networks is that they represent cellular states, such as *proliferation* (cell cycle), *apoptosis* (programmed cell death), and *differentiation* (execution of tissue-specific tasks). This highly appealing view was expounded by Huang (1999, 2001) with substantial biological justification. Such an interpretation can provide new insights into cellular homeostasis and cancer progression, the latter being characterized by a disbalance between these cellular states. For instance, if a (structural) mutation occurs, resulting in a very small probability of the network entering the apoptosis attractor(s), then the cells will be unable to undergo apoptosis and will exhibit uncontrolled growth. Similarly, large basins of attraction for the proliferation attractor would result in hyperproliferation, typical of tumorigenesis. Such an interpretation need not be at odds with the interpretation that attractors represent cellular types. To the contrary, these views are complementary to each other, since for a given cell type, different cellular functional states must exist and be determined by the collective behavior of gene activity. Thus, one cell type can comprise several "neighboring" attractors each of which corresponds to different cellular functional states.

Let us now discuss the various ways the network properties, such as connectivity, bias, and classes of functions, can affect the dynamical behavior. One way to gain insight into such relationships between structure and dynamics is to construct, simulate, and analytically study networks within an ensemble framework. This entails constructing many random networks with a given property (ensemble) and studying the ensuing statistical behavior of the quantity of interest as regards the networks' dynamics or using probabilistic tools to obtain analytical results for typical members of such an ensemble.

## 2.2.2 Network Properties and Dynamics

Let us consider the wiring (input-output relationships between genes) in a Boolean network in terms of the corresponding directed graph. The in-degree of a gene is the number of inputs to the Boolean function governing its behavior. The simplest scenario is when each gene has the same in-degree $K$. This implies that the in-degree distribution (probability mass function) has all of its mass at $K$ and zero elsewhere (a delta function). This is the original definition proposed by Kauffman (1969a). Just as we discussed for graphs in section 2.1, other distributions are also possible, such as the power law distribution, which may be more reflective of the heterogeneous nature of biological systems (Oosawa and Savageau, 2002; Aldana and Cluzel, 2003).

First, let us examine the simplest case, where $k_i = K$ for all $i = 1, \ldots, n$. The $K$ inputs for each Boolean function are chosen randomly from among the $n$ genes. In a random Boolean network, each function is also chosen randomly. An important parameter characterizing each random function $f_i$ is the bias $p$, which is the probability that the function takes on the value 1. A $K$-input random Boolean function with bias $p$ can be generated by flipping a $p$-biased coin $2^K$ times and filling in its truth table. If $p = 0.5$, then the function is said to be *unbiased*. We can also refer to an entire RBN as being $p$-biased or unbiased, referring to the corresponding property of each of its randomly generated functions.

Boolean networks can undergo a phase transition between the ordered and chaotic regimes as $n \to \infty$ (the thermodynamic limit). In the ordered regime, two trajectories through the state space that start in nearby states tend to converge (i.e., differences tend to die out). By contrast, in the chaotic regime, such trajectories tend to diverge (i.e., differences are amplified). Let us make this notion more formal.

Let $\mathbf{x}^{(1)}(t) = (x_1^{(1)}(t), \ldots, x_n^{(1)}(t))$ and $\mathbf{x}^{(2)}(t) = (x_1^{(2)}(t), \ldots, x_n^{(2)}(t))$ be two randomly chosen states of the network at time $t$. The normalized Hamming distance (henceforth, simply Hamming distance), which represents the proportion of bits (genes) that differ between the two states, is defined as

$$\rho(t) = \frac{1}{n} \sum_{i=1}^{n} (x_i^{(1)}(t) \oplus x_i^{(2)}(t)),$$

where $\oplus$ is addition modulo 2 (exclusive OR). Given a realization of an RBN (i.e., a specific Boolean network), let $\mathbf{x}^{(1)}(t+1)$ and $\mathbf{x}^{(2)}(t+1)$ be the successor states of $\mathbf{x}^{(1)}(t)$ and $\mathbf{x}^{(2)}(t)$, respectively. Similarly, let $\rho(t+1)$ be the Hamming distance between these successor states.

The Derrida curve consists of plotting $\rho(t+1)$ versus $\rho(t)$ (Derrida and Pomeau, 1986). If the network is operating in the chaotic regime, then the divergence of small Hamming distances implies that the Derrida curve lies above the main diagonal for small initial Hamming distances. That is, its slope close to the origin is greater than 1. This also implies that small gene perturbations (i.e., nearby states) tend to spread further apart and that networks are sensitive to initial condition—a hallmark of chaotic behavior. On the other hand, networks in the ordered regime exhibit convergence for nearby states, with the Derrida curve lying below the main

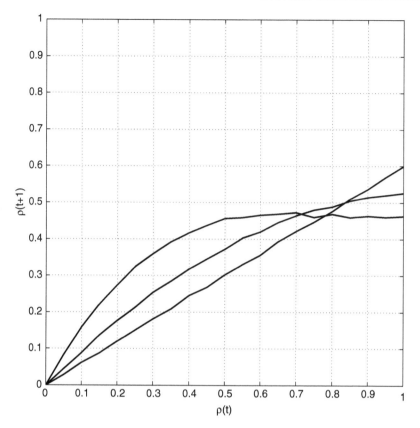

Figure 2.3. An example of three Derrida curves corresponding to ordered, critical, and chaotic dynamics. The slope of the Derrida curve at the origin indicates the regime ($<1$, ordered; $=1$, critical; $>1$, chaotic).

diagonal. The farther the Derrida curve lies above the main diagonal for small values of $\rho(t)$, the more chaotic the network. Figure 2.3 shows an example of three Derrida curves corresponding to ordered, critical, and chaotic dynamics.

Since the slope of the Derrida curve at the origin is a measure of the network's behavior in terms of order/chaos, let us derive this quantity for RBNs in terms of the bias $p$ and the connectivity $K$. But first, we need some definitions. In a Boolean function, some variables have a greater influence on the output of the function than other variables. To formalize this concept, let $f : \{0, 1\}^K \to \{0, 1\}$ be a Boolean function of $K$ variables $x_1, \ldots, x_K$. Let

$$\frac{\partial f(\mathbf{x})}{\partial x_j} = f\left(\mathbf{x}^{(j,0)}\right) \oplus f\left(\mathbf{x}^{(j,1)}\right)$$

be the partial derivative of $f$ with respect to $x_j$, where

$$\mathbf{x}^{(j,k)} = \left(x_1, \ldots, x_{j-1}, k, x_{j+1}, \ldots x_K\right), \qquad k = 0, 1.$$

Clearly, the partial derivative is a Boolean function itself that specifies whether a change in the $j$th input causes a change in the original function $f$.

Now, the activity of variable $x_j$ in function $f$ can be defined as

$$\alpha_j^f = \frac{1}{2^K} \sum_{\mathbf{x} \in \{0,1\}^K} \frac{\partial f(\mathbf{x})}{\partial x_j}.$$

Note that although the vector $\mathbf{x}$ consists of $K$ components (variables), the $j$th variable is fictitious in $\partial f(\mathbf{x})/\partial x_j$. Recall that a variable $x_j$ is fictitious in $f$ if $f(\mathbf{x}^{(j,0)}) = f(\mathbf{x}^{(j,1)})$ for all $\mathbf{x}^{(j,0)}$ and $\mathbf{x}^{(j,1)}$. For a $K$-variable Boolean function $f$, we can form its activity vector $\alpha^f = [\alpha_1^f, \ldots, \alpha_K^f]$. It is easy to see that $0 \leq \alpha_j^f \leq 1$ for any $j = 1, \ldots, K$. In fact, we can consider $\alpha_j^f$ to be the probability that toggling the $j$th input bit changes the function value when the input vectors $\mathbf{x}$ are distributed uniformly over $\{0, 1\}^K$. Since we are in the binary setting, the activity is also the expectation of the partial derivative with respect to the uniform distribution: $\alpha_j^f = E[\partial f(\mathbf{x})/\partial x_j]$. Under an arbitrary (not necessarily uniform) distribution, $\alpha_j^f$ is sometimes referred to as the *influence* of variable $x_j$ on the function $f$ (Kahn et al., 1988).

Another important quantity is the *sensitivity* of a Boolean function $f$, which measures how sensitive the output of the function is to changes in the inputs (this was introduced by Cook et al. (1986) as "critical complexity"). The sensitivity $s^f(\mathbf{x})$ of $f$ on vector $\mathbf{x}$ is defined as the number of Hamming neighbors of $\mathbf{x}$ on which the function value is different than on $\mathbf{x}$ (two vectors are Hamming neighbors if they differ in only one component; that is, the Hamming distance between them is equal to 1). That is,

$$s^f(\mathbf{x}) = |\{i \in \{1, \ldots, K\} : f(\mathbf{x} \oplus e_i) \neq f(\mathbf{x})\}|$$

$$= \sum_{i=1}^{K} \chi\left[f(\mathbf{x} \oplus e_i) \neq f(\mathbf{x})\right],$$

where $e_i$ is the unit vector with 1 in the $i$th position and 0s everywhere else and $\chi[A]$ is an indicator function that is equal to 1 if and only if $A$ is true. The *average sensitivity* $s^f$ is defined by taking the expectation of $s^f(\mathbf{x})$ with respect to the distribution of $\mathbf{x}$. Under the uniform distribution, the average sensitivity is equal to the sum of the activities:

$$s^f = E\left[s^f(\mathbf{x})\right] = \sum_{i=1}^{K} E\left[\chi\left[f(\mathbf{x} \oplus e_i) \neq f(\mathbf{x})\right]\right] \qquad (2.2)$$

$$= \sum_{i=1}^{K} \alpha_i^f.$$

Therefore, $s^f$ is a number between 0 and $K$.

Intuition tells us that highly biased functions (i.e., those with $p$ far away from 0.5) should have low average sensitivity since such a function with $K$ inputs would be unlikely to change much between neighboring vectors on the $K$-dimensional hypercube. Indeed, consider a random Boolean function with bias $p$. The truth table of such a function is a $2^K$-length vector of independent and identically distributed Bernoulli ($p$) random variables. Therefore, the probability that two Hamming neighbors are different is equal to $2p(1-p)$ since one can be 1 (with probability $p$) and the other 0 (with probability $1-p$), and vice versa. Since this is the same for all Hamming neighbors, all expected activities should be equal. That is, for each $i = 1, \ldots, K$,

$$E[\alpha_i^f] = 2p(1-p),$$

where the expectation is taken with respect to the distribution of the truth table for the function $f$. Thus, using equation (2.2), the expectation of the average sensitivity is

$$E\left[s^f\right] = \sum_{i=1}^{K} E[\alpha_i^f] = \sum_{i=1}^{K} 2p(1-p) \tag{2.3}$$

$$= K2p(1-p).$$

This confirms our intuition about the relationship between bias and (expected) average sensitivity. Note that an unbiased function ($p = 0.5$) has expected average sensitivity equal to $K/2$.

Let us now consider a random Boolean network with $n$ genes and with connectivity $K$ and bias $p$. Two states that are Hamming neighbors should, on average, be succeeded at the next time moment by two states that are separated by a Hamming distance equal to $K2p(1-p)$, by the definition of average sensitivity and equation (2.3). It is straightforward to see that as $n \to \infty$, the average sensitivity determines the slope of the Derrida curve at its origin. Consequently, if $K2p(1-p) > 1$, the system is chaotic, and if $K2p(1-p) < 1$, the system is ordered. Solving this equation for $K$ at the phase transition, we obtain the critical connectivity $K_c$ in terms of the bias $p$:

$$K_c = [2p(1-p)]^{-1}. \tag{2.4}$$

Similarly, we can solve for the critical bias $p_c$ in terms of $K$. The critical curve given in equation (2.4) is shown in figure 2.4. All values of $p$ and $K$ above the critical curve yield chaotic network behavior, while values below the curve result in ordered behavior.

Equation (2.4) was derived by Derrida and Pomeau (1986) using annealed approximation, where at each time step a new set of Boolean functions is selected randomly. Other approaches based on damage spreading (Luque and Solé, 1997), Lyapunov exponents (Luque and Solé, 2000), a stable core (Flyvbjerg, 1988), and others have been used. The reader can consult (Aldana et al., 2002) for a good review.

From figure 2.4, it is clear that the parameter space is dominated by the chaotic regime. Thus, for most values of $p$, the critical connectivity $K_c$ is quite small. Also, for higher values of $K$, we are required to "tune" $p$ far away from 0.5 if we

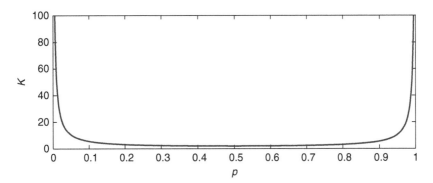

Figure 2.4. The critical curve relating the bias $p$ to the connectivity $K$ in a random Boolean network. All values of $p$ and $K$ above the curve result in chaotic network behavior, whereas all values below the curve yield ordered behavior.

demand that the network remain in the ordered regime. Aldana (2003) and Aldana and Cluzel (2003) have studied the dynamical behavior of Boolean networks, in particular as regards the phase transition from order to chaos, under the power law distribution of the inputs. This distribution is given by

$$P(k) = \left[ \zeta(\gamma) k^\gamma \right]^{-1},$$

where $\gamma > 1$ is the scale-free exponent and $\zeta(\gamma) = \sum_{k=1}^{\infty} k^{-\gamma}$ is the Riemann zeta function that normalizes the distribution. It is interesting that the phase transition in this case is also given by the same equation, $K2p(1-p) = 1$, but here $K = \sum_{k=1}^{\infty} kP(k)$ is the mean of the distribution. It can be shown (Aldana, 2003) that for $\gamma > 2$, the mean is equal to

$$K_\gamma = \frac{\zeta(\gamma - 1)}{\zeta(\gamma)}.$$

Consequently, the phase transition is determined by the equation

$$2p(1-p) \frac{\zeta(\gamma_c - 1)}{\zeta(\gamma_c)} = 1, \tag{2.5}$$

where $\gamma_c$ denotes the critical value of the scale-free exponent. The solution to equation (2.5) is shown in figure 2.5. All parameter values $(p, \gamma)$ to the left of the curve result in chaotic behavior, while all values to the right of the curve yield ordered behavior. The maximum value of $\gamma_c$ is approximately 2.47875 (at $p = 0.5$), which implies that for all values of $\gamma$ higher than this value, ordered behavior is guaranteed regardless of the bias $p$. It is also quite remarkable that most real scale-free networks that have been analyzed have $\gamma \in [2, 2.5]$ (Aldana, 2003).

So far we have focused on the bias and the connectivity of random Boolean networks in studying their dynamical behavior. We have assumed that all Boolean functions are generated randomly from the set of all possible Boolean functions.

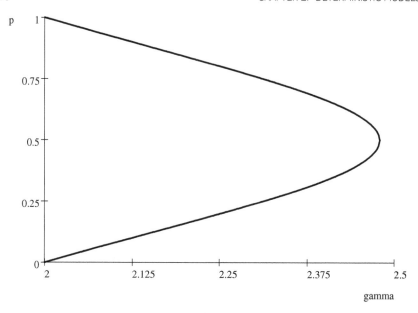

Figure 2.5. The critical curve for Boolean networks with a power-law input distribution. The curve represents a solution of equation (2.5), where the bias $p$ is on the $y$-axis and the scale-free exponent $\gamma$ is on the $x$-axis. The chaotic regime corresponds to values of the parameters to the left of the curve, and the ordered regime results from all parameter values to the right of the curve.

Another important aspect is the class of Boolean functions from which random functions are selected. As we will now discuss, properly constraining the set of Boolean functions in a network can not only significantly affect its dynamical behavior but can also be justifiable on the basis of evolutionary and inferential considerations, where by the latter we mean that it confers statistical advantages in terms of learning the model structure from experimental data.

First, we will consider the class of Boolean functions called *canalizing* functions. Recall that a canalizing function is a type of Boolean function in which at least one of the input variables, called a canalizing variable, is able to determine the function output regardless of the values of the other variables. For example, the function $f(x_1, x_2, x_3) = x_1 \vee (x_2 \wedge x_3)$, where the symbols $\vee$ and $\wedge$ denote the Boolean disjunction and conjunction, respectively, is a canalizing function since setting $x_1$ to 1 guarantees that the function value is 1 regardless of the value of $x_2$ or $x_3$. On the other hand, the exclusive OR function $f(x_1, x_2) = x_1 \oplus x_2$ is not a canalizing function since the values of both variables always need to be known in order to determine the function output.

It has been known for quite some time that canalizing functions play a role in preventing chaotic behavior (Kauffman, 1969a, 1993; Stauffer, 1987; Lynch, 1995). By increasing the percentage of canalizing functions in a Boolean network, one can move closer toward the ordered regime and, depending on the connectivity and the distribution of the number of canalizing variables, cross the phase transition boundary (Kauffman, 2000). In fact, there is strong evidence that canalizing functions are abundantly utilized in higher-vertebrate gene regulatory systems (Kauffman, 1993). A recent large-scale study of the literature on transcriptional regulation in eukaryotes demonstrated a high bias toward canalizing rules (Harris et al., 2002).

One of the reasons why an abundance of canalizing functions in a Boolean network tends to prevent chaotic behavior is that canalizing functions "prefer" to be biased (Stauffer, 1987). Consider the plots in figure 2.6a and 2.6c. These plots show histograms of the number of canalizing functions of $K = 4$ and $K = 5$ variables, respectively, versus the number of 1s in their truth tables—also called the *Hamming weight*. So, for example, for $K = 4$, there are only 8 canalizing functions (out of 3514) that have exactly eight 1s in their truth table. The plots in Figure 2.6b and 2.6d show the probability that a randomly generated Boolean function with bias $p$ will be a canalizing function (for $K = 4$ and $K = 5$, respectively). Again, it can be seen that for $p = 0.5$ and $K = 4$, this probability is only about 0.0536. This phenomenon lends support to the observed role of canalizing functions in the formation of ordered behavior.

Let us look more closely at the reasons why canalizing functions promote ordered behavior. We have already discussed the relationship between average sensitivity and the slope of the Derrida curve at the origin. Therefore, we would expect canalizing functions to have lower average sensitivity than noncanalizing functions. Moreover, we would expect that for canalizing functions, the importance of canalizing variables is greater than that of noncanalizing variables, as measured by their activities. The following discussion, adapted from (Shmulevich and Kauffman, 2004), illustrates this.

Function $f : \{0, 1\}^K \to \{0, 1\}$ is said to be canalizing if there exists an $i \in \{1, \ldots, K\}$ and $u, v \in \{0, 1\}$ such that for all $x_1, \ldots, x_K \in \{0, 1\}$, if $x_i = u$, then $f(x_1, \ldots, x_K) = v$. The input variable $x_i$ is called the canalizing variable with canalizing value $u$ and canalized value $v$. A canalizing function $f$ can be written as either

$$f(x_1, \ldots, x_K) = x_i^q \vee g(x_1, \ldots, x_{i-1}, x_{i+1}, \ldots, x_K)$$

or

$$f(x_1, \ldots, x_K) = x_i^q \wedge g(x_1, \ldots, x_{i-1}, x_{i+1}, \ldots, x_K),$$

where $q \in \{0, 1\}$. Here, $x_i^1 = x_i$ and $x_i^0 = x_i'$, where $x_i'$ is the complement or negation of $x_i$. Also, recall that $\wedge$ takes precedence over $\vee$, so that $a \vee b \wedge c$ is the same as $a \vee (b \wedge c)$.

Let $f(x_1, \ldots, x_K)$ be a random canalizing function of the form

$$f(x_1, \ldots, x_K) = x_1 \vee g(x_2, \ldots, x_K),$$

where $g$ is chosen randomly from the set of all $2^{2^{K-1}}$ Boolean functions. Without loss of generality, we suppose that the first variable, $x_1$, is a canalizing variable. Furthermore, the discussion for other types of canalizing functions (e.g., $f(x_1, \ldots, x_K) = x_1 \wedge g(x_2, \ldots, x_K)$) would be nearly identical. Our first aim is

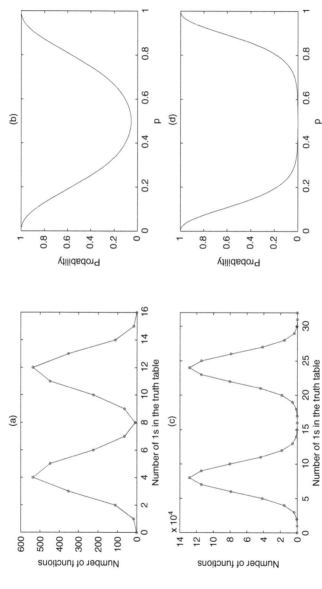

Figure 2.6. Subplots (a) and (c), with $K = 4$ and $K = 5$, respectively, show histograms of the number of canalizing functions versus the number of 1s in their truth tables (Hamming weight). Subplots (b) and (d), with $K = 4$ and $K = 5$, respectively, show the probability that a randomly chosen Boolean function with bias $p$ is a canalizing function.

to characterize the activities of each of the variables. The activities are also random variables themselves by virtue of $f$ being random. It is clear that the activity of variables $x_2, \ldots, x_K$ should be identical in the probabilistic sense if $g(x_2, \ldots, x_K)$ is a random unbiased function. Consequently, it will suffice to examine the activity of variable $x_2$, with the other variables behaving identically.

Let us first compute $\alpha_1^f$—the activity of $x_1$ in $f$. First, we have

$$\frac{\partial f}{\partial x_1} = (0 \vee g(x_2, \ldots, x_K)) \oplus (1 \vee g(x_2, \ldots, x_K))$$

$$= g(x_2, \ldots, x_K) \oplus 1$$

$$= g'(x_2, \ldots, x_K).$$

Now, since $g$ is a random unbiased function (i.e., $p = 1/2$), the expected activity of the canalizing variable $x_1$ is equal to

$$E\left[\alpha_1^f\right] = E[2^{-(K-1)} \cdot \sum_{\mathbf{x} \in \{0,1\}^{K-1}} g'(x_2, \ldots, x_K)]$$

$$= 2^{-(K-1)} \cdot \sum_{\mathbf{x} \in \{0,1\}^{K-1}} E\left[g'(x_2, \ldots, x_K)\right]$$

$$= 2^{-(K-1)} \cdot \sum_{\mathbf{x} \in \{0,1\}^{K-1}} \frac{1}{2}$$

$$= \frac{1}{2}.$$

Now let us consider the expected activity of variable $x_2$. We have

$$\frac{\partial f}{\partial x_2} = \left(x_1 \vee g\left(\mathbf{x}^{(2,0)}\right)\right) \oplus \left(x_1 \vee g\left(\mathbf{x}^{(2,1)}\right)\right)$$

$$= \left(x_1 \vee g\left(\mathbf{x}^{(2,0)}\right)\right) \wedge \left(x_1 \vee g\left(\mathbf{x}^{(2,1)}\right)\right)'$$

$$\vee \left(x_1 \vee g\left(\mathbf{x}^{(2,0)}\right)\right)' \wedge \left(x_1 \vee g\left(\mathbf{x}^{(2,1)}\right)\right)$$

$$= \left(x_1 \vee g\left(\mathbf{x}^{(2,0)}\right)\right) \wedge \left(x_1' \wedge g'\left(\mathbf{x}^{(2,1)}\right)\right)$$

$$\vee \left(x_1' \wedge g'\left(\mathbf{x}^{(2,0)}\right)\right) \wedge \left(x_1 \vee g\left(\mathbf{x}^{(2,1)}\right)\right)$$

$$= x_1' \wedge g\left(\mathbf{x}^{(2,0)}\right) \wedge g'\left(\mathbf{x}^{(2,1)}\right) \vee x_1' \wedge g'\left(\mathbf{x}^{(2,0)}\right) \wedge g\left(\mathbf{x}^{(2,1)}\right)$$

$$= x_1' \wedge \left(g\left(\mathbf{x}^{(2,0)}\right) \oplus g\left(\mathbf{x}^{(2,1)}\right)\right)$$

$$= x_1' \wedge \frac{\partial g}{\partial x_2},$$

where in the second equality we used the fact that $a \oplus b = a \wedge b' \vee a' \wedge b$, in the third equality we used de Morgan's identity $(a \vee b)' = a' \wedge b'$, in the fifth equality we again used the definition of $\oplus$, and in the last equality we used the definition of partial derivative. The expected activity of variable $x_2$ is now equal to

$$E\left[\alpha_2^f\right] = E\left[2^{-(K-1)} \cdot \sum_{\mathbf{x} \in \{0,1\}^{K-1}} x_1' \wedge \frac{\partial g}{\partial x_2}\right].$$

Note that $\partial g(x_2, \ldots, x_K)/\partial x_2$ is a Boolean function of $(K-2)$ variables and that the above summation is taken over all $\mathbf{x} = (x_1, x_3, \ldots, x_K)$. Let us break up this summation into parts corresponding to $x_1 = 0$ and $x_1 = 1$. We have

$$E\left[\alpha_2^f\right] = 2^{-(K-1)} \cdot \left[\sum_{\mathbf{x}^{(1,0)} \in \{0,1\}^{K-1}} E\left[1 \wedge \frac{\partial g}{\partial x_2}\right] + \sum_{\mathbf{x}^{(1,1)} \in \{0,1\}^{K-1}} E\left[0 \wedge \frac{\partial g}{\partial x_2}\right]\right]$$

$$= 2^{-(K-1)} \cdot \sum_{\mathbf{x}^{(1,0)} \in \{0,1\}^{K-1}} E\left[\frac{\partial g}{\partial x_2}\right].$$

Since $g$ is a random unbiased function, so is $\partial g/\partial x_2$. This essentially means that the probability that a random function $g$ differs on $\mathbf{x}^{(j,0)}$ and $\mathbf{x}^{(j,1)}$ is equal to $1/2$. Thus,

$$E\left[\alpha_2^f\right] = 2^{-(K-1)} \cdot \sum_{\mathbf{x}^{(1,0)} \in \{0,1\}^{K-1}} \frac{1}{2},$$

and since there are exactly $2^{K-2}$ different vectors $\mathbf{x}^{(1,0)} = (0, x_3, \ldots, x_K)$,

$$E\left[\alpha_2^f\right] = 2^{-(K-1)} \cdot \frac{1}{2} \cdot 2^{K-2} = \frac{1}{4}.$$

Thus, the expected activity of all noncanalizing variables is equal to $1/4$. The expected activity vector is then equal to $E[\alpha^f] = (\frac{1}{2}, \frac{1}{4}, \ldots, \frac{1}{4})$, and the expected sensitivity, by equation (2.2), is equal to $E[s(f)] = \frac{1}{2} + \frac{1}{4} \cdot (K-1) = (K+1)/4$. The situation for two or more canalizing variables is analogous. Note that a random unbiased function is expected to have each activity equal to $1/2$, which means the expected average sensitivity is equal to $K/2$. Therefore, for $K > 1$ a canalizing function with just one canalizing variable is expected to have a lower average sensitivity, implying that the network constructed from such functions should be more ordered than an unbiased random network.

Recently, another class of Boolean functions has been found to be highly plausible from the points of view of evolution, noise resilience, network dynamical behavior, and existing knowledge about regulatory control rules gathered from the published literature (Shmulevich et al., 2003). We will now discuss some of the properties of this class of functions, especially as they relate to the behavior of Boolean networks.

In 1921 the American mathematician Emil Post characterized all classes of Boolean functions that are closed under composition (Post, 1921, 1941). Several well-known classes of Boolean functions, such as monotone, linear, and self-dual functions, fall into this category and are referred to as *Post classes*. For example, it is relatively straightforward to show that a composition of an arbitrary number of monotone Boolean functions results in a monotone Boolean function. Of course, the class of all Boolean functions is trivially a Post class as well. However, the class that we will consider is of a somewhat more mysterious nature and has a connection with the class of canalizing functions (which is not a Post class).

**DEFINITION 2.2** *A Boolean function f belongs to class $A^\infty$ if all vectors on which the function takes on the value 1 have a common component equal to 1. Similarly, we define the Post class $a^\infty$ by replacing 1s by 0s in the previous sentence.*

For example, the function $f(x_1, x_2, x_3) = x_1 \vee (x_2 \wedge x_3)$ belongs to $a^\infty$ since all the vectors on which the function is equal to 0, namely, (000), (001), and (010), have a common component equal to 0 (the first one). In this example, the first variable is also canalizing since if $x_1 = 1$, then $f(x_1, x_2, x_3) = 1$. For simplicity, we will also say "$f$ is $a^\infty$," referring to the corresponding property of the function. If we denote the set of canalizing functions by $\mathcal{C}$, then it is clear from the above definition that $A^\infty \cup a^\infty \subseteq \mathcal{C}$. We can also generalize this definition as follows.

**DEFINITION 2.3** *A Boolean function f belongs to class $A^\mu$, $\mu \geq 2$, if any $\mu$ vectors on which the function takes on the value 1 have a common component equal to 1 (some of these $\mu$ vectors may be repeated). The class $a^\mu$ is defined by replacing 1s by 0s in the previous sentence.*

As an example, the function

$$f(x_1, x_2, x_3) = (x_1 \wedge x_2) \vee (x_2 \wedge x_3) \vee (x_1 \wedge x_3) \tag{2.6}$$

is $A^2$ since any two vectors on which the function is equal to 1 have a common unity component. The function in equation (2.6) is the well-known majority or three-point median operation. It is easy to see, by definition, that if $f$ is $A^\mu$, then it is also $A^\nu$ for any $\nu$ satisfying $2 \leq \nu \leq \mu$. A similar result holds for $a^\mu$. Thus, these classes are nested, with $A^2$ (respectively, $a^2$) being the largest and $A^\infty$ (respectively, $a^\infty$) being the smallest. Another interesting result is that if the function $f : \{0, 1\}^n \rightarrow \{0, 1\}$ is $A^\mu$, then for $\mu \geq n$ it is also $A^\infty$ (Yablonsky, 1966). It is easy to show that if $f$ is $A^\mu$, then any collection of $m \leq \mu$ terms in the disjunctive normal form (DNF) must contain a variable that belongs to each one of these terms. Thus, there is once again an interesting relationship to canalizing functions in the sense that the functions defined by these terms are themselves canalizing functions. For example, if we take any two terms from the function in equation (2.6), such as $(x_1 \wedge x_2) \vee (x_2 \wedge x_3)$, then the Boolean function corresponding to these terms is canalizing, with canalizing variable $x_2$.

Since the classes $a^\infty$, $A^\infty$, $a^\mu$, $A^\mu$ are all Post classes, they are closed under the operation of composition by definition. This implies that a Boolean network constructed from one of these classes has the property that any $t$-step transition function (outputting the gene value $t$ time steps into the future) also belongs to this class. A remarkable implication of this closure property relates to the connection between "model time" and actual physical time. In the context of genetic regulatory networks, Boolean networks, which are discrete-time systems, are coarsely quantized models of the real continuous-time systems that they represent. For instance, to fit the models to real gene expression data, which we will discuss later, we would like to sample the cell system at consecutive points in time (either uniformly or nonuniformly) and use the time-series data to infer the Boolean functions in the network. Therefore, there is an implicit albeit unspecified mapping between the discrete model time and the actual physical time of the cells. If genes are regulated by rules belonging to the considered Post classes, then regardless of how many discrete time steps correspond to the actual physical time interval (we emphasize that this mapping need not even be time-invariant), the overall rule specifying the gene value at the end of the physical time interval still belongs to the same Post class. Thus, the plausibility of the Post classes representing gene regulatory mechanisms does not depend on the relationship between model time and physical time.

It can be shown that functions from $A^\mu$ or $a^\mu$ also prefer to be biased. Consider figure 2.7, which is similar to figure 2.6, but instead of canalizing functions we use functions from the class $A^2 \cup a^2$ for the same number of variables. We notice a striking similarity between the two figures and can readily conclude that functions from $A^2$ or $a^2$ are also more likely to be biased, as suggested by the two-peaked histograms in figure 2.7a and 2.7c. Similarly, figure 2.7b and 2.7d show a marked tendency for a $p$-biased random Boolean function to be either $A^2$ or $a^2$ with increasing $p$. A similar relationship exists for $A^\mu \cup a^\mu$, $\mu > 2$. Thus, it appears that the functions in $A^2$ or $a^2$ behave very similarly to canalizing functions in that they both prefer to be biased. This is promising from the point of view of network behavior in terms of order and chaos. Had the situation been the opposite, namely, if the functions in $A^2$ or $a^2$ were unlikely to be biased, then networks constructed from them would have been more likely to behave chaotically, especially for $K > 2$.

Let us look at the dynamical behavior of networks constructed from these Post classes using Derrida curves. In figure 2.8, the dash-dot plot corresponds to random (and unbiased) Boolean networks for $K = 4$. As can be seen, the Derrida plot is significantly above the main diagonal for small $\rho(t)$, indicating that $K = 4$ random Boolean networks behave chaotically. In the same plot, the solid line corresponds to networks constructed from canalizing functions and the dashed line corresponds to networks constructed from $A^2$ and $a^2$ functions, also for $K = 4$. In both cases, these networks behave much less chaotically than their random unbiased counterparts.

There is evidence that the classes $A^\mu$ and $a^\mu$ ($\mu = 2, \ldots, \infty$) are also equipped with a natural mechanism to ensure robustness against noise. The problem of reliable computation with unreliable components in the Boolean circuit setting dates back to the work of von Neumann (1956). The relevant characteristic of a reliable circuit is redundancy. Intuitively, higher redundancy ensures reliability. The aim is to construct circuits that, with high probability, can compute the correct function even when their components are noisy.

In the 1960s it was discovered that the Post classes $A^\mu$ and $a^\mu$ play a significant role in the synthesis and reliability of control systems and self-correcting circuits

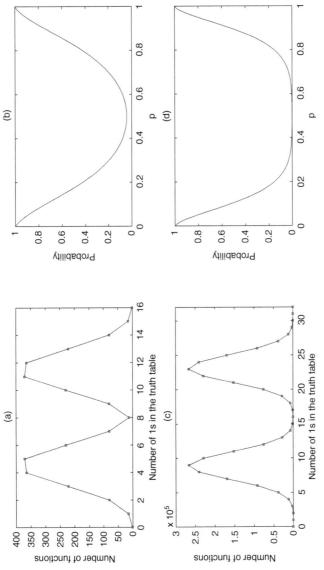

Figure 2.7. Subplots (a) and (c), with $K = 4$ and $K = 5$, respectively, show histograms of the number of $A^2 \cup a^2$ functions versus the number of 1s in their truth tables. Subplots (b) and (d), with $K = 4$ and $K = 5$, respectively, show the probability that a randomly chosen Boolean function with bias $p$ is an $A^2 \cup a^2$ function.

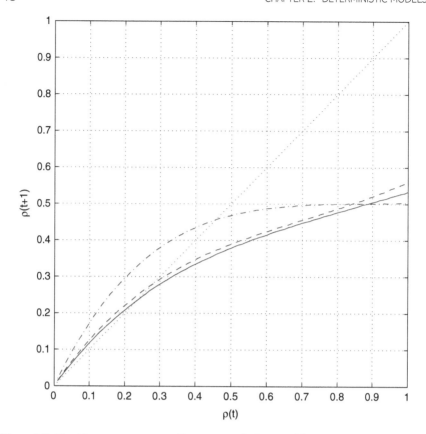

Figure 2.8. Derrida curves corresponding to several classes with $K = 4$: canalizing functions (solid), the class $A^2 \cup a^2$ (dashed), random Boolean functions (dash-dot). (Shmulevich et al., 2003.)

(Muchnik and Gindikin, 1962; Kirienko, 1964). These results rest on the concept of *basis functions*. Any closed class of Boolean functions contains a set of basis functions such that any function in this class can be constructed by iterative composition of these basis functions. For example, since the class of all Boolean functions is a closed class (Post class), one possible set of basis functions is $\{x \wedge y, x'\}$ (i.e., conjunction and negation). Thus, any Boolean function can be constructed by repeated composition of these two functions. The basis set for the class $A^2$ is (Yablonsky et al., 1966)

$$\{x \wedge y', (x \wedge y) \vee (x \wedge z) \vee (y \wedge z)\}.$$

The second function is again the majority function in equation (2.6). These functions are essentially building blocks from which any function in $A^2$ can be constructed. By cascading such building blocks a sufficient number of times in a treelike fashion, it can be shown (Kirienko, 1964) that any Boolean function can be realized in a reliable manner, meaning that the correct output is guaranteed even when some of the components are faulty. Thus, Post classes $A^\mu$ and $a^\mu$ are strongly implicated in fault-tolerant Boolean circuit synthesis. One might speculate that evolution employed a similar mechanism to give rise to highly resilient and robust genetic networks.

Finally, it was shown by Shmulevich et al. (2003) that these Post classes, in particular $A^2$ and its dual class $a^2$, are significantly larger than the class of canalizing functions for biologically realistic values of $K$. Thus, a class that is more abundant in number than canalizing functions, while exhibiting high fitness in terms of robustness and globally ordered network behavior, may be more plausible based on evolutionary considerations.

### 2.2.3 Network Inference

So far, we have discussed ensemble properties of Boolean networks. To the extent that this model represents certain characteristics of real genetic networks, the ensemble approach can be useful for gaining insight into their dynamical behavior and organization. The ensemble approach can also be used in conjunction with experimental data by measuring a global property from data and matching it to the corresponding ensemble property in the model network. For example, Serra et al. (2004) studied the distribution of avalanches, which measures the size of the perturbation generated by knocking out a single gene, in random Boolean network models. It was found that these distributions are close to those observed in actual experiments performed with *S. cerevisiae* using DNA microarrays. Shmulevich et al. (2005) used ensembles of Boolean networks operating in the ordered, critical, and chaotic regimes in conjunction with information-theoretic measures of complexity to gain insight into the dynamical behavior of real genetic circuits in HeLa cells.

In order to make progress in understanding the genetic regulation in specific organisms and develop tools for rational therapeutic intervention in diseases such as cancer, it is necessary to be able to identify the networks from experimental measurement data. In the context of Boolean networks, this entails inferring the Boolean function for each gene along with its inputs (i.e., wiring). This task generally requires the application of techniques and methods from the fields of statistical inference and computational learning theory. A number of authors have addressed the problem of Boolean network inference (Liang et al., 1998; Akutsu et al., 1998, 1999; Ideker et al., 2000; Maki et al., 2001; Shmulevich et al., 2002; Lähdesmäki et al ., 2003).

*Consistent and Best-Fit Extensions*

Perhaps the simplest approach is the well-known *consistency problem*, which entails a search for a rule from examples (Pitt and Valiant, 1988; Boros et al., 1998; Shmulevich et al., 2001). The problem is formulated as follows. Given some sets $T$ and $F$ of "true" and "false" vectors, respectively, the aim is to discover a Boolean function $f$ that takes on the value 1 for all vectors in $T$ and the value 0 for all vectors in $F$. It is also commonly assumed that the target function $f$ is chosen from some class of possible target functions. In the context of Boolean networks, such a class could be the class of canalizing functions, the $A^{\mu}$ or $a^{\mu}$ functions discussed above, or simply functions with a fixed number $K$ of essential (nonfictitious) variables.

More formally, let

$$T(f) = \left\{ v \in \{0, 1\}^n : f(v) = 1 \right\}$$

be the *on-set* of function $f$ and let

$$F(f) = \left\{ v \in \{0, 1\}^n : f(v) = 0 \right\}$$

be the *off-set* of $f$. The sets $T, F \subseteq \{0, 1\}^n$, $T \cap F = \emptyset$ define a *partially defined* Boolean function $g_{T,F}$ as

$$g_{T,F}(v) = \begin{cases} 1 & v \in T, \\ 0 & v \in F, \\ * & \text{otherwise.} \end{cases}$$

A function $f$ is called an *extension* of $g_{T,F}$ if $T \subseteq T(f)$ and $F \subseteq F(f)$. The consistency problem (also called the extension problem) can be posed as: Given a class $C$ of functions and two sets $T$ and $F$, is there an extension $f \in C$ of $g_{T,F}$?

It is easy to see that if $C$ is the class of all Boolean functions, an extension $f \in C$ exists if and only if $T \cap F = \emptyset$. This condition can be checked in time $O(|T| \cdot |F| \cdot \text{poly}(n))$ by comparing all examples pairwise, where $\text{poly}(n)$ is the time needed to examine one pair of examples (Boros et al., 1998). Sorting algorithms can be used to provide a more efficient solution to the consistency problem (Lähdesmäki et al., 2003). Suppose that there are $|T| + |F| = m$ examples. By encoding all vectors $x \in T \cup F$ by positive integers, sorting them, and comparing consecutive elements in the sorted list to determine whether or not one belongs to $T$ and the other to $F$, the problem can be solved in time $O(m \cdot \log m \cdot \text{poly}(n))$.

In reality, a consistent extension may not exist either because of errors or, more likely, because of a number of underlying latent factors. This is no doubt the case for gene expression profiles as measured from microarrays. In this case, we may have to give up our goal of establishing a consistent extension and settle for a Boolean formula that minimizes the number of misclassifications. This problem is known as the *best-fit extension problem* (Boros et al., 1998) and is formulated as follows. Suppose we are given positive weights $w(x)$ for all vectors $x \in T \cup F$ and define $w(S) = \sum_{x \in S} w(x)$ for a subset $S \subseteq T \cup F$. Then, the *error size* of function $f$ is defined as

$$\varepsilon(f) = w(T \cap F(f)) + w(F \cap T(f)). \tag{2.7}$$

If $w(x) = 1$ for all $x \in T \cup F$, then the error size is just the number of misclassifications. The goal is then to output subsets $T^*$ and $F^*$ such that $T^* \cap F^* = \emptyset$ and $T^* \cup F^* = T \cup F$ for which the partially defined Boolean function $g_{T^*, F^*}$ has an extension in some class of functions $C$ and so that $w\,(T^* \cap F) + w\,(F^* \cap T)$ is a minimum. Consequently, any extension $f \in C$ of $g_{T^*, F^*}$ has a minimum error size.

It is clear that the best-fit extension problem is computationally more difficult than the consistency problem since the latter is a special case of the former, that is, when $\varepsilon\,(f) = 0$. Nonetheless, it has been shown that for many function classes, including the class of all Boolean functions, the best-fit extension problem for Boolean networks is polynomial-time-solvable (Shmulevich et al., 2002). Further, Lähdesmäki et al. (2003) have introduced a method that can be used to significantly accelerate the search for consistent gene regulatory networks (not just decide whether one exists) under the Boolean network model. The same method, with minor changes, also applies to learning gene regulatory networks under the best-fit extension problem.

### Coefficient of Determination

Another approach that has been used to infer rules (functions) in genetic regulatory networks is based on the *coefficient of determination* (CoD). The CoD, long used in the context of linear regression, was introduced in the context of optimal non-linear filter design (Dougherty et al., 2000) but since then has been used to infer multivariate relationships between genes (Kim et al., 2000a, b). Such relationships, referred to as *predictors*, are the basic building blocks of a rule-based network. For the case of Boolean networks, a predictor for each gene is just a Boolean function. The CoD measures the degree to which the expression levels of an observed gene set can be used to improve prediction of the expression of a target gene relative to the best possible prediction in the absence of observations. The method allows the incorporation of knowledge of other conditions relevant to the prediction, such as the application of particular stimuli or the presence of inactivating gene mutations, as predictive elements affecting the expression level of a given gene. Using the CoD, one can find sets of genes related multivariately to a given target gene.

Let us briefly discuss the CoD in the context of Boolean networks. Let $x_i$ be a *target* gene that we wish to predict by observing some other genes $x_{i_1}, x_{i_2}, \ldots, x_{i_k}$. Also, suppose $f(x_{i_1}, x_{i_2}, \ldots, x_{i_k})$ is an optimal predictor of $x_i$ relative to some error measure $\varepsilon$. For example, in the case of mean-square-error (MSE) estimation, it is well known that the optimal predictor is the conditional expectation of $x_i$ given $x_{i_1}, x_{i_2}, \ldots, x_{i_k}$ (Scharf, 1991). Let $\varepsilon_{\text{opt}}$ be the optimal error achieved by $f$. Then, the CoD for $x_i$ relative to $x_{i_1}, x_{i_2}, \ldots, x_{i_k}$ is defined as

$$\theta = \frac{\varepsilon_i - \varepsilon_{\text{opt}}}{\varepsilon_i}, \tag{2.8}$$

where $\varepsilon_i$ is the error of the best (constant) estimate of $x_i$ in the absence of any conditional variables. It is easily seen that the CoD must be between 0 and 1 and measures the relative decrease in error by estimating $x_i$ via $f$ rather than by just the best constant estimate.

In practice, the CoD must be estimated from training data with designed approximations being used in place of $f$. The sets of (predictive) genes that yield the highest CoD compared to all other sets of genes are the ones used to construct the optimal predictor of the target gene. This procedure is applied to all target genes, thus estimating all the functions in a Boolean network. The method is computationally intensive, and massively parallel architectures have been employed to handle large gene sets (Suh et al., 2002).

*Constrained Prediction*

Given limited amounts of training data, it is prudent to constrain the complexity of the predictor by limiting the number of possible predictive genes that can be used. This corresponds to limiting the connectivity $K$ of the Boolean network. Constraining the predictor to some subclass (e.g., canalizing functions or Post classes) is another viable approach. Indeed, inference typically involves designing an estimator from some predefined class of estimators that minimizes the error of estimation among all estimators in the class. An important role in predictor design is played by these predefined classes or constraints. Although it seems that imposing such constraints can only result in degradation of the performance (larger error) relative to the optimal predictor with no imposed constraints, constraining may have certain advantages. These include prior knowledge of the likely class of functions, such as canalizing functions, tractability of the predictor design, and precision of the estimation procedure by which the optimal predictor is estimated from observations. For example, in digital filter design—a surprisingly related area—it is often the case that a certain class of filters provides a very good suboptimal filter while lessening the data requirements for its estimation.

To quantify the advantage (or disadvantage) of predictor constraint, suppose that data from a sample of size $n$ are used to design an estimate $\psi_n$ of the optimal predictor $\psi_{\mathrm{opt}}$. The error $\varepsilon_n$ of $\psi_n$ cannot be less than the error $\varepsilon_{\mathrm{opt}}$ of the optimal predictor. The corresponding design cost is $\Delta_n = \varepsilon_n - \varepsilon_{\mathrm{opt}}$, and the error of the designed predictor is decomposed as $\varepsilon_n = \varepsilon_{\mathrm{opt}} + \Delta_n$. Hence, the expected error of the designed predictor is

$$E[\varepsilon_n] = \varepsilon_{\mathrm{opt}} + E[\Delta_n].$$

For the predictors being considered, $E[\Delta_n] \to 0$ as $n \to \infty$. The essential problem is that if many input variables are required for estimation of the optimal predictor function (large $K$), it is often impossible to obtain large enough samples to sufficiently reduce $E[\Delta_n]$. Thus, optimization can be constrained to some subclass $C$ of predictors such as the class of canalizing functions. If $\psi_C$ is an optimal predictor in $C$ with error $\varepsilon_C$, then typically $\varepsilon_C \geq \varepsilon_{\mathrm{opt}}$ unless $C$ happens to include the optimal predictor $\psi_{\mathrm{opt}}$. The *cost of constraint* is given by $\Delta_C = \varepsilon_C - \varepsilon_{\mathrm{opt}}$.

The error $\varepsilon_{n,C}$ of the designed constrained predictor, $\psi_{n,C} \in C$, satisfies $\varepsilon_{n,C} \geq \varepsilon_C \geq \varepsilon_{\mathrm{opt}}$ with the associated design error $\Delta_{n,C} = \varepsilon_{n,C} - \varepsilon_C$. Hence, $\varepsilon_{n,C} = \varepsilon_{\mathrm{opt}} + \Delta_C + \Delta_{n,C}$ and

$$E[\varepsilon_{n,C}] = \varepsilon_{\mathrm{opt}} + \Delta_C + E[\Delta_{n,C}].$$

Constraint is statistically beneficial if and only if $E[\varepsilon_{n,C}] \le E[\varepsilon_n]$, which is true if and only if

$$\Delta_C \le E[\Delta_n] - E[\Delta_{n,C}]. \tag{2.9}$$

In other words, if the cost of constraint is less than the decrease in the expected design cost, the expected error of the designed constrained predictor $\psi_{n,C}$ is less than that of the designed unconstrained predictor $\psi_n$. A fundamental problem is to find constraints for which equation (2.9) is satisfied.

Another approach that has been used for estimating predictors in genetic networks is based on the well-known minimum-description-length (MDL) principle (Rissanen, 1978, 1986; Barron et al., 1998). This principle is based on an intuitive but profound idea: the best model class for a set of observed data is the one whose representation permits the most compact (minimal) encoding of the data. This idea is predicated on the belief that the better a theory can compress the data, the better it can predict future data. This general paradigm is a form of Occam's razor. Using the MDL principle, one can characterize the goodness of a model in terms of code length via a two-part measure: one for the description of the data using the optimal model class, and another for the description of the model class itself. Thus, it provides a tradeoff between goodness of fit and model complexity. Tabus and Astola (2001) used the MDL principle for the gene prediction problem. They showed that it not only can be naturally used to find the optimal predictor along with the optimal number of input variables but is also well suited for comparing predictors having different complexities.

## 2.3 GENERALIZATIONS OF BOOLEAN NETWORKS

There are two main aspects of Boolean network models that can be naturally generalized. First is the assumption that each gene is restricted to be in one of two possible states, namely, on or off. The natural extension of this idea would be to allow genes to take on more than two discrete values. The other assumption that can be relaxed is that at each "clock tick," all genes are updated synchronously. The obvious generalization would be having genes update themselves asynchronously. Here there is more flexibility in how the asynchronous updating schedule can take place. Much of the work in this regard was done by René Thomas and colleagues and reported in a book by Thomas and D'Ari (1990).

### 2.3.1 Asynchrony

In his work, Thomas distinguishes a gene as being on or off, denoted by uppercase letters (e.g., $X = 1$ or $X = 0$), from the gene product (protein) being absent or present, denoted by lowercase letters (e.g., $x = 1$ or $x = 0$). For example, the logical expression $X = y \wedge z'$ means that if protein $y$ is present and protein $z$ is not present, gene $X$ is activated. To capture the fact that proteins require time to be produced and to degrade, time delays can be introduced in the logical relationships to give rise to system dynamics. For example, suppose that at some particular time

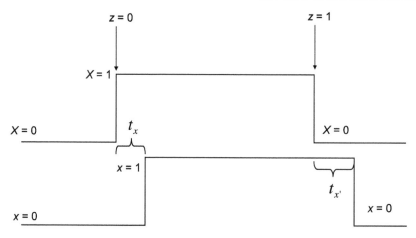

Figure 2.9. A timing diagram illustrating the behavior of gene $X$ and its product $x$. Gene
$X$ switches on or off in response to the absence or presence of protein $z$. Gene
product $x$ becomes present and absent after time delays $t_x$ and $t_{x'}$, respectively.

gene $X=0$ because protein $y$ is present, but so is protein $z$. Then, suppose that $z$
suddenly becomes absent. Then, gene $X$ immediately switches on ($X=1$), but its
product $x$ goes on only after a time delay $t_x$. Suppose further that some time after
$x=1$, protein $z$ appears again. Then, gene $X$ switches off, but protein $x$ switches
off only after another time delay $t_{x'}$. Note that typically $t_x \neq t_{x'}$ since $t_x$ depends on
the rate of synthesis of $x$, whereas $t_{x'}$ depends on the stability of $x$. In general, a
different time delay corresponds to each logical transition. The above is illustrated
in figure 2.9.

Generally speaking, if a gene and its product have the same logical value (e.g.,
$X=0$ and $x=0$), no change should take place, implying a logical steady state.
However, if they disagree, such as when the gene is turned on but the product
is still absent, one or more changes will take place after appropriate time delays.
Hence, such situations are transient.

Consider a simple two-element system where product $x$ activates gene $Y$ and its
product $y$ inactivates gene $X$ (Thomas and D'Ari, 1990). The logical description is
given by

$$X = y',$$

$$Y = x.$$

A state of the system is given by the values of the products $x$ and $y$, as well as
by the gene values $X$ and $Y$, which can be written as $xy/XY$. Consider, for example, 00/10. This implies that products $x$ and $y$ are absent but gene $X$ is switched
on, which further implies that product $x$ is expected to eventually become present.
Thus, the next state is expected to be 10/10. But since product $x$ is present, gene
$Y$ will eventually be switched on, so the next state should be 10/11. Note that
the notation is somewhat redundant and that the state 00/10 can be described with
the notation $\overline{0}0$. The bar over the 0 signifies that product $x$ is committed to changing

from being absent to being present. Thus, the system goes through the following state transitions:

$$\begin{array}{ccc} \overline{0}0 & \longrightarrow & 1\overline{0} \\ \uparrow & & \downarrow \\ 0\overline{1} & \longleftarrow & \overline{1}1 \end{array}$$

In this simple example, the state sequence is independent of the values of the time delays. In general, this is not the case.

In the above example, there is no logical steady state because the values for which $xy = XY$ must change. This is why periodic behavior is observed. Consider now the system

$$X = y',$$

$$Y = x',$$

representing the situation where two genes repress each other. Consider the state 01. Since gene product $x$ is absent, gene $Y$ should be on. At the same time, since product $y$ is present, gene $X$ should be off. Clearly, $xy = XY$ and 01 is a logical steady state; similarly for state 10. Now let us consider state $\overline{00}$. We can see that since both gene products are absent, genes $XY$ should both be turned on. Since two variables need to change their values, depending on which changes faster, the state $\overline{00}$ will be followed by either 10 or 01. Of course, it is also possible that $\overline{00} \longrightarrow \overline{11}$ if both events occur simultaneously. The state transition diagram is

$$\begin{array}{ccc} \overline{0}\overline{0} & \longrightarrow & 10 \\ \downarrow & & \uparrow \\ 01 & \longleftarrow & \overline{1}\overline{1} \end{array}$$

It is interesting that even such a simple system can be in more than one steady state, a property called *multistationarity*. Additionally, because of the asynchrony, each state can be succeeded by more than one state. The trajectory of states that the system takes, therefore, is dependent on the relationships between the time delays corresponding to all transitions. In cases when there are several successor states, the relationships between time delays, expressed by inequalities such as $t_{x'} \gtrless t_y + t_z$, determine which of the states will succeed a given state. If all the time delays for all the genes are known and fixed, the system is deterministic.

It is convenient to represent inequalities between time delays using Boolean variables, where the value 1 corresponds to $<$ and the value 0 corresponds to $>$. For example, $m = 1$ if $t_x < t_y$ and $m = 0$ if $t_x > t_y$. Thus, the trajectory that the system takes toward a steady state, when there is more than one trajectory, can be represented by Boolean expressions involving variables corresponding to these inequalities. In a simulation, the time delays can be selected from some underlying distribution so that the Boolean variables representing the inequalities become random variables. Since the state transitions become probabilistic, the process can be described by a Markov chain. We will return to this topic later when we discuss probabilistic Boolean networks.

Another fascinating property of such asynchronous systems is that, depending on the relationships between time delays, two kinds of cycles can emerge. *Stable cycles* depend on the inequalities between (sums of) time delays, and if the actual values of the time delays fluctuate slightly, as long as the inequality remains unchanged, the system will persist in that cycle. *Unstable cycles*, on the other hand, can occur if specific values of time delays satisfy some equality, such as $t_x + t_y = t_z + t_{y'}$, but as soon as one of the time delays fluctuates away from the value satisfying the equality, the system immediately escapes from the unstable cycle. Thus, in practice, where rates of protein synthesis and degradation are described by distributions, unstable cycles are not expected to occur. This is particularly true when a population of cells is considered since it is well known that even clonal cell populations exhibit significant variation in the rates of concentrations of molecular species (McAdams and Arkin, 1999). It was demonstrated by Thomas and Kaufman (2001a,b) that the asynchronous description fits extremely well with the differential description, which will be discussed later.

There have been other proposed generalizations of Boolean networks that relax the assumption of gene synchrony. For example, Öktem et al. (2003) introduced continuous-time Boolean networks, which are special cases of the class of Boolean delay equations. They showed that this model class can exhibit a wider range of dynamical behavior than classical Boolean networks. In particular, whereas Boolean networks are necessarily periodic, Boolean delay equations can exhibit aperiodic behavior. Silvescu and Honavar (1997) proposed temporal Boolean networks, which allow temporal dependencies between genes spanning several time steps. That is, instead of a gene being functionally dependent on the values of some other genes at the previous time step, it can be dependent on the values of these genes during a certain time window, reminiscent of higher-order Markov chains. Thus, a gene at time $t + 1$ is dependent on the values of other genes at times $t, t - 1, \ldots, t - T + 1$.

### 2.3.2 Multivalued Networks

As mentioned earlier, another obvious way to generalize Boolean networks is to allow each gene to take on more than just two values. Thus, unlike the all-or-none nature of Boolean networks, *generalized logical networks* can capture intermediate levels of expression (Thomas and D'Ari, 1990). There are a number of ways in which one could formulate the interactions between genes in a multilevel domain. For instance, suppose that instead of a gene being 0 or 1 it can take on one of three different values: 0, 1, or 2. One approach for describing the interactions between genes is to represent the functions with multilevel truth tables. For instance, suppose that genes $A$ and $B$ control gene $C$ and that each of them can be 0, 1, or 2. Thus, a truth table for gene $C$ might look like the one shown in table 2.2. Such ternary-valued functions were used by Kim et al. (2000a,b) in the context of inferring multivariate relationships from cDNA microarray data. Ternary networks arise naturally in the context of cDNA microarrays owing to the use of ratios, which result in the quantization $+1$ (up-regulated), $-1$ (down-regulated), and 0 (invariant) based on a significant expression differential (Chen et al., 1997).

Table 2.2 Example of a Truth Table with three Values

| A | B | C |
|---|---|---|
| 0 | 0 | 0 |
| 0 | 1 | 1 |
| 0 | 2 | 0 |
| 1 | 0 | 0 |
| 1 | 1 | 1 |
| 1 | 2 | 2 |
| 2 | 0 | 2 |
| 2 | 1 | 1 |
| 2 | 2 | 2 |

Such cDNA microarray–based ternary networks have been used extensively. There is a conceptual difference between multivalued logical networks and discrete-valued numerical networks arising from quantization since the latter incorporate order in the usual real-valued measurement sense; nonetheless, from a structural perspective (connectivity and predictors), they are equivalent.

In the context of quantization, Snoussi (1989) suggested the following framework of which the Boolean formalization is a special case. Suppose a gene product $x$ can act in different circumstances with different characteristic thresholds. The threshold refers to the limiting case of a sigmoidal curve (to be discussed later) whose critical transition point can vary depending on these circumstances. It is then natural to associate a multilevel discrete variable $\mathbf{x}$ with this gene product such that each value of the discrete variable corresponds to where $x$ falls in relation to the threshold. Thus, given thresholds $\tau_1 < \tau_2 < \cdots$, the discrete variable is defined as $\mathbf{x} = Q(x) = r_k$ for $\tau_k < x \leq \tau_{k+1}$. The values of the thresholds $\tau_k$ can either come from prior knowledge or be chosen automatically using the measurement data.

In the field of signal processing, the operation $Q(x)$ is referred to as (scalar) *quantization*. This process of representing continuous-valued signals by a set of finitely many values results in approximating the data at the expense of precision. However, it can also have beneficial effects by enhancing inference precision on account of reduced model complexity and mitigating the effects of noise. Quantization has been shown to significantly improve error rates in microarray-based classification (Mircean et al., 2004). Since quantization is a many-to-one mapping, it is *lossy* in the sense that the original signal cannot be recovered.

There are several ways one could determine the thresholds $\tau_k$ in an optimal way. The most popular method is the Lloyd-Max quantizer (Gersho and Gray, 1992), also known as $k$-means in the pattern recognition community. Given the number of quantization levels $L$, the optimality consists of selecting the $\tau_k$ and $r_k$, $k = 1, \ldots, L$, in order to minimize the mean-square quantization error

$$D = E[(x - r_k)^2] \tag{2.10}$$

$$= \sum_{k=1}^{L} \int_{\tau_{k-1}}^{\tau_k} (x - r_k)^2 p(x) \, dx$$

for a given even input probability density function $p(x)$. Differentiating the error in equation (2.10) with respect to $r_k$ and $\tau_k$ gives us the necessary conditions that must be satisfied by the optimal quantizer. Indeed, $\partial D/\partial r_k = 0$ gives us

$$r_k = \frac{\displaystyle\int_{\tau_{k-1}}^{\tau_k} x p(x)\, dx}{\displaystyle\int_{\tau_{k-1}}^{\tau_k} p(x)\, dx}, \tag{2.11}$$

which can be interpreted as the expectation of $x$, or the center of mass of the density function, given that we are in the interval $(\tau_{k-1}, \tau_k]$. Similarly, $\partial D/\partial \tau_k = 0$ gives

$$\tau_k = \frac{r_k + r_{k+1}}{2}, \tag{2.12}$$

which means that the optimal decision levels must be midway between the quantization levels. In general, equations (2.11) and (2.12) do not yield closed-form solutions but can be solved by the following iterative algorithm. The algorithm starts with arbitrary positions of the quantization values and then iterates the following two steps until convergence:

1. For each data point, find and associate with it the nearest quantization value.
2. For each quantization interval, find the average of the data points associated with it and assign this average to be the new quantization value.

This iterative algorithm is expected to reach a local optimum. Unfortunately, it can be sensitive to the initial quantization values (seeds). To address this problem, one approach is to run the algorithm a fixed number of times and return the solution with the lowest quantization error $D$. Another approach is to run a preliminary quantization on 10 percent of the data and use the solution of that run as an initial condition.

Having established the threshold levels, we can return to the multivariate formulation of gene interactions. Consider the Boolean expression $X = y \vee z$. The meaning of this expression is that gene $X$ is switched on when $y$, $z$, or both are present. However, these three cases cannot be discriminated. To do so, we can associate positive real weights $K_y$ and $K_z$ with the gene products $y$ and $z$, respectively, and form the expression $K_y \cdot y + K_z \cdot z$, where $+$ and $\cdot$ are the usual addition and multiplication operations and the binary-valued variables $y$ and $z$ are treated as real variables. This results in the following four possibilities:

| $y$ | $z$ | $K_y \cdot y + K_z \cdot z$ |
|---|---|---|
| 0 | 0 | 0 |
| 0 | 1 | $K_z$ |
| 1 | 0 | $K_y$ |
| 1 | 1 | $K_y + K_z$ |

which can then be quantized. Thus, the multivalued generalization of the expression $X = y \vee z$ is

$$X = Q(K_y \cdot y + K_z \cdot z). \tag{2.13}$$

It is easy to see that, depending on the relationships between the weights $K_y$ and $K_z$, very different values of $X$ are possible. Such functions can be combined to construct multivalued networks. The kinetic behavior of such multivalued networks has much in common with differential equation formalism (Thomas and D'Ari, 1990), which we discuss in the next section. Finally, we should point out that these methods have been implemented by Thieffry et al. (1993) and applied to several systems such as $\lambda$ phage (Thieffry and Thomas, 1995), *Drosophila* (Sánchez and Thieffry, 2001, 2003), and *Arabidopsis thaliana* (Mendoza et al., 1999).

## 2.4 DIFFERENTIAL EQUATION MODELS

A model of a genetic network based on differential equations typically expresses the rates of change of an element, such as a gene product, in terms of the levels of other elements of the network and possibly external inputs. Typically, as there are a number of variables involved in the network, a system of differential equations is required. In general, a nonlinear time-dependent differential equation has the form

$$\dot{x} = f(x, u, t),$$

where $x$ is a state vector denoting the values of the physical variables in the system, $\dot{x} = dx/dt$ is the elementwise derivative of $x$, $u$ is a vector of external inputs, and $t$ is time.

If time is discretized and the functional dependency specified by $f$ does not depend on time, then the system is said to be *time-invariant*. For example, the equations

$$\begin{bmatrix} \dot{x}_1 \\ \dot{x}_2 \end{bmatrix} = \begin{bmatrix} x_2 \\ -4\sin(x_1) \end{bmatrix}$$

describe the motion of an undamped pendulum, with $x_1$ and $x_2$ representing the angular position and the angular velocity, respectively. Here, the nonlinear function $f$ is time-invariant and does not depend on any external inputs $u$. Thus, for any given initial state $x(t_0) = [x_1(t_0), x_2(t_0)]^T$, these equations allow us to determine the state of the system, represented by $x$, at any time $t > t_0$. Finally, if $f$ is linear and time-invariant, then it can be expressed as

$$\dot{x} = \mathbf{A}x + \mathbf{B}u,$$

where $\mathbf{A}$ and $\mathbf{B}$ are constant matrices.

When $\dot{x} = 0$, the variables no longer change with time and thus define the *steady state* of the system. Consider the simple case of a gene product $x$ (a scalar) whose rate of synthesis is proportional, with kinetic constant $k_1$, to the abundance of another protein $a$ that is sufficiently abundant such that the overall concentration of $a$ is not significantly changed by the reaction (Thomas and D'Ari, 1990). However,

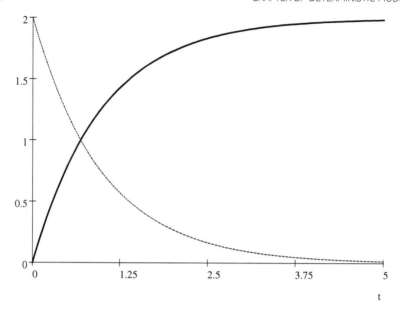

Figure 2.10. The behavior of the solution to $\dot{x} = k_1 a - k_2 x$, $x(0) = 0$. We have used $k_1 = 2$, $k_2 = 1$, and $a = 1$. As can be seen, the gene product $x$, shown with a solid plot, tends toward its steady state value given in equation (2.14). The time derivative $\dot{x}$, which starts at initial value of $k_1 a$ and tends toward zero, is shown with a dashed plot.

$x$ is also subject to degradation, the rate of which is proportional, with constant $k_2$, to the concentration of $x$ itself. This situation can be expressed as

$$\dot{x} = k_1 a - k_2 x \qquad a, x > 0.$$

Let us analyze the behavior of this simple system. If initially $x = 0$, then the decay term is also zero and $\dot{x} = k_1 a$. However, as $x$ is produced, the decay term $k_2 x$ also increases, thereby decreasing the rate $\dot{x}$ toward zero and stabilizing $x$ at some steady-state value $\bar{x}$. It is easy to determine this value since setting $\dot{x} = 0$ and solving for $x$ yields

$$\bar{x} = \frac{k_1 a}{k_2}. \tag{2.14}$$

This behavior is shown in figure 2.10, where $x$ starts off at $x = 0$ and approaches the value in equation (2.14). The exact form of the kinetics is

$$x(t) = \frac{k_1 a}{k_2} \left(1 - e^{-k_2 t}\right).$$

Similarly, the derivative $\dot{x}$, also shown in figure 2.10, starts off at the initial value of $k_1 a$ and thereafter tends toward zero.

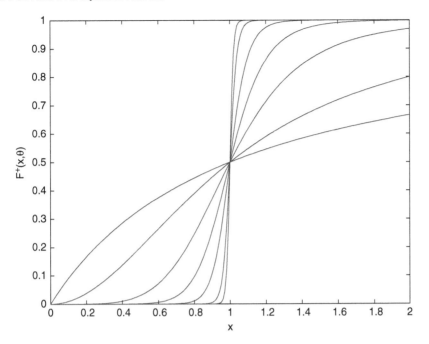

Figure 2.11. The function $F^+(x, \theta)$ for $\theta = 1$ and $n = 1$, 2, 5, 10, 20, 50, and 100. As $n$ gets large, $F^+(x, \theta)$ approaches an ideal step function.

Now suppose that $a$ is suddenly removed after the steady-state value $\bar{x}$ is reached. Since $a = 0$, we have $\dot{x} = -k_2 x$, and since the initial condition is $x = k_1 a / k_2$, $\dot{x} = -k_1 a$ initially. The solution of this equation is

$$x(t) = \frac{k_1 a}{k_2} e^{-k_2 t},$$

and it can be seen that it will eventually approach zero.

This example describes a linear relationship between $a$ and $\dot{x}$. However, most gene interactions are highly nonlinear. When the regulator is below some critical value, it has very little effect on the regulated gene. When it is above the critical value, it has virtually a full effect that cannot be significantly amplified by increased concentrations of the regulator. This nonlinear behavior is typically described by *sigmoid* functions, which can be either monotonically increasing or decreasing. A common form is the Hill functions given by

$$F^+(x, \theta) = \frac{x^n}{\theta^n + x^n}$$

$$F^-(x, \theta) = \frac{\theta^n}{\theta^n + x^n} = 1 - F^+(x, \theta).$$

The function $F^+(x, 1)$ is illustrated in figure 2.11 for $n = 1, 2, 5, 10, 20, 50$, and 100. It can be seen that it approaches an ideal step function with increasing $n$. This behavior is the basis for using the threshold functions for the multivalued networks we discussed earlier. In fact, the parameter $\theta$ essentially plays the role of the threshold value. Glass (1975) used step functions in place of sigmoidal functions in differential equation models, resulting in piecewise linear differential equations. Glass and Kauffman (1973) also showed that many systems exhibit the same qualitative behavior for a wide range of sigmoidal steepnesses, parameterized by $n$. Given that gene regulation is nonlinear, differential equation models can incorporate the Hill functions into their synthesis and decay terms. For simulating and analyzing such dynamical systems using a variety of methods and algorithms (Lambert, 1991), there are many available computer tools such as DBsolve (Goryanin et al., 1999), GEPASI (Mendes, 1993), and Dizzy (Ramsey et al., 2005). Dizzy is also capable of modeling stochastic kinetics of large-scale genetic, metabolic, and signaling networks and simulating multistep reaction processes and spatial compartmentalization.

### 2.4.1 A Differential Equation Model Incorporating Transcription and Translation

Having described differential equation models in a general form, we now discuss a biologically detailed model whose essentials have appeared in various forms (de Jong, 2002; Mestle et al., 1995; Smolen et al., 2000; Wolf and Beckman, 1998). The model takes into account the interaction of genes and regulatory proteins (transcription factors) and assumes a large population of genetically identical cells expressing the same set of genes. We follow the presentation in (Goutsias and Kim, 2004).

Let the concentrations of mRNAs and regulatory proteins at time $t$ be described by the vectors

$$\mathbf{r}(t) = \begin{pmatrix} r_1(t) \\ r_2(t) \\ \vdots \\ r_m(t) \end{pmatrix} \qquad (2.15)$$

and

$$\mathbf{p}(t) = \begin{pmatrix} p_1(t) \\ p_2(t) \\ \vdots \\ p_m(t) \end{pmatrix}, \qquad (2.16)$$

respectively. Given a target gene $x_k$ regulated by genes $x_i$ and $x_j$, according to the central dogma, regulation follows the program illustrated in figure 2.12.

The transcriptional regulation model is oversimplified in a number of ways, for instance, in assuming that there are no external factors or posttranslational regulation. Here we are concerned with accounting for two simplifications. First, transcription follows cis-regulation (chapter 1). The second assumption is that the

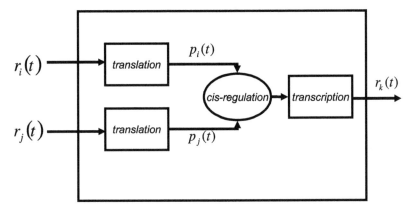

Figure 2.12. A model for transcriptional regulation of a target gene $k$ involving translation, cis-regulation, and transcription. See (Goutsias and Kim, 2004).

various reactions are instantaneous. Translation, transcription, and cis-regulation involve time delays, which we denote by $\tau_{p,i}$, $\tau_{r,i}$, and $\tau_{c,i}$, respectively.

To formulate the model mathematically, define the following terms: $T$ is the absolute temperature in kelvins; $R = 1.9872$ cal mol$^{-1}$ K$^{-1}$ is the gas constant; $U_{\text{tr},i}$ is the activation energy of translation of the $i$th mRNA, $U_{\text{dg},i}$ is the activation energy of degradation of the $i$th regulatory protein, $r_i(t|U > U_{\text{tr},i})$ is the concentration at time $t$ of the $i$th mRNA with energy greater than $U_{\text{tr},i}$; and $p_i(t|U > U_{\text{dg},i})$ is the concentration at time $t$ of the $i$th regulatory protein with energy greater than $U_{\text{dg},i}$. The rate of protein synthesis is assumed to be proportional to the concentration of mRNAs possessing energy greater than $U_{\text{tr},i}$, with proportionality constant $\alpha_{\text{tr},i}$, and the rate of protein degradation is assumed to be proportional to the concentration of proteins possessing energy greater than $U_{\text{dg},i}$, with proportionality constant $\alpha_{\text{dg},i}$.

To analyze the protein concentration, let $\xi_i(\Delta t)$ be the concentration of the $i$th regulatory protein produced by translation during the time interval $(t, t + \Delta t]$ and let $\zeta_i(\Delta t)$ be the concentration of the $i$th regulatory protein degraded during the time interval $(t, t + \Delta t]$. Then,

$$\xi_i(\Delta t) = \alpha_{\text{tr},i} \int_t^{t+\Delta t} r_i(\tau - \tau_{p,i}|U > U_{\text{tr},i}) \, d\tau \tag{2.17}$$

and

$$\zeta_i(\Delta t) = \alpha_{\text{dg},i} \int_t^{t+\Delta t} p_i(\tau|U > U_{\text{dg},i}) \, d\tau. \tag{2.18}$$

According to the theory of statistical mechanics, in a large population of identical molecules at statistical equilibrium with concentration $\eta$, the concentration of molecules with kinetic energy $U$ is given by the Boltzmann distribution,

$$\eta(U) = \frac{\eta}{RT} e^{-U/RT}. \tag{2.19}$$

Letting $\eta(U > U_0)$ be the concentration of molecules in the population with kinetic energy exceeding $U_0$,

$$\eta(U > U_0) = \int_{U_0}^{\infty} \eta(U)\, dU = \eta e^{-U_0/RT}. \tag{2.20}$$

Hence,

$$\xi_i(\Delta t) = \alpha_{\mathrm{tr},i}\, e^{-U_{\mathrm{tr},i}/RT} \int_{t}^{t+\Delta t} r_i(\tau - \tau_{p,i})\, d\tau \tag{2.21}$$

$$\zeta_i(\Delta t) = \alpha_{\mathrm{dg},i}\, e^{-U_{\mathrm{dg},i}/RT} \int_{t}^{t+\Delta t} p_i(\tau). \tag{2.22}$$

Since

$$p_i(t + \Delta t) = p_i(t) + \xi_i(\Delta t) - \zeta_i(\Delta t), \tag{2.23}$$

the difference quotient for regulatory protein concentration is given by

$$\frac{p_i(t + \Delta t) - p_i(t)}{\Delta t} = \alpha_{\mathrm{tr},i}\, e^{-U_{\mathrm{tr},i}/RT} \frac{1}{\Delta t} \int_{t}^{t+\Delta t} r_i(\tau - \tau_{p,i})\, d\tau$$
$$- \alpha_{\mathrm{dg},i}\, e^{-U_{\mathrm{dg},i}/RT} \frac{1}{\Delta t} \int_{t}^{t+\Delta t} p_i(\tau)\, d\tau. \tag{2.24}$$

Letting $\Delta t \to 0$ yields

$$\frac{dp_i(t)}{dt} = \lambda_i r_i(t - \tau_{p,i}) - \gamma_i p_i(t), \tag{2.25}$$

where

$$\lambda_i = \alpha_{\mathrm{tr},i}\, e^{-U_{\mathrm{tr},i}/RT} \tag{2.26}$$

and

$$\gamma_i = \alpha_{\mathrm{dg},i}\, e^{-U_{\mathrm{dg},i}/RT}. \tag{2.27}$$

Equation (2.25) is quite intuitive. It states that the regulatory protein rate of change is proportional to the mRNA level at time $t - \tau_{p,i}$ and that the protein is degrading proportionally to its concentration.

A similar analysis applies to the mRNA rate of change. To do so, we define the vector

$$\mathbf{c}(t) = \begin{pmatrix} c_1(t) \\ c_2(t) \\ \vdots \\ c_m(t) \end{pmatrix}, \tag{2.28}$$

where $c_i$, $0 \le c_i \le 1$, is the fraction, of DNA templates in the population that are *committed* to transcription of the $i$th gene, that is, when the transcription initiation complex is attached to the promoter of the gene. Then,

$$\frac{dr_i(t)}{dt} = \kappa_i c_i(t - \tau_{r,i}) - \beta_i r_i(t), \tag{2.29}$$

where $\kappa_i > 0$ is the transcription rate of gene $i$ and $\beta_i > 0$ is the degradation rate of the $i$th gene.

Solving equations (2.25) and (2.29) yields

$$p_i(t) = \lambda_i e^{-\gamma_i t} \int_{\tau_{p,i}}^{t} e^{\gamma_i \tau} r_i(\tau - \tau_{p,i}) \, d\tau + e^{-\gamma_i(t - \tau_{p,i})} p_i(\tau_{p,i}) \qquad t \ge \tau_{p,i}. \tag{2.30}$$

$$r_i(t) = \kappa_i e^{-\beta_i t} \int_{\tau_{r,i}}^{t} e^{\beta_i \tau} c_i(\tau - \tau_{r,i}) \, d\tau + e^{-\beta_i(t - \tau_{r,i})} r_i(\tau_{r,i}) \qquad t \ge \tau_{r,i}. \tag{2.31}$$

The model is completed by characterizing cis-regulation in terms of the committed fraction $c_i$. Specifically, cis-regulation is modeled by a (typically) nonlinear function $\phi_i$, a regulatory set of genes $R_i$, and the functional relation

$$c_i(t) = \phi_i[p_j(t - \tau_{c,j}), \, j \in R_i], \tag{2.32}$$

where $R_i$ is the set of genes that produce proteins that regulate the transcription of the $i$th gene and $\phi_i$ is called the *cis-regulatory function* for the $i$th gene.

## 2.4.2 Discretization of the Continuous Differential Equation Model

To apply computational techniques to the preceding continuous differential equation model, it is useful to discretize it. Discretization can also be viewed as a natural step based on conceiving transcription and translation as discrete processes. Goutsias and Kim (2004) derived a discrete system under the assumption that it models average transcriptional activity between discrete time steps $\delta t, 2\delta t, \ldots$, so that $c_i$, $r_i$, and $p_i$ are constant in each time interval. This leads to discrete forms of equations (2.30)–(2.32) based on the time-step parameter $n$, and, under the assumption that the time delays are independent of the genes, these discrete equations take the matrix form

$$\mathbf{r}(n) = \mathbf{D}_b \mathbf{r}(n-1) + \mathbf{KS}_b(n-1)\mathbf{\Phi}[\mathbf{p}(n - \nu n_p - 1)], \tag{2.33}$$

$$\mathbf{p}(n) = \mathbf{D}_c \mathbf{p}(n-1) + \mathbf{LS}_c(n-1)\mathbf{r}(n - n_p - 1), \tag{2.34}$$

for $n = 1, 2, \ldots$, where $v = (\tau_r + \tau_c)/\tau_p$, $n_p = \tau_p/\delta t$, and $\mathbf{\Phi}[\mathbf{p}(n - vn_p - 1)]$ is an $m \times 1$ vector-valued cis-regulatory functional whose $i$th element is $\phi_i[p_j(n - vn_p - 1), j \in R_i]$. Further,

$$
\mathbf{D}_b = \begin{pmatrix}
e^{-\beta_1 \delta t} & 0 & \cdots & 0 \\
0 & e^{-\beta_2 \delta t} & \cdots & 0 \\
\vdots & \vdots & \ddots & \vdots \\
0 & 0 & \cdots & e^{-\beta_m \delta t}
\end{pmatrix}, \tag{2.35}
$$

$$
\mathbf{D}_c = \begin{pmatrix}
e^{-\gamma_1 \delta t} & 0 & \cdots & 0 \\
0 & e^{-\gamma_2 \delta t} & \cdots & 0 \\
\vdots & \vdots & \ddots & \vdots \\
0 & 0 & \cdots & e^{-\gamma_m \delta t}
\end{pmatrix}, \tag{2.36}
$$

$$
\mathbf{K} = \begin{pmatrix}
\kappa_1 & 0 & \cdots & 0 \\
0 & \kappa_2 & \cdots & 0 \\
\vdots & \vdots & \ddots & \vdots \\
0 & 0 & \cdots & \kappa_m
\end{pmatrix}, \tag{2.37}
$$

$$
\mathbf{L} = \begin{pmatrix}
\lambda_1 & 0 & \cdots & 0 \\
0 & \lambda_2 & \cdots & 0 \\
\vdots & \vdots & \ddots & \vdots \\
0 & 0 & \cdots & \lambda_m
\end{pmatrix}, \tag{2.38}
$$

$$
\mathbf{S}_b(n) = u(n - vn_p) \begin{pmatrix}
s(\beta_1, \delta t) & 0 & \cdots & 0 \\
0 & s(\beta_2, \delta t) & \cdots & 0 \\
\vdots & \vdots & \ddots & \vdots \\
0 & 0 & \cdots & s(\beta_m, \delta t)
\end{pmatrix}, \tag{2.39}
$$

$$
\mathbf{S}_c(n) = u(n - n_p) \begin{pmatrix}
s(\gamma_1, \delta t) & 0 & \cdots & 0 \\
0 & s(\gamma_2, \delta t) & \cdots & 0 \\
\vdots & \vdots & \ddots & \vdots \\
0 & 0 & \cdots & s(\gamma_m, \delta t)
\end{pmatrix}, \tag{2.40}
$$

$$
s(x, y) = \frac{1 - e^{-xy}}{x}, \tag{2.41}
$$

and

$$u(n) = \begin{cases} 1 & \text{for } n \geq 0, \\ 0 & \text{otherwise.} \end{cases}$$

We now consider steady-state behavior, focusing our attention solely on fixed-point attractors (there also possibly being limit-cycle attractors). For a fixed-point attractor $(\mathbf{r}_\infty, \mathbf{p}_\infty)$, the concentrations are no longer a function of $n$ in equations (2.33) and (2.34), so these reduce to

$$\mathbf{r}_\infty = \mathbf{D}_b \mathbf{r}_\infty + \mathbf{KS}_b \mathbf{\Phi}[\mathbf{p}_\infty], \tag{2.42}$$

$$\mathbf{p}_\infty = \mathbf{D}_c \mathbf{p}_\infty + \mathbf{LS}_c \mathbf{r}_\infty. \tag{2.43}$$

Straightforward matrix algebra shows that $\mathbf{r}_\infty$ and $\mathbf{p}_\infty$ satisfy these equations if

$$\mathbf{r}_\infty = \mathbf{BK}\mathbf{\Phi}[\mathbf{CL}\mathbf{r}_\infty], \tag{2.44}$$

$$\mathbf{p}_\infty = \mathbf{CLBK}\mathbf{\Phi}[\mathbf{p}_\infty], \tag{2.45}$$

where

$$\mathbf{B} = \begin{pmatrix} \beta_1^{-1} & 0 & \cdots & 0 \\ 0 & \beta_2^{-1} & \cdots & 0 \\ \vdots & \vdots & \ddots & \vdots \\ 0 & 0 & \cdots & \beta_m^{-1} \end{pmatrix}, \tag{2.46}$$

$$\mathbf{C} = \begin{pmatrix} \gamma_1^{-1} & 0 & \cdots & 0 \\ 0 & \gamma_2^{-1} & \cdots & 0 \\ \vdots & \vdots & \ddots & \vdots \\ 0 & 0 & \cdots & \gamma_m^{-1} \end{pmatrix}. \tag{2.47}$$

According to equations (2.44) and (2.45), $\mathbf{r}_\infty$ and $\mathbf{p}_\infty$ are individually fixed-point attractors for the *genomic regulatory functional* $\mathbf{\Psi}_r$ and *proteomic regulatory functional* $\mathbf{\Psi}_p$ defined by

$$\mathbf{\Psi}_r[\mathbf{r}_\infty] = \mathbf{BK}\mathbf{\Phi}[\mathbf{CL}\mathbf{r}_\infty] \tag{2.48}$$

and

$$\mathbf{\Psi}_p[\mathbf{p}_\infty] = \mathbf{CLBK}\mathbf{\Phi}[\mathbf{p}_\infty], \tag{2.49}$$

respectively. Specifically, $\mathbf{\Psi}_r[\mathbf{r}_\infty] = \mathbf{r}_\infty$ and $\mathbf{\Psi}_p[\mathbf{p}_\infty] = \mathbf{p}_\infty$.

Equations (2.44) and (2.45) show that the steady-state mRNA concentration vector $\mathbf{r}_\infty$ can be found independently of the protein concentration vector $\mathbf{p}_\infty$ via solution of the nonlinear system in equation (2.48), and that $\mathbf{p}_\infty$ can be found independently of $\mathbf{r}_\infty$ via solution of equation (2.49). This decoupling of the mRNA and protein concentrations occurs only in the steady state, not in transient states.

From equations (2.42) and (2.43), we see that

$$\mathbf{r}_\infty = \mathbf{BK}\Phi[\mathbf{p}_\infty], \tag{2.50}$$

$$\mathbf{p}_\infty = \mathbf{CLr}_\infty. \tag{2.51}$$

Since $\mathbf{CL}$ is nonsingular, this means that there is a one-to-one correspondence between the fixed-point-attractor mRNA concentration vectors and the fixed-point-attractor protein concentration vectors, which in turn implies that the fixed-point mRNA concentration vectors can be deduced from the fixed-point protein concentration vectors, and vice versa. Consequently, at least with respect to the model under consideration, in the steady state, mRNA measurements on expression microarrays are sufficient to characterize the fixed-point attractors of regulatory protein concentrations in the sense of equation (2.51). This is useful to know because microarrays yield mRNA measurements, not protein measurements, and a common assumption is that static (non-time-course) mRNA measurements are taken from the steady state. Of course, to apply equation (2.51), one would have to know the parameters involved in $\mathbf{C}$ and $\mathbf{L}$.

It is informative to compare the foregoing transcriptional regulatory system (tRS) model with the kind of gene regulatory modeling involved in discrete function-based networks like Boolean networks, where there is a time-invariant function system $\Psi$ governing network transitions according to

$$\mathbf{r}(n) = \Psi[\mathbf{r}(n-1)]. \tag{2.52}$$

As pointed out by Goutisas and Kim (2004), equations (2.33) and (2.34) cannot be reduced to equation (2.52). Even if we set the time delay to zero ($n_p = v = 0$ in equations (2.33) and (2.34)), the resulting solution for $\mathbf{r}(n)$ becomes

$$\mathbf{r}(n) = \mathbf{D}_b\mathbf{r}(n-1) + \mathbf{KS}_b\Phi[\mathbf{LS}_c\mathbf{r}(n-2) + \mathbf{D}_c\mathbf{p}(n-2)], \tag{2.53}$$

so that $\mathbf{r}(n)$ depends on $\mathbf{r}(n-1)$, $\mathbf{r}(n-2)$, and $\mathbf{p}(n-2)$. Not only can $\mathbf{r}(n)$ not be characterized by a one-step transition, as in equation (2.52), but it also cannot be decoupled from the regulatory protein concentrations. Hence, gene regulatory models built on equation (2.52) cannot model the behavior of the tRS model, at least insofar as transient behavior is concerned. However, in the steady state, for fixed-point attractors, equation (2.52) becomes $\Psi[\mathbf{r}_\infty] = \mathbf{r}_\infty$, which is precisely the fixed-point relation satisfied by the genomic regulatory functional. Hence, fixed-point steady-state analysis relative to equation (2.52) is compatible with such steady-state analysis in the tRS model.

Two points need to be made regarding these observations. First, it is not surprising, or disheartening, that the gene-only model of equation (2.52) cannot capture the behavior of the more complex tRS model. If we are interested in gene behavior, then the salient issue is which is better for predicting gene behavior. The model-complexity issue that occurs throughout genomic signal processing comes immediately to the fore: there are far more parameters that require estimation in the tRS model, and its performance could be worse than that of the simpler model if there are insufficient data to estimate these parameters. Second, the steady-state conclusions are important from a practical perspective. If we are interested in gene expression and not in protein expression, which is often the case simply owing to the availability of gene expression measurements, then the ability of the gene-only system in equation (2.52) to capture the fixed-point-attractor structure of the more complex tRS system for mRNA concentration means that the more complex system is not necessary in this regard. From the point of view of system dynamics, this conclusion may not be overly salutary; however, given the often used hypothesis that non-time-course expression data are from the steady state and that one of the most important aspects of a gene regulatory network is its ability to capture steady-state behavior, the steady-state decoupling of the mRNA and protein concentrations in the tRS model is of consequence. Even if the tRS model requires more data to estimate its parameters than are available, it nevertheless incorporates a substantial amount of biological understanding in a mathematical framework, and therefore it is encouraging when conclusions derived from it are concordant with simpler models or experimental methodologies.

# Bibliography

Akutsu T, Miyano S, Kuhara S. (1999) Identification of genetic networks from a small number of gene expression patterns under the Boolean network model. *Pac Symp Biocomput* 4:17–28.

Albert R, Barabási AL. (2002) Statistical mechanics of complex networks. *Rev Mod Phys* 74:47–97.

Albert R, Jeong H, Barabási AL. (2000) Error and attack tolerance of complex networks. *Nature* 406:378–82.

Aldana M, Cluzel P. (2003) A natural class of robust networks. *Proc Natl Acad Sci USA* 100(15):8710–4.

Aldana M. (2003) Boolean dynamics of networks with scale-free topology. *Phys D* 185:45–66.

Aldana M, Coppersmith S, Kadanoff LP. (2002) Boolean dynamics with random couplings. In *Perspectives and Problems in Nonlinear Science*, Kaplan E, Marsden JE, Sreenivasan KR, eds., Springer, New York, pp. 23–89.

Ananko EA, Podkolodny NL, Stepanenko IL, Ignatieva EV, Podkolodnaya OA, Kolchanov NA. (2002) GeneNet: a database on structure and functional organisation of gene networks. *Nucleic Acids Res* 30(1):398–401.

Bang-Jensen J, Gutin G. (2001) *Digraphs: Theory, Algorithms and Applications.* Springer-Verlag, London.

Barabási AL, Albert R. (1999) Emergence of scaling in random networks. *Science* 286(5439):509–12.

Barron A, Rissanen J, Yu B. (1998) The minimum description length principle in coding and modeling. *IEEE Trans Inform Theory* 44(6):2743–60.

Boros E, Ibaraki T, Makino K. (1998) Error-free and best-fit extensions of partially defined Boolean functions. *Inform Comput* 140:254–83.

Bu D, Zhao Y, Cai L, Xue H, Zhu X, Lu H, Zhang J, Sun S, Ling L, Zhang N, Li G, Chen R. (2003) Topological structure analysis of the protein-protein interaction network in budding yeast. *Nucleic Acids Res* 31(9):2443–50.

Chen Y, Dougherty ER, Bittner ML (1997) Ratio-based decisions and the quantitative analysis of cDNA microarray images. *J Biomed Opt* 2(4):364–74.

Cook S, Dwork C, Reischuk R. (1986) Upper and lower time bounds for parallel RAMS without simultaneous writes. *SIAM J Comput* 15:87–97.

de Jong H. (2002) Modeling and simulation of genetic regulatory systems: a literature review. *J Comput Biol* 9:69–105.

Derrida B, Pomeau Y. (1986) Random networks of automata: a simple annealed approximation. *Europhys Lett* 1:45–9.

Dougherty ER, Kim S, Chen Y. (2000) Coefficient of determination in nonlinear signal processing. *Signal Process* 80:2219–35.

Flyvbjerg H. (1988) An order parameter for networks of automata. *J Phys A: Math Gen* 21:L955–60.

Fraser AG, Marcotte EM. (2004) A probabilistic view of gene function. *Nat Genet* 36(6):559–64.

Gersho A, Gray RM. (1992) *Vector Quantization and Signal Compression*, Kluwer Academic Publishers, Boston.

Glass L. (1975) Classification of biological networks by their qualitative dynamics. *J Theor Biol* 54:85–107.

Glass L, Kauffman SA. (1973) The logical analysis of continuous, nonlinear biochemical control networks. *J Theor Biol* 39:103–29.

Goodwin BC. (1963) *Temporal Organization in Cells*, Academic Press, New York.

Goodwin BC. (1965). Oscillatory behavior in enzymatic control processes. In *Advances in Enzyme Regulation*, Weber, G. ed., Pergamon Press, Oxford, UK, pp. 425–38.

Goryanin I, Hodgman TC, Selkov E. (1999) Mathematical simulation and analysis of cellular metabolism and regulation. *Bioinformatics* 15(9):749–58.

Han K, Ju BH, Jung H. (2004) WebInterViewer: visualizing and analyzing molecular interaction networks. *Nucleic Acids Res* 32(web server issue):W89–95.

Harris SE, Sawhill BK, Wuensche A, Kauffman, SA. (2002) A model of transcriptional regulatory networks based on biases in the observed regulation rules. *Complexity* 7(4),23–40.

Holland JH. (1995) *Hidden Order: How Adaptation Builds Complexity*, Helix Books, Reading, MA.

Huang S. (1999) Gene expression profiling, genetic networks, and cellular states: an integrating concept for tumorigenesis and drug discovery. *J Mol Med* 77(6): 469–80.

Huang S. (2001) Genomics, complexity and drug discovery: insights from Boolean network models of cellular regulation. *Pharmacogenomics* 2(3):203–22.

Huang S, Ingber DE. (2000) Shape-dependent control of cell growth, differentiation, and apoptosis: switching between attractors in cell regulatory networks. *Exp Cell Res* 261(1):91–103.

Ideker TE, Thorsson V, Karp RM. (2000) Discovery of regulatory interactions through perturbation: inference and experimental design. *Pac Symp Biocomput* 5:305–16.

Ito T, Chiba T, Ozawa R, Yoshida M, Hattori M, Sakaki Y. (2001) A comprehensive two-hybrid analysis to explore the yeast protein interactome. *Proc Natl Acad Sci USA* 98(8):4569–74.

Jeong H, Tombor B, Albert R, Oltvai ZN, Barabasi AL. (2000) The large-scale organization of metabolic networks. *Nature* 407(6804):651–4.

Jeong H, Mason SP, Barabasi AL, Oltvai ZN. (2001) Lethality and centrality in protein networks. *Nature* 411(6833):41–2.

Kahn J, Kalai G, Linial N. (1988) The influence of variables on Boolean functions. In *29th Annual Symposium on Foundations of Computer Science*, IEEE, Washington, DC, pp. 68–80.

Kanehisa M, Goto S. (2000) KEGG: Kyoto encyclopedia of genes and genomes. *Nucleic Acids Res* 28(1):27–30.

Kauffman SA. (1969a) Metabolic stability and epigenesis in randomly constructed genetic nets. *J Theor Biol* 22:437–67.

Kauffman SA. (1969b) Homeostasis and differentiation in random genetic control networks. *Nature* 224:177–8.

Kauffman SA. (1974) The large scale structure and dynamics of genetic control circuits: an ensemble approach. *J Theor Biol* 44:167–90.

Kauffman SA. (1993) *The Origins of Order: Self-organization and Selection in Evolution*. Oxford University Press, New York.

Kauffman SA. (1995) *At Home in the Universe*, Oxford University Press, New York.

Kauffman SA. (2000) *Investigations*. Oxford University Press, New York.

Kim S, Dougherty ER, Chen Y, Sivakumar K, Meltzer P, Trent JM, Bittner M. (2000a) Multivariate measurement of gene expression relationships. *Genomics* 67(2):201–9.

Kim S, Dougherty ER, Bittner ML, Chen Y, Sivakumar K, Meltzer P, Trent JM. (2000b) General nonlinear framework for the analysis of gene interaction via multivariate expression arrays. *J Biomed Opt* 5(4):411–24.

Kirienko GI. (1964) On self-correcting schemes from functional elements. *Problemy Kibernet* 12:29–37 (in Russian).

Lähdesmäki H, Shmulevich I, Yli-Harja O. (2003) On learning gene regulatory networks under the Boolean network model. *Mach Learning* 52:147–67.

Lambert JD. (1991) *Numerical Methods for Ordinary Differential Equations*, John Wiley & Sons, Chichester, UK.

Liang S, Fuhrman S, Somogyi R. (1998) REVEAL, a general reverse engineering algorithm for inference of genetic network architectures. *Pac Symp Biocomput* 3:18–29.

Luque B, Solé RV. (1997) Phase transitions in random networks: simple analytic determination of critical points. *Phys Rev E* 55:257–60.

Luque B, Solé RV. (2000) Lyapunov exponents in random Boolean networks. *Phys A* 284:33–45.

Lynch JF. (1995) On the threshold of chaos in random Boolean cellular automata. *Random Structures Algorithms* 6:239–60.

Ma HW, Zeng AP. (2003) The connectivity structure, giant strong component and centrality of metabolic networks. *Bioinformatics* 19(11):1423–30.

MacLeod MC. (1996) A possible role in chemical carcinogenesis for epigenetic, heritable changes in gene expression. *Mol Carcinog* 15(4):241–50.

Maki Y, Tominaga D, Okamoto M, Watanabe S, Eguchi Y. (2001) Development of a system for the inference of large scale genetic networks. *Pac Symp Biocomput* 6:446–58.

McAdams HH, Arkin A. (1999) It's a noisy business! Genetic regulation at the nanomolar scale. *Trends Genet* 15(2):65–9.

Mendes P. (1993) GEPASI: a software package for modelling the dynamics, steady states and control of biochemical and other systems. *Comput Appl Biosci* 9(5): 563–71.

Mendoza L, Thieffry D, Alvarez-Buylla ER. (1999) Genetic control of flower morphogenesis in *Arabidopsis thaliana*: a logical analysis. *Bioinformatics* 15(7–8): 593–606.

Mestl T, Plahte E, Omholt SW. (1995) A mathematical framework for describing and analysing gene regulatory networks. *J Theor Biol* 176(2):291–300.

Milgram S. (1967) The small world problem. *Psychology Today*, May 1967, 60–7.

Mircean C, Tabus I, Astola J, Kobayashi T, Shiku H, Yamaguchi M, Shmulevich I, Zhang W. (2004) Quantization and similarity measure selection for discrimination of lymphoma subtypes under *k*-nearest neighbor classification. *BIOS 2004*, San Jose, CA, January 2004.

Moriyama T, Shinohara A, Takeda M, Maruyama O, Goto T, Miyano S, Kuhara S. (1999) A system to find genetic networks using a weighted network model. *Genome Inform Ser Workshop Genome Inform* 10:186–95.

Muchnik AA, Gindikin SG. (1962) On the completeness of systems of unreliable elements which realize functions of the algebra of logic. *Dokl Akad Nauk SSSR* 144:1007–10 (in Russian).

Noda K, Shinohara A, Takeda M, Matsumoto S, Miyano S, Kuhara S. (1998) Finding a genetic network from experiments by a weighted network model. *Genome Inform Ser Workshop Genome Inform* 9:141–50.

Öktem H, Pearson R, Egiazarian K. (2003) An adjustable aperiodic model class of genomic interactions using continuous time Boolean networks (Boolean delay equations). *Chaos* 13(4):1167–74.

Oosawa C, Savageau MA. (2002) Effects of alternative connectivity on the behavior of randomly constructed Boolean networks. *Phys D* 170:143–61.

Pitt L, Valiant LG. (1988) Computational limitations on learning from examples. *J ACM* 35:965–84.

Post E. (1921) Introduction to a general theory of elementary propositions. *Am J Math* 43:163–85.

Post E. (1941) *Two-valued Iterative Systems of Mathematical Logic*, Princeton University Press, Princeton, NJ.

Przulj N, Wigle DA, Jurisica I. (2004) Functional topology in a network of protein interactions. *Bioinformatics* 20(3):340–8.

Ramsey S, Orrell D, Bolouri H. (2005) Dizzy: stochastic simulations of large-scale genetic regulatory networks. *J Bioinform Comput Biol* 3(2):1–21.

Rissanen J. (1978) Modelling by shortest data description. *Automatica* 14:465–71.

Rissanen J. (1986) Stochastic complexity and modelling. *Ann Statist* 14:1080–100.

Samsonova MG, Savostyanova EG, Serov VN, Spirov AV, Reinitz J. (1998) GeNet, a database of genetic networks. *Proceedings of the 1st International Conference on Bioinformatics of Genome Regulation and Structure, BGRS'98*, Novosibirsk, pp. 105–10.

Sánchez L, Thieffry D. (2001) A logical analysis of the *Drosophila* gap-gene system. *J Theor Biol* 211(2):115-41.

Sánchez L, Thieffry D. (2003) Segmenting the fly embryo: a logical analysis of the pair-rule cross-regulatory module. *J Theor Biol* 224(4):517-37.

Scharf LL. (1991) *Statistical Signal Processing*, Addison-Wesley, Reading, MA.

Serov VN, Spirov AV, Samsonova MG. (1998) Graphical interface to the genetic network database GeNet. *Bioinformatics* 14(6):546–7.

Serra R, Villani M, Semeria A. (2004) Genetic network models and statistical properties of gene expression data in knock-out experiments. *J Theor Biol* 227(1): 149–57.

Shmulevich I, Gabbouj M, Astola J. (2001) Complexity of the consistency problem for certain Post classes. *IEEE Trans Systems Man Cybern B* 31(2):251-3.

Shmulevich I, Saarinen A, Yli-Harja O, Astola J. (2002) Inference of genetic regulatory networks via best-fit extensions. In *Computational and Statistical Approaches to Genomics*, Zhang W, Shmulevich I, eds., Kluwer Academic Publishers, Boston, pp. 197–210.

Shmulevich I, Lähdesmäki H, Dougherty ER, Astola J, Zhang W. (2003) The role of certain Post classes in Boolean network models of genetic networks. *Proc Natl Acad Sci USA* 100(19):10734–9.

Shmulevich I, Kauffman SA. (2004) Activities and sensitivities in Boolean network models. *Phys Rev Lett* 93(4):048701.

Shmulevich I, Kauffman SA, Aldana M. (2005) Eukaryotic cells are dynamically ordered or critical but not chaotic. *Proc Natl Acad Sci USA* 102(38):13439–44.

Silvescu A, Honavar V. (1997) Temporal Boolean network models of genetic networks and their inference from gene expression time series. *Complex Syst* 13:54–75.

Smolen P, Baxter DA, Byrne JH. (2000) Modeling transcriptional control in gene networks—methods, recent results, and future directions. *Bull Math Biol* 62(2):247–92.

Snoussi EH. (1989) Qualitative dynamics of piecewise-linear differential equations: a discrete mapping approach. *Dyn Stability Syst* 4:189–207.

Spellman PT, Sherlock G, Zhang MQ, Iyer VR, Anders K, Eisen MB, Brown PO, Botstein D, Futcher B. (1998) Comprehensive identification of cell cycle-regulated genes of the yeast *Saccharomyces cerevisiae* by microarray hybridization. *Mol Biol Cell* 9(12):3273–97.

Stauffer D. (1987) On forcing functions in Kauffman's random Boolean networks. *J Stat Phys* 46, 789–94.

Suh EB, Dougherty ER, Kim S, Bittner ML, Chen Y, Russ DE, Martino R. (2002) Parallel computation and visualization tools for codetermination analysis of multivariate gene-expression relations. In *Computational and Statistical Approaches to Genomics*, Zhang W, Shmulevich I, eds., Kluwer Academic Publishers, Boston.

Tabus I, Astola J. (2001) On the use of MDL principle in gene expression prediction. *J Appl Signal Process* 2001(4):297–303.

Thieffry D, Colet M, Thomas R. (1993) Formalisation of regulatory nets: a logical method and its automatization. *Math Model Sci Comput* 2:144–51.

Thieffry D, Thomas R. (1995) Dynamical behaviour of biological networks: II. Immunity control in bacteriophage lambda. *Bull Math Biol* 57(2):277–97.

Thomas R, D'Ari R. (1990) *Biological Feedback* CRC Press, Boca Raton, FL.

Thomas R, Kaufman M. (2001a) Multistationarity, the basis of cell differentiation and memory. I. Structural conditions of multistationarity and other nontrivial behavior. *Chaos* 11(1):170–9.

Thomas R, Kaufman M. (2001b) Multistationarity, the basis of cell differentiation and memory. II. Logical analysis of regulatory networks in terms of feedback circuits. *Chaos* 11(1):180–95.

Uetz P, Giot L, Cagney G, Mansfield TA, Judson RS, Knight JR, Lockshon D, Narayan V, Srinivasan M, Pochart P, Qureshi-Emili A, Li Y, Godwin B, Conover D, Kalbfleisch T, Vijayadamodar G, Yang M, Johnston M, Fields S, Rothberg JM. (2000) A comprehensive analysis of protein-protein interactions in *Saccharomyces cerevisiae. Nature* 403(6770):623–7.

von Neumann J. (1956) Probabilistic logics and the synthesis of reliable organisms from unreliable components. In *Automata Studies*, Shannon CE, McCarthy J, eds., Princeton University Press, Princeton, NJ, pp. 43–98.

Wagner A. (2001) The yeast protein interaction network evolves rapidly and contains few redundant duplicate genes. *Mol Biol Evol* 18(7):1283–92.

Watts DJ, Strogatz SH. (1998) Collective dynamics of "small-world" networks. *Nature* 393(6684):440–2.

Whitfield ML, Sherlock G, Saldanha AJ, Murray JI, Ball CA, Alexander KE, Matese JC, Perou CM, Hurt MM, Brown PO, Botstein D. (2002) Identification of genes periodically expressed in the human cell cycle and their expression in tumors. *Mol Biol Cell* 13(6):1977–2000.

Wolf DM, Eeckman FH. (1998) On the relationship between genomic regulatory element organization and gene regulatory dynamics. *J Theor Biol* 195(2):167–86.

Yablonsky SV, Gavrilov GP, Kudryavtsev VB. (1966) *Functions of the Algebra of Logic and Post Classes*, Nauka, Moscow (in Russian).

Yook SH, Oltvai ZN, Barabasi AL. (2004) Functional and topological characterization of protein interaction networks. *Proteomics* 4(4):928–42.

Zhou X, Kao MC, Wong WH. (2002) Transitive functional annotation by shortest-path analysis of gene expression data. *Proc Natl Acad Sci USA* 99(20):12783–8.

# Chapter Three

## Stochastic Models of Gene Networks

Stochastic models of genetic regulatory networks differ from their deterministic counterparts by incoporating randomness or uncertainty. Most deterministic models can be generalized such that we associate probabilities with particular components or aspects of the model. For example, in an undirected graph model, instead of specifying whether an edge between two vertices is present or absent, we can associate a probability with that event. Thus, an edge can be present with probability 0.8, in essence making it a probabilistic edge. For example, because of the uncertainty inherent in making inferences from data such as from yeast two-hybrid experiments (Ito et al., 2001; Uetz et al., 2000), we may find that protein A interacts with protein B with a probability of 0.8 and with protein C with a probability of 0.2. Indeed, the data are known to include up to 50 percent false-positive interactions (Von Mering et al., 2002). At the same time, the probabilities may reflect actual biological uncertainty in that A may sometimes interact with B and sometimes with C.

Similarly, models that are capable of representing dynamical processes can be extended into a probabilistic framework as well. Thus, the values of the variables in the model can be described by probability distributions, typically incorporating multivariate relationships between several variables in terms of joint probability distributions. Dynamic Bayesian networks (DBNs) have this property. As we shall see later, the model structure can also be endowed with a probability distribution, as in the case of probabilistic Boolean networks. Finally, models such as ordinary differential equations can be extended by adding noise terms, as in Langevin equations.

## 3.1 BAYESIAN NETWORKS

A Bayesian network is essentially a compact representation of a joint probability distribution (Pearl, 1988; Jensen, 2001; Lauritzen, 1996). This representation takes the form of a directed acyclic graph (DAG) in which the nodes (vertices) of the graph represent random variables and the directed edges or lack thereof represent conditional dependencies or independencies. The other ingredient in this model is the specification of a conditional probability distribution for each of the random variables (nodes). This distribution, which in the case of discrete-valued variables takes the form of a conditional probability table, specifies the probability of a *child* node taking on a particular value given the values of its *parent* nodes. For example, in the case of genetic networks, the values of the nodes can correspond to

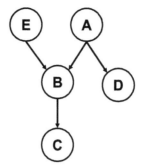

Figure 3.1.  An example of a Bayesian network consisting of five nodes.

(possibly discretized) gene expression levels or other measurable events, including external conditions.

Specifically, consider a random vector $\mathbf{X} = (X_1, \ldots, X_n)$ representing some measurements, such as expressions of genes. A Bayesian network $B = (G, \Theta)$ is defined by (1) a DAG $G$ whose vertices correspond to $X_1, \ldots, X_n$ and (2) a set $\Theta$ of local conditional probability distributions for each vertex $X_i$ given its parents in the graph. The graph encodes the *Markov assumption*, which states that each variable $X_i$ is independent of its nondescendants given its parents in $G$. Using the chain rule of probabilities, any joint distribution satisfying the Markov assumption can be decomposed into a product of the local conditional probabilities. Let us denote by $\mathrm{Pa}(X_i)$ the Markovian parents of $X_i$ in the graph $G$. Then,

$$P\{x_1, \ldots, x_n\} = \prod_{i=1}^{n} P\{x_i | \mathrm{Pa}(x_i)\}. \tag{3.1}$$

Consider the following example used by Friedman et al. (2000), which is shown in figure 3.1. The graph, consisting of five nodes, implies a number of conditional independencies. For example, the only nondescendant of $B$ is $D$, and the parents of $B$ are $A$ and $E$. Thus, according to the Markov assumption, $B$ is independent of $D$, given both $A$ and $E$. In other words,

$$P\{B|D, A, E\} = P\{B|A, E\}.$$

Similarly,

$$P\{D|B, C, E, A\} = P\{D|A\}$$

since $B$, $C$, and $E$ are nondescendants of $D$ and $A$ is the only parent of $D$. There are other such independencies implied by this graph. The joint distribution, then, factors as

$$P\{A, B, C, D, E\} = P\{A\}P\{E\}P\{B|A, E\}P\{D|A\}P\{C|B\}. \tag{3.2}$$

Thus, a Bayesian network graphically represents a joint probability distribution in a compact and intuitive way.

To see why such a representation is more compact, consider the simplest case where nodes are binary-valued. Then, to represent the joint distribution of $A$, $B$, $C$, $D$, and $E$ in the example above, one would need to store a conditional probability table (CPT) containing $2^5 = 32$ entries, one for each combination of the 5 variables. However, with the factored form given in equation (3.2), we need only 1 entry to store $P\{A\}$, 1 entry to store $P\{E\}$, 4 entries to store $P\{B|A, E\}$, and 2 entries each to store $P\{D|A\}$ and $P\{C|B\}$, for a total of 10 entries. As an arbitrary example, the CPT for $P\{B|A, E\}$ might look something like this.

| $A$ | $E$ | $P\{B=1|A, E\}$ |
|-----|-----|------------------|
| 0 | 0 | 0.3 |
| 0 | 1 | 0.6 |
| 1 | 0 | 0.6 |
| 1 | 1 | 0.8 |

For large numbers of variables, the savings become much more appreciable. In general, for a binary-valued network with $n$ nodes in which there are no more than $k$ parents of any given node, the representation requires $O(n2^k)$ space as opposed to $O(2^n)$ for storing the entire joint probability distribution.

For continuous-valued random variables, a CPT clearly cannot be used. One common choice is to use linear Gaussian conditional densities,

$$P\{x|y_1, \ldots, y_k\} \sim N\left(a_0 + \sum_i a_i \cdot y_i, \sigma^2\right),$$

where the random variable corresponding to a node is normally distributed around a mean that is a linear function of the parents of this node (Friedman et al., 2000). Here the variance $\sigma^2$ is assumed to be constant and independent of the parents. If all the nodes have this form, the joint distribution is multivariate Gaussian.

As we discussed above, each Bayesian network implies a set of independence statements. It turns out that two different Bayesian networks can imply identical sets of such independence statements. As a trivial example, consider $X \to Y$ and $Y \to X$. Clearly, in each case, there are no independencies, and so both networks, although different, are said to be *equivalent* because they imply the same independencies. This is an important issue because when we infer the network from measurement data, as we will discuss below, we cannot distinguish between equivalent networks. One way around this difficulty is to perform targeted intervention experiments where genes are manipulated. This can help to narrow down the range of possible networks. An effective approach in selecting the targeted interventions is to choose those that tend to split large and highly likely equivalence classes into many small ones (Pournara and Wernisch, 2004).

*Inference*

One of the most common and important tasks concerning Bayesian networks is their probabilistic inference from measurement data. Let us pose the learning problem as follows. Given a set of data $D$, find the Bayesian network $B = (G, \Theta)$, meaning the graph structure along with its family of local conditional probability distributions, that best explains $D$. In view of the above discussion concerning equivalence classes, we are really finding the optimal equivalence class of networks. A detailed tutorial on learning Bayesian networks can be found in (Heckerman, 1998). Most approaches employ a *scoring function* that associates a score with a candidate network that reflects the network's predictive accuracy in light of the data. For example, the Bayesian score reflects the posterior probability of the graph $G$ given the data $D$. That is,

$$\text{BayesianScore}(G|D) = \log P\{G|D\} \tag{3.3}$$
$$= \log P\{D|G\} + \log P\{G\} + c$$

and

$$P\{D|G\} = \int P\{D|G, \Theta\} P\{\Theta|G\} \, d\Theta. \tag{3.4}$$

Equation (3.4) is known as the marginal likelihood, which essentially averages the probability of the data relative to the distribution of the parameters (local conditional probabilities). The term $P\{G\}$ in equation (3.3) is the prior probability of the network (model prior), and the term $P\{\Theta|G\}$ in equation (3.4) is the parameter prior. We can see that maximizing the Bayesian score is equivalent to maximum a posteriori (MAP) estimation and that the model prior $P\{G\}$ is effectively used to penalize model complexity, thus discouraging overfitting of the model to the data.

Exhaustively searching for the optimal network is problematic because as the number of nodes (genes) increases, the set of possible networks grows exponentially. In fact, it is known to be an NP-hard problem (Chickering, 1996). Thus, heuristic search methods are typically employed. For example, one could use a greedy approach whereby an edge is added, deleted, or reversed such that the score is maximally increased. Such an approach requires one to assume a useful property, namely, the decomposability of the score, which means that the overall score of the network is a sum of individual score contributions from each node $X_i$ and its parents $\text{Pa}(X_i)$ (Friedman et al., 2000). Thus, by adding or deleting an edge, one can immediately evaluate the effect on the overall score. One could also use simulated annealing or Markov chain Monte Carlo sampling (Friedman and Koller, 2000). Hartemink et al. (2002) observed that simulated annealing consistently found the highest scoring models among other approaches.

Another useful approach is to restrict the search space of the networks to facilitate efficient learning. One way to do this is to focus the search on relevant regions of the search space using the sparse candidate algorithm developed by Friedman et al. (1999). This approach restricts the set of possible parents of each node by using simple local statistics such as correlation. Thus, only those genes that are sufficiently correlated with a given gene can be its parents, thereby dramatically reducing the search space and yielding a solution much quicker. On the other hand, one of the dangers of such an approach is that restricting the parents in such a way during the early stages of the search can significantly reduce the search space and produce highly suboptimal solutions. Friedman et al. (2000) proposed an iterative algorithm that adapts the candidate parent sets during the search.

One of the most critical challenges in inferring the Bayesian network structure and parameters is doing so reliably and accurately in the face of the small sample sizes available for learning, especially relative to the number of genes and parameters to be inferred. This challenge pertains not only to Bayesian networks but also to all other network models. The number of microarrays, for instance, is typically in the tens or, in rare cases, in the several hundreds. However, the number of genes (nodes) in the network can be in the thousands. This is often referred to as the "curse of dimensionality" and can result in very poor predictive accuracy of the inferred model. A priori information can often help mitigate this problem. One common assumption is the sparsity of the regulatory network, which essentially means that the number of parent genes is limited. Pe'er et al. (2002) used a method that examines only those networks in which a small number of gene regulators explain the expression of all the other genes. Such a constraint not only forces the learning procedure to identify the most pronounced regulators but also makes the problem computationally simpler.

Another approach introduced by Segal et al. (2003) examines networks composed of modules of coregulated genes, where each module is controlled by a shared regulator. The learning procedure simultaneously constructs the modules and finds the regulators for each module that best predict the expression of its genes. The advantage of this approach is that far fewer parameters are required and the robustness of the inferred model is increased.

One may also be interested in calculating a statistical confidence measure of a certain feature of a network. A feature in this case, as defined by Friedman et al. (2000), is an indicator function whose input is a network (DAG) and whose output is either 1 or 0, depending on whether the network satisfies a given criterion. For instance, we might be interested in knowing whether there is an edge between $X$ and $Y$ or perhaps whether $X$ is an ancestor of $Y$. In a given network $G$, if this is so, then the feature $f(G) = 1$; otherwise, $f(G) = 0$. The posterior probability of a feature is

$$P\{f(G)|D\} = \sum_G f(G)P\{G|D\},$$

which is also the expectation of the feature since it is binary-valued. Note that $P\{G|D\} = \exp(\text{BayesianScore}(G|D))$. One approach in estimating the confidence of a feature is to use the bootstrap method (Efron and Tibshirani, 1993). Briefly, this entails generating $m$ data sets $D_1, \ldots, D_m$ by sampling $D$ (which is of size $n$) with replacement and then inferring the optimal network from each data set $D_i$, $i = 1, \ldots, m$, resulting in $m$ networks $G_1, \ldots, G_m$. Then, the confidence of the feature is defined as

$$\text{conf}(f) = \frac{1}{m} \sum_{i=1}^{m} f(G_i),$$

which can be interpreted as a probability: the closer this value is to 1, the more confidence we have in this feature.

Friedman et al. (2000) applied the sparse candidate algorithm and a 200-fold bootstrap procedure to infer a Bayesian network from the Spellman et al. (1998) data set consisting of 76 gene expression measurements of 6177 genes in yeast, with the inference applied to 800 genes whose expression level varied over the cell cycle. The data consist of six time series under different cell cycle synchronization methods. Despite this, each measurement was treated as an independent sample without taking into account the temporal nature of the data. To compensate for this, an additional variable (node) was introduced that denoted the cell cycle phase. Moreover, this node was forced to be the root node in all inferred networks. This design allows one to monitor the dependency of gene expression levels on the particular cell cycle phase. Some gene relations that were found were not revealed by the cluster analysis of Spellman et al. (1998), indicating a potentially more powerful methodology afforded by the multivariate probabilistic relationships represented by a Bayesian network in contrast to clustering.

Bayesian networks are also naturally well suited for capturing relationships between variables from multiple data sources. For example, Hartemink et al. (2002) used microarray expression data in conjunction with location data, where the latter were gathered from chromatin immunoprecipitation assays—a methodology that allows identification of the sites in a genome that are occupied by a gene regulatory protein. Such location data can be used to influence the model prior $P\{G\}$, for example, by requiring that certain edges be present (or absent). Hartemink et al. (2002) used such an approach to infer genetic regulatory models involved in yeast pheromone response. Imoto et al. (2004) used a similar approach to integrate microarray gene expression data, protein-protein interactions, protein-DNA interactions, binding site information, and existing literature.

### Dynamic Bayesian Networks

Although we have focused on Bayesian networks as representations of static probabilistic relationships, they can naturally be extended to the temporal domain and used to model stochastic processes, resulting in *dynamic Bayesian networks* (DBNs) (Murphy and Mian, 1999). Thus, DBNs generalize hidden Markov models and linear dynamical systems by representing the conditional dependencies and independencies between (state) variables over time. This added flexibility allows us to model feedback relationships, which we know are ubiquitous in genetic regulation, whereas standard Bayesian networks whose structure is a directed acyclic graph cannot. It is typically assumed that the DBN model is time-invariant (graph structure and parameters do not change over time). Let us restrict our attention to first-order Markov processes in $X$, that is, to processes whose transition probability obeys

$$P\{X(t)|X(0), X(1), \ldots, X(t-1)\} = P\{X(t)|X(t-1)\}.$$

A DBN that represents the joint probability distribution over all possible time series of variables in $\mathbf{X}$ consists of two parts: (1) an initial Bayesian network $B_0 = (G_0, \Theta_0)$ that defines the joint distribution of the variables in $\mathbf{X}(0)$, and (2) a transition Bayesian network $B_1 = (G_1, \Theta_1)$ that specifies the transition probabilities $P\{\mathbf{X}(t)|\mathbf{X}(t-1)\}$ for all $t$. Here boldface indicates vector-valued random variables. Thus, a DBN is defined by a pair $(B_0, B_1)$. The structure of DBNs is restricted in two ways. First, the directed acyclic graph $G_0$ in the initial

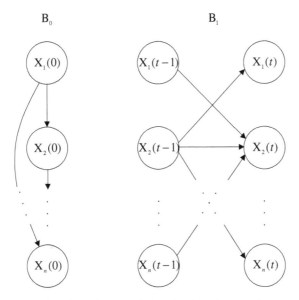

Figure 3.2. An example of the basic building blocks of dynamic Bayesian networks $(B_0, B_1)$. $B_0$ and $B_1$ are the initial and transition Bayesian networks, respectively. (Lähdesmäki et al., 2006.)

Bayesian network $B_0$ is assumed to have only "within-slice" connections; that is, $\text{Pa}(X_i(0)) \subseteq \mathbf{X}(0)$ for all $1 \leq i \leq n$. Second, because of the first-order Markovian assumption, we also constrain the variables in time slice $\mathbf{X}(t)$, $t > 0$, to have all their parents in slice $t - 1$; that is, $\text{Pa}(X_i(t)) \subseteq \mathbf{X}(t - 1)$ for all $1 \leq i \leq n$ and $t > 0$. An example of the basic building blocks of DBNs, $B_0$ and $B_1$, is shown in figure 3.2.

Using equation (3.1) and the assumptions on the initial and transition Bayesian networks discussed above, the joint probability over a finite set of random variables $\mathbf{X}(0) \cup \mathbf{X}(1) \cup \cdots \cup \mathbf{X}(T)$ can be expressed as (Friedman et al., 1998)

$$P\{\mathbf{X}(0), \mathbf{X}(1), \ldots, \mathbf{X}(T)\} = P\{\mathbf{X}(0)\} \prod_{t=1}^{T} P\{\mathbf{X}(t)|\mathbf{X}(t - 1)\}$$

$$= \prod_{i=1}^{n} P\{X_i(0)|\text{Pa}(X_i(0))\} \prod_{t=1}^{T} \prod_{j=1}^{n} P\{X_j(t)|\text{Pa}(X_j(t))\}.$$

DBNs have also been used to infer genetic regulatory networks from gene expression data (Murphy and Mian, 1999; Kim et al., 2003; Perrin et al., 2003; Zou and Conzen, 2005). Bernard and Hartemink (2005) have used dynamic Bayesian networks for learning dynamical models of transcriptional regulatory networks from gene expression data and transcription factor binding location data.

## 3.2 PROBABILISTIC BOOLEAN NETWORKS

As already discussed, recent research indicates that many biological questions can be tackled within the seemingly simplistic Boolean formalism, which emphasizes high-level relations between observable variables, not biochemical details (Huang, 1999, 2001; Huang and Ingber, 2000; Somogyi and Sniegoski, 1996). For example, Albert and Othmer (2003) constructed the wiring diagram of the interactions seven segment polarity genes in *Drosophila* and formulated a Boolean description of this network in which the state of each mRNA and protein is either 1 (on) or 0 (off). The topology and Boolean functions in the network were based on work published by von Dassow et al. (2000), who used 13 nonlinear differential equations with 48 kinetic parameters.

The Boolean model was validated by showing that it reproduces the "wild type" or normal behavior of the system. Starting from the known initial state of the segment polarity genes, the network dynamics were simulated, and it was found that the gene expression pattern stabilized into patterns that are observed experimentally. Furthermore, by simulating the effects of gene mutations, it was found that the spatial pattern of expression predicted by the model was in perfect agreement with experiments (Albert and Othmer, 2003). These results indicated that a detailed kinetic model may not be needed at this level of description and that the topology and nature of the connections (activating or repressing) is more important to account for the dynamics.

However, despite the success of Boolean networks in reproducing the behavior of genetic networks, when the model structure needs to be inferred from measurement data, it is unlikely that the determinism of the Boolean network model will be concordant with the data. As discussed in the previous chapter, one could pick the predictor set (input genes) with the highest coefficient of determination, although because of a typically small number of measurements, there may be a number of almost equally performing predictor sets and the CoDs we have for them are only estimates from the data. By associating several predictor sets with each target gene, it is not possible to obtain with certainty the transcriptional status of the target gene at the next time point; however, one can compute the probability that the target gene will be transcriptionally active at time $t + 1$ based on the gene activity profile at time $t$.

The time evolution of the gene activity profile then defines a stochastic dynamical system. Since the gene activity profile at a particular time point depends only on the profile at the immediately preceding time point, the dynamical system is Markovian. Such systems can be studied in the established framework of Markov chains and Markov decision processes. These ideas are mathematically formalized in probabilistic Boolean networks (PBNs), which we now discuss (Shmulevich et al., 2002a,b).

As mentioned above, from an empirical point of view, the assumption of only one logical rule per gene (as is the case in Boolean networks) may lead to incorrect conclusions when inferring these rules from gene expression measurements, as the latter are typically noisy and the number of samples is small relative to the number of parameters to be inferred. PBNs were originally introduced to cope with this kind of uncertainty.

There are various reasons for utilizing a probabilistic network. A model incorporates only a partial description of a physical system. This means that a Boolean function giving the next state of a variable is likely to be only partially accurate.

There are occasions when different Boolean functions may actually describe the transition, but these are outside the scope of the conventional Boolean model. If, consequently, we are uncertain as to which transition rule should be used, then a probabilistic Boolean network involving a set of possible Boolean functions for each variable may be more suitable than a network in which there is only a single function for each variable.

Even if one is fairly confident that a model is sufficiently robust that other variables can be ignored without significant impact, there remains the problem of inferring the Boolean functions from sample data. In the case of gene expression microarrays, the data are severely limited relative to the number of variables in the system. Should it happen that a particular Boolean function has even a moderately large number of essential variables, then its design from the data is likely to be imprecise because the number of possible input states will be too large for precise estimation. This situation is exacerbated if some essential variables are either unknown or unobservable. As a consequence of the inability to observe sufficient examples to design the transition rule, it is necessary to restrict the number of variables over which a function is defined. For each subset of the full set of essential variables, there may be an optimal function in the sense that the prediction error is minimized for that function given the variables in the subset. These optimal functions must be designed from sample data. Owing to inherent imprecision in the design process, it may be prudent to allow a random selection from among several functions, with the weight of selection based on a probabilistic measure of worth such as the CoD.

A final consideration is that one may wish to model an open system rather than a closed system. An open system has inputs (stimuli). Depending upon a particular external condition at a given moment in time, the system may transition differently than it would in the absence of this condition. Such effects have been considered in the framework of using the coefficient of determination in the presence of external stresses (Kim et al., 2000). Under the assumption that external stimuli affect the transition rules, it is prudent to allow uncertainty among the transition rules and weight their likelihood accordingly. It may be that the probability of applying a Boolean function corresponding to an unlikely condition is low; however, system behavior might be seriously misunderstood if the possibility of such a transition is ignored. The concept of an open system plays a key role in distinguishing probabilistic approaches to Boolean networks.

To set the stage, let us first reexamine standard Boolean network models and their dynamics in the probabilistic context of Markov chains. PBNs represent an interface between the absolute determinism of Boolean networks and the probabilistic nature of Bayesian networks in that they incorporate rule-based uncertainty. This compromise is important because rule-based dependencies between genes are biologically meaningful, while mechanisms for handling uncertainty are conceptually and empirically necessary.

Since Boolean networks are completely deterministic, so are their dynamics. This can be captured by considering the joint probability distribution of all the genes. In order for us to have a useful probabilistic description of the dynamics of such systems, it is necessary to consider the joint probabilities of all Boolean functions corresponding to all the nodes because even after one step of the network, the nodes become dependent, regardless of assumptions on the initial distribution.

To this end, suppose we are given a Boolean network containing $n$ nodes (genes) $x_1, \ldots, x_n$ and an initial joint probability distribution $D(\mathbf{x})$, $\mathbf{x} \in \{0, 1\}^n$, over the $n$-dimensional hypercube. We are interested in computing the joint probability of every node after one step of the network. It is easy to see that

$$P\{f_1(\mathbf{x}) = i_1, f_2(\mathbf{x}) = i_2, \ldots, f_n(\mathbf{x}) = i_n\} = \sum_{\mathbf{x} \in \{0,1\}^n : f_k(\mathbf{x}) = i_k, k = 1, \ldots, n} D(\mathbf{x}), \quad (3.5)$$

where $i_k \in \{0, 1\}$. Equation (3.5) can then be used in an iterative fashion, following the dynamical nature of the system. In other words, the computed joint distribution $P\{f_k(\mathbf{x}) = i_k, k = 1, \ldots, n\}$ can be used in place of $D(\mathbf{x})$ to compute the joint distribution at the next time point. This defines the iterative system

$$D^{t+1} = \Psi(D^t), \quad (3.6)$$

where the mapping $\Psi : [0, 1]^{2^n} \to [0, 1]^{2^n}$ is implicitly defined by equation (3.5). In fact, $\Psi$ is an affine mapping. To see this, let $\mathbf{D}^t$ and $\mathbf{D}^{t+1}$ be $1 \times 2^n$ vectors containing the joint probabilities in $D^t$ and $D^{t+1}$ and let the matrix $\mathbf{A}$ be a $2^n \times 2^n$ binary matrix defined as

$$\mathbf{A}_{ij} = \left. \begin{cases} 1 & \text{if } \exists\, \mathbf{x} \in \{0, 1\}^n, \ C(f_1(\mathbf{x}), \ldots, f_n(\mathbf{x})) = j, \ C(x_1, \ldots, x_n) = i, \\ 0 & \text{otherwise,} \end{cases} \right\}$$
$$(3.7)$$

where $C(i_1, \ldots, i_n) = 1 + \sum_{j=1}^{n} 2^{n-j} \cdot i_j$ and each $i_j \in \{0, 1\}$. Thus, $i$ and $j$ are simply indices such that $i$ is the integer representation of the binary vector $\mathbf{x} = (x_1, \ldots, x_n)$ and $j$ encodes the binary vector $(f_1(\mathbf{x}), \ldots, f_n(\mathbf{x}))$. This representation is efficient because the matrix $\mathbf{A}$ contains exactly one nonzero entry in each row. Thus, equation (3.6) can be written as

$$\mathbf{D}^{t+1} = \mathbf{D}^t \cdot \mathbf{A} \quad (3.8)$$
$$= \mathbf{D}^0 \cdot \mathbf{A}^{t+1},$$

where $\mathbf{D}^0$ is the vector corresponding to the starting (prior) joint distribution $D(\mathbf{x})$. Equation (3.8) is the familiar Markov chain representation, where the state transition matrix $\mathbf{A}$ is binary. This is to be expected since the state transitions are completely specified by the Boolean functions and the probability of transition can be either 0 or 1.

### 3.2.1 Definitions

The basic idea behind PBNs is to extend the Boolean network model to accommodate more than one possible function for each node. Thus, to every node $x_i$, there corresponds a set

$$F_i = \left\{ f_j^{(i)} \right\}_{j=1,\ldots,l(i)}, \quad (3.9)$$

where each $f_j^{(i)}$ is a possible function determining the value of gene $x_i$ and $l(i)$ is the number of possible functions assigned to gene $x_i$. We will also refer to the functions $f_j^{(i)}$ as *predictors* since the process of inferring these functions from measurements or, equivalently, the process of producing a minimum-error estimate of the value of a gene at the next time point is known as prediction in estimation theory. A realization of the PBN at a given instant in time is determined by a vector of Boolean functions. If there are $s_0$ possible realizations, then there are $s_0$ vector functions $\mathbf{f}_1, \mathbf{f}_2, \ldots, \mathbf{f}_{s_0}$ of the form $\mathbf{f}_k = (f_{k_1}^{(1)}, f_{k_2}^{(2)}, \ldots, f_{k_n}^{(n)})$ for $k = 1, 2, \ldots, s_0$, $1 \le k_i \le l(i)$ and where $f_{k_i}^{(i)} \in F_i$ $(i = 1, \ldots, n)$. In other words, the vector function $\mathbf{f}_k : \{0, 1\}^n \to \{0, 1\}^n$ acts as a transition function (mapping) representing a possible realization of the entire PBN. Thus, given the values of all the genes $(x_1, \ldots, x_n)$, $\mathbf{f}_k(x_1, \ldots, x_n) = (x_1', \ldots, x_n')$ gives us the state of the genes after one step in the network given by the realization $\mathbf{f}_k$.

Now, let $\mathbf{F} = (f^{(1)}, \ldots, f^{(n)})$ be a random vector taking values in $F_1 \times \cdots \times F_n$. That is, $\mathbf{F}$ can take on all possible realizations of the PBN. Then, the probability that predictor $f_j^{(i)}$ is used to predict gene $i$ $(1 \le j \le l(i))$ is equal to

$$c_j^{(i)} = P\{f^{(i)} = f_j^{(i)}\} = \sum_{k: f_{k_i}^{(i)} = f_j^{(i)}} P\{\mathbf{F} = \mathbf{f}_k\}. \tag{3.10}$$

Since the $c_j^{(i)}$ are probabilities, they must satisfy

$$\sum_{j=1}^{l(i)} c_j^{(i)} = 1. \tag{3.11}$$

It is not necessary that the selection of the Boolean functions composing a specific network be independent. This means that it is not necessarily the case that

$$P\{f^{(i)} = f_j^{(i)}, f^{(l)} = f_k^{(l)}\} = P\{f^{(i)} = f_j^{(i)}\} \cdot P\{f^{(l)} = f_k^{(l)}\}.$$

A PBN is said to be *independent* if the random variables $f^{(1)}, f^{(2)}, \ldots, f^{(n)}$ are independent. In the dependent case, product expansions such as the one given in the preceding equation, as well as ones involving more functions, require conditional probabilities. Figure 3.3 illustrates the basic building block of a PBN. A number of predictors share common inputs, while their outputs are synthesized, in this case by random selection, into a single output. This type of structure is known as a *synthesis filter bank* in the digital signal processing literature. The wiring diagram for the entire PBN consists of $n$ such building blocks.

The above definitions refer to *instantaneously random* PBNs. An alternative approach is to take the view that the data on the microarrays come from distinct sources, each representing a context of the cell. That is, the data derive from a family of deterministic networks, and if we were able to separate the samples according to context, there would in fact be CoDs with value 1, indicating deterministic biochemical activity for the wiring of a particular constituent network. From this perspective, the only reason that it is not possible to find predictor sets with a CoD

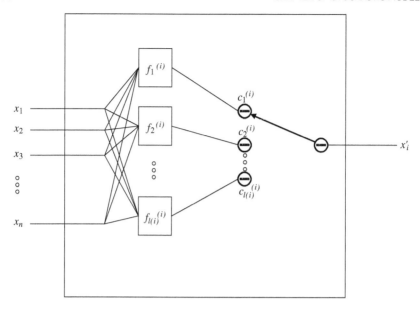

Figure 3.3.  A basic building block of a probabilistic Boolean network. (Shmulevich et al., 2002a.)

equal (or very close) to 1 is that they represent averages across the various cellular contexts with their correspondingly various wirings.

This perspective leads to the view that a PBN is a collection of Boolean networks in which one constituent network governs gene activity for a random period of time before another randomly chosen constituent network takes over, possibly in response to some random event such as an external stimulus. Since the latter is not part of the model, network switching is random. This model defines a *context-sensitive* PBN. The probabilistic nature of the constituent choice reflects the fact that the system is open, not closed. Context sensitivity results not only from the effects of variables outside the genome but also from the pragmatic condition that a genetic regulatory model represents a very small number of genes and is therefore susceptible to the effects of latent genomic variables. The context-sensitive model reduces to the instantaneously random model by having network switching at every time point.

An instantaneously random probabilistic Boolean network $G(V, F)$ is defined by a set of nodes $V = \{x_1, \ldots, x_n\}$ and the list $F = (F_1, \ldots, F_n)$, where the latter is defined in equation (3.9). Assuming independence, $s_0 = \prod_{i=1}^{n} l(i)$ is the number of possible PBN realizations. If $l(i) = 1$ for all $i = 1, \ldots, n$, then $s_0 = 1$ and the PBN reduces to the standard Boolean network.

The dynamics of an instantaneously random PBN are essentially the same as in Boolean networks, but at any given point in time, the value of each node is determined by one of the possible predictors chosen according to its corresponding probability. This can be interpreted by saying that at any point in time, we have one out of $s_0$ possible networks. In order to calculate the probability that a particular network is selected, let us define the matrix

$$
\mathbf{K} =
\begin{bmatrix}
1 & 1 & \cdots & 1 & 1 \\
1 & 1 & \cdots & 1 & 2 \\
\vdots & \vdots & \ddots & \vdots & \vdots \\
1 & 1 & \cdots & 1 & l(n) \\
1 & 1 & \cdots & 2 & 1 \\
1 & 1 & \cdots & 2 & 2 \\
\vdots & \vdots & \ddots & \vdots & \vdots \\
1 & 1 & \cdots & 2 & l(n) \\
\vdots & \vdots & \ddots & \vdots & \vdots \\
l(1) & l(2) & \cdots & l(n-1) & l(n)
\end{bmatrix}
$$

containing lexicographically ordered rows, each one corresponding to a possible network configuration. That is, row $m$ corresponds to network $m$, and the entry $j$ in the $i$th column specifies that predictor $f_j^{(i)}$ should be used for gene $x_i$. It is easy to see that $\mathbf{K}$ is an $s_0 \times n$ matrix. Using this matrix, the probability that network $i$ is selected is

$$
p_i = P\{\text{network } i \text{ is selected}\} = \prod_{j=1}^{n} c_{\mathbf{K}_{ij}}^{(j)}, \tag{3.12}
$$

where $\mathbf{K}_{ij}$ is the $ij$th entry in matrix $\mathbf{K}$. It can easily be checked that $\sum_{i=1}^{s_0} p_i = 1$ by noting that

$$
\sum_{i=1}^{s_0} p_i = \sum_{i=1}^{s_0} \prod_{j=1}^{n} c_{\mathbf{K}_{ij}}^{(j)} = \prod_{j=1}^{n} \sum_{i=1}^{l(j)} c_i^{(j)} = \prod_{j=1}^{n} 1 = 1, \tag{3.13}
$$

where we have used equation (3.11).

Now, let us consider the state space of the PBN, also consisting of $2^n$ states. As in the case of equation (3.7), we are interested in establishing the state transition matrix $\mathbf{A}$. In this case, however, the network may transition from a state to a number of other possible states, hence defining a random process. The probability of transitioning from state $(x_1, \ldots, x_n)$ to state $(x_1', \ldots, x_n')$ can be obtained as

$$
P\{(x_1, \ldots, x_n) \to (x_1', \ldots, x_n')\} \tag{3.14}
$$

$$
= \sum_{i: f_{\mathbf{K}_{i1}}^{(1)}(x_1, \ldots, x_n) = x_1', f_{\mathbf{K}_{i2}}^{(2)}(x_1, \ldots, x_n) = x_2', \ldots, f_{\mathbf{K}_{in}}^{(n)}(x_1, \ldots, x_n) = x_n'} p_i
$$

$$
= \sum_{i=1}^{s_0} p_i \underbrace{\left[ \prod_{j=1}^{n} \left( 1 - \left| f_{\mathbf{K}_{ij}}^{(j)}(x_1, \ldots, x_n) - x_j' \right| \right) \right]}_{\in \{0,1\}},
$$

where in the last expression binary values are treated as real values. Since $p_i = P\{\text{network } i \text{ is selected}\}$, equation (3.14) can be interpreted as

$$P\{(x_1, \ldots, x_n) \rightarrow (x_1', \ldots, x_n')\}$$

$$= \sum_{i=1}^{s_0} P\{(x_1, \ldots, x_n) \rightarrow (x_1', \ldots, x_n') \mid \text{network } i \text{ is selected}\} \cdot p_i$$

and $P\{(x_1, \ldots, x_n) \rightarrow (x_1', \ldots, x_n') \mid \text{network } i \text{ is selected}\} \in \{0, 1\}$, as is the case for standard Boolean networks. By using equations (3.13) and (3.14) and the fact that for any $(x_1, \ldots, x_n)$ there always exists an $(x_1', \ldots, x_n')$ and $i$ such that

$$\prod_{j=1}^{n} \left(1 - \left| f_{\mathbf{K}_{ij}}^{(j)}(x_1, \ldots, x_n) - x_j' \right| \right) = 1,$$

we have that

$$\sum_{j=1}^{2^n} \mathbf{A}_{ij} = 1$$

for any $i = 1, \ldots, 2^n$. Thus, $\mathbf{A}$ is also a Markov matrix and the PBN is a homogeneous Markov process, meaning that the transition probabilities do not change with time. It also follows that $\mathbf{A}$ has at most $s_0 \cdot 2^n$ nonzero entries and reduces to the binary state transition matrix when $s_0 = 1$, given in equation (3.7). Let us consider an example PBN illustrating the above constructions.

**Example 3.1**  *Suppose we are given a PBN consisting of three genes $V = (x_1, x_2, x_3)$ and the function sets $F = (F_1, F_2, F_3)$, where $F_1 = \{f_1^{(1)}, f_2^{(1)}\}$, $F_2 = \{f_1^{(2)}\}$, and $F_3 = \{f_1^{(3)}, f_2^{(3)}\}$. Let the functions be given by the following truth tables.*

| $x_1 x_2 x_3$ | $f_1^{(1)}$ | $f_2^{(1)}$ | $f_1^{(2)}$ | $f_1^{(3)}$ | $f_2^{(3)}$ |
|---|---|---|---|---|---|
| 000 | 0 | 0 | 0 | 0 | 0 |
| 001 | 1 | 1 | 1 | 0 | 0 |
| 010 | 1 | 1 | 1 | 0 | 0 |
| 011 | 1 | 0 | 0 | 1 | 0 |
| 100 | 0 | 0 | 1 | 0 | 0 |
| 101 | 1 | 1 | 1 | 1 | 0 |
| 110 | 1 | 1 | 0 | 1 | 0 |
| 111 | 1 | 1 | 1 | 1 | 1 |
| $c_j^{(i)}$ | 0.6 | 0.4 | 1 | 0.5 | 0.5 |

*Since there are 2 functions for node $x_1$, 1 function for node $x_2$, and 2 functions for node $x_3$, there are $s_0 = 4$ possible networks and matrix $\mathbf{K}$ is equal to*

$$\mathbf{K} = \begin{bmatrix} 1 & 1 & 1 \\ 1 & 1 & 2 \\ 2 & 1 & 1 \\ 2 & 1 & 2 \end{bmatrix}.$$

*For example, the second row of* **K** *containing* $(1, 1, 2)$ *means that the predictors* $(f_1^{(1)}, f_1^{(2)}, f_2^{(3)})$ *will be used. Finally, by using equation* (3.14), *the state transition matrix* **A** *is given by*

$$\mathbf{A} = \begin{bmatrix} 1 & 0 & 0 & 0 & 0 & 0 & 0 & 0 \\ 0 & 0 & 0 & 0 & 0 & 0 & 1 & 0 \\ 0 & 0 & 0 & 0 & 0 & 0 & 1 & 0 \\ p_4 & p_3 & 0 & 0 & p_2 & p_1 & 0 & 0 \\ 0 & 0 & 1 & 0 & 0 & 0 & 0 & 0 \\ 0 & 0 & 0 & 0 & 0 & 0 & p_2 + p_4 & p_1 + p_3 \\ 0 & 0 & 0 & 0 & p_2 + p_4 & p_1 + p_3 & 0 & 0 \\ 0 & 0 & 0 & 0 & 0 & 0 & 0 & 1 \end{bmatrix}.$$

*Let us consider one of the entries in matrix* **A** *to clarify its construction. Suppose we wish to compute the transition probability* $P\{(1, 1, 0) \rightarrow (1, 0, 0)\}$, *which corresponds to the entry* $\mathbf{A}_{7,5}$ *(the indexing starts with 1). To do this, we need to use the row corresponding to* $(x_1, x_2, x_3) = (1, 1, 0)$ *in the network truth table given above. Then we look for possible combinations of the predictors for each of the three genes that will give us the values* $(1, 0, 0)$. *By direct inspection, we can see that either* $(f_1^{(1)}, f_1^{(2)}, f_2^{(3)})$ *or* $(f_2^{(1)}, f_1^{(2)}, f_2^{(3)})$ *results in* $(1, 0, 0)$. *The two possible combinations correspond to the second and fourth rows of matrix* **K**. *That is why this transition probability is equal to* $p_2 + p_4$. *All the other entries in* **A** *are computed similarly. The state transition diagram corresponding to this matrix is shown in figure* 3.5. *For example, the seventh row of matrix* **A** *corresponds to* $(1, 1, 0)$, *and it can be seen that the only possible transitions are to* $(1, 0, 0)$ *or* $(1, 0, 1)$, *corresponding to columns 5 and 6, respectively.*

*Influences of Genes in PBNs*

Given a gene $x_i$ and a predictor $f_j^{(i)}$ for this gene, along with the genes used to make the prediction, it is important to distinguish those genes that have a major impact on the predictor from those that have only a minor impact. In other words, some (parent) genes are more "important" than others in determining the value of a target gene. Many examples of such biased regulation of gene expression are known to biologists. For example, the cell cycle regulator gene p21/WAF1/cip1 can be transcriptionally activated by a series of genes p53, smad4, AP2, BRCA1, and so on (Gartel and Tyner, 1999). Among these genes, p53 has the most potent effect.

Recall that we have already defined the *activity* of a gene $x_j$ to be the expectation of the partial derivative of the function with respect to the uniform distribution: $\alpha_j^f = E[\partial f(\mathbf{x})/\partial x_j]$. We have also mentioned that the *influence* of the variable $x_j$ on the function $f$ is the expectation of the partial derivative with respect to an arbitrary distribution $D(\mathbf{x})$:

$$I_j(f) = E_D \left[ \frac{\partial f(\mathbf{x})}{\partial x_j} \right] = P \left\{ \frac{\partial f(\mathbf{x})}{\partial x_j} = 1 \right\} = P\{f(\mathbf{x}) \neq f(\mathbf{x}^{(j)})\}, \qquad (3.15)$$

where $\mathbf{x}^{(j)}$ is the same as $\mathbf{x}$ except that the $j$th component is toggled (from 0 to 1 or from 1 to 0). The last expression gives the influence as the probability (under $D$) that a toggle of the $j$th variable changes the value of the function (Kahn et al., 1988).

Recall that in (standard) Boolean networks, a predictor function $f_i$ is assigned to node $x_i$. Thus, instead of thinking of a variable as influencing the function, we think of it as influencing the target node or gene. Furthermore, the partial derivative itself is a Boolean function, and thus equation (3.15) can be interpreted as a conditional probability that the value of node $x_i$ is equal to 1, given that the partial derivative was used as a predictor of the value of node $x_i$:

$$I_j(x_i) = P\left\{X_i = 1 \middle| \frac{\partial f_i(\mathbf{x})}{\partial x_j} \text{ is used as a predictor}\right\}, \qquad (3.16)$$

where the notation $I_j(x_i)$ represents the influence of gene $x_j$ on gene $x_i$ given $f_i$ as the predictor. The point of this observation is to show that the same methods and framework used for gene predictors in Boolean networks can be used for influences as well simply by replacing predictors by their partial derivatives.

Let us go one step further and consider probabilistic Boolean networks. In this case, we have a number of predictors for each gene along with their probabilities. As before, let $F_i$ be the set of predictors for gene $x_i$ with corresponding probabilities $c_1^{(i)}, \ldots, c_{l(i)}^{(i)}$. Let $I_k(f_j^{(i)})$ be the influence of variable $x_k$ on the predictor $f_j^{(i)}$. Since many possible predictors can be used for gene $x_i$, we would like to determine the overall influence of gene $x_k$ on gene $x_i$. Thus, by unconditioning (i.e., "integrating out") all the partial derivatives of the predictors, the influence of gene $x_k$ on gene $x_i$ becomes

$$I_k(x_i) = \sum_{j=1}^{l(i)} I_k(f_j^{(i)}) \cdot c_j^{(i)}. \qquad (3.17)$$

This calculation can be performed between all pairs of variables, and an $n \times n$ matrix $\Gamma$ of influences can be constructed. That is, $\Gamma_{ij} = I_i(x_j)$. We call $\Gamma$ the *influence matrix* and will show later how it can be used to construct subnetworks. Let us consider an example illustrating the computation of the influence of one variable on another in the context of PBNs.

**Example 3.2**   *Consider the same PBN as in example* 3.1. *Again, let* $c_1^{(1)} = 0.6$, $c_2^{(1)} = 0.4$, $c_1^{(2)} = 1$, $c_1^{(3)} = 0.5$, $c_2^{(3)} = 0.5$. *Suppose we would like to compute the influence of variable* $x_2$ *on variable* $x_1$. *Therefore, we will need to use both of the predictors* $f_1^{(1)}$ *and* $f_2^{(1)}$ *given in the table in example* 3.1. *Further, suppose* $D$ *is the uniform distribution; that is,* $D(\mathbf{x}) = 1/8$ *for all* $\mathbf{x} \in \{0, 1\}^3$. *First, we have*

$$I_2\left(f_1^{(1)}\right) = E_D\left[\frac{\partial f_1^{(1)}(\mathbf{x})}{\partial x_2}\right] = 0.5,$$

$$I_2\left(f_2^{(1)}\right) = E_D\left[\frac{\partial f_2^{(1)}(\mathbf{x})}{\partial x_2}\right] = 0.75.$$

*Putting these two influences together, we obtain*

$$I_2(x_1) = 0.5 \cdot 0.6 + 0.75 \cdot 0.4 = 0.6.$$

*If we repeat these calculations between all pairs of variables, we will obtain the influence matrix*

$$\mathbf{\Gamma} = \begin{bmatrix} 0.1 & 0.75 & 0.375 \\ 0.6 & 0.75 & 0.375 \\ 0.6 & 0.75 & 0.375 \end{bmatrix}.$$

*It is not surprising that $\mathbf{\Gamma}$ is not symmetric since the influence of variable $x_i$ on variable $x_j$ may be stronger than the influence of variable $x_j$ on variable $x_i$, or vice versa.*

Recall also that the average sensitivity $s^f$ of a function $f$ is the sum of the influences of all the variables on $f$ (under the uniform distribution, it is the sum of the activities). An interpretation of $s^f$ is how much, on average, the function $f$ changes between Hamming neighbors. Since for probabilistic Boolean networks we have several predictors for each gene, we will again use the notion of the influence of a gene on another gene, using equation (3.17). In other words, the average sensitivity of gene $x_i$ can be expressed as $s(x_i) = \sum_{j=1}^{n} I_j(x_i)$. Given the influence matrix, this can be computed as

$$s(x_i) = \sum_{k=1}^{n} \Gamma_{ki}.$$

Biologically, the sensitivity of a gene represents its stability or, in some sense, its autonomy. If the sensitivity of a gene is low, this implies that other genes have little effect on it. The "housekeeping" genes that encode structural proteins in cells may fall into this category. In example 3.2, gene $x_2$ is the most sensitive.

Since we are dealing with probabilistic Boolean networks, we always have at least as many predictors as genes. Hence, another informative measure would be the collective effect of a gene on all the other genes. This can simply be called the *influence* of gene $x_i$, denoted by $r(x_i)$, and can also be obtained from the influence matrix as

$$r(x_i) = \sum_{k=1}^{n} \Gamma_{ik}.$$

Biologically, a gene with a high influence factor has a high collective impact on the other genes. It is precisely these genes that have the potential to regulate the dynamics of the network, as their perturbation can lead to significant "downstream" effects, possibly forcing the system to transition to a different basin of attraction. We will consider this when we discuss gene intervention in PBNs. Many transcriptional factor genes likely fall into this category. In example 3.2, genes $x_2$ and $x_3$ are equally more important than gene $x_1$.

*Random Gene Perturbations*

Suppose that any gene, out of $n$ possible genes, can become perturbed with probability $p$, independently of other genes. In the Boolean setting, this is represented by a flip of value from 1 to 0, or vice versa, and directly corresponds to the bit-flipping mutation operator in $NK$ landscapes (Kauffman and Levin, 1987; Kauffman, 1993) as well as in genetic algorithms (Goldberg, 1989). This type of "randomization," namely, allowing genes to randomly flip value, is biologically meaningful. Since the genome is not a closed system but rather has inputs from the outside, it is known that genes can become either activated or inhibited because of external stimuli such as mutagens, heat stress, and so on. Thus, a network model should be able to capture this phenomenon. If $p = 0$, then the model is reduced to the standard PBN described above. If $p > 0$, then we have the following situation. With probability $(1 - p)^n$, the transition from one state to another occurs as usual by one of the randomly selected network realizations, while with probability $1 - (1 - p)^n$, the state changes because of random bit perturbation(s). Note that we are assuming instantaneously random PBNs here.

We can frame the random gene perturbations as follows. Suppose that at every step in the network we have the realization of a random *perturbation vector* $\gamma \in \{0, 1\}^n$. If the $i$th component of $\gamma$ is equal to 1, then the $i$th gene is flipped; otherwise it is not. In general, $\gamma$ need not be independent and identically distributed (i.i.d.), but we will assume this for simplicity. The generalization to the non-i.i.d. case is conceptually straightforward. Thus, we suppose that $P\{\gamma_i = 1\} = E[\gamma_i] = p$ for all $i = 1, \ldots, n$. Clearly,

$$P\{\gamma = (0, \ldots, 0)\} = (1 - p)^n.$$

Let $\mathbf{x} = (x_1, \ldots, x_n)$ be the state of the network (i.e., the values of all the genes) at some given time. Then, the next state $\mathbf{x}'$ is given by

$$\mathbf{x}' = \begin{cases} \mathbf{x} \oplus \gamma & \text{with probability } 1 - (1 - p)^n, \\ \mathbf{f}_k(x_1, \ldots, x_n) & \text{with probability } (1 - p)^n, \end{cases} \tag{3.18}$$

where $\oplus$ is componentwise addition modulo 2 and $\mathbf{f}_k(x_1, \ldots, x_n)$, $k = 1, 2, \ldots, s_0$, is the transition function representing a possible realization of the entire PBN. In other words, equation (3.18) states that if no genes are perturbed, the standard network transition function will be used, whereas if at least one perturbation occurs, the next state will be determined according to the genes that are perturbed.

An important observation to make here is that for $p > 0$, any state of the network in principle becomes accessible from any other state because of the possibility of any combination of random gene perturbations. In fact, we can say the following.

**Proposition 3.3**     For $p > 0$, the Markov chain corresponding to the PBN is ergodic.

*Proof.* Since there are only a finite number of states, ergodicity is equivalent to the chain being aperiodic and irreducible. First, by virtue of equation (3.18), we note that the Markov state transition matrix has no zero entries except possibly on the diagonal, the latter corresponding to the case where there does not exist a network transition function $\mathbf{f}_k$ ($k = 1, 2, \ldots, s_0$) such that $\mathbf{f}_k(\mathbf{x}) = \mathbf{x}$. This immediately implies that the chain is irreducible since all the states communicate. Indeed, let $\mathbf{x}$ be such a state (i.e., a state for which $\mathbf{f}_k(\mathbf{x}) \neq \mathbf{x}$ for all $k = 1, 2, \ldots, s_0$) and let $\mathbf{y} \neq \mathbf{x}$ be any other state. The probability of transitioning from $\mathbf{x}$ to $\mathbf{y}$ is positive, as is the probability of going from $\mathbf{y}$ back to $\mathbf{x}$. Therefore, there is a positive probability that $\mathbf{x}$ is accessible from itself in just two steps. Using the same reasoning, the process can return to the same state after any number of steps except possibly after one step, implying that the chain is also aperiodic.                         □

This result extends to the context-sensitive PBN. The fact that the Markov chain is ergodic implies that it possesses a steady-state distribution equal to the stationary distribution, which can be estimated empirically simply by running the network for a sufficiently long time and by collecting information about the proportion of time that the process spends in each state. The convergence rate, however, will surely depend on the parameter $p$ and will be discussed later.

A simulation-based analysis of the network involving gene perturbation may require one to compute the transition probabilities,

$$A(\mathbf{x}, \mathbf{x}') = P\{(x_1, \ldots, x_n) \to (x'_1, \ldots, x'_n)\},$$

between any two arbitrary states of the network, which can be stored in the matrix $\mathbf{A}$. We turn to this next.

**THEOREM 3.4** *Given a PBN* $G(V, F)$ *with genes* $V = \{x_1, \ldots, x_n\}$ *and a list* $F = (F_1, \ldots, F_n)$ *of sets* $F_i = \{f_1^{(i)}, \ldots, f_{l(i)}^{(i)}\}$ *of Boolean predictors, as well as a gene perturbation probability* $p > 0$,

$$A(\mathbf{x}, \mathbf{x}') = \left( \sum_{i=1}^{s_0} p_i \left[ \prod_{j=1}^{n} \left( 1 - \left| f_{\mathbf{K}_{ij}}^{(j)}(x_1, \ldots, x_n) - x'_j \right| \right) \right] \right) \times (1 - p)^n +$$

$$+ p^{\eta(\mathbf{x}, \mathbf{x}')} \times (1 - p)^{n - \eta(\mathbf{x}, \mathbf{x}')} \times 1_{[\mathbf{x} \neq \mathbf{x}']},$$

*where* $\eta(\mathbf{x}, \mathbf{x}') = \sum_{i=1}^{n}(x_i \oplus x'_i)$ *is the (nonnormalized) Hamming distance between vectors* $\mathbf{x}$ *and* $\mathbf{x}'$, $p_i$ *is given in equation (3.12), and* $1_{[\mathbf{x} \neq \mathbf{x}']}$ *is an indicator function that is equal to 1 only when* $\mathbf{x} \neq \mathbf{x}'$.

*Proof.* The two terms in theorem 3.4 essentially correspond to the two cases in equation (3.18). First, consider the case when no gene is perturbed or, equivalently, $\gamma = (0, \ldots, 0)$. This occurs with probability $(1 - p)^n$. Thus, the next state is determined via the Boolean functions selected at that time step. The probability of transitioning from $\mathbf{x} = (x_1, \ldots, x_n)$ to $\mathbf{x}' = (x'_1, \ldots, x'_n)$, then, is equal to the sum of the probabilities of all network realizations $\mathbf{f}_k$ such that $\mathbf{f}_k(x_1, \ldots, x_n) = (x'_1, \ldots, x'_n)$, $k = 1, 2, \ldots, s_0$. Thus, given that no perturbation occurred,

$$A(\mathbf{x}, \mathbf{x}') = \sum_{i: \mathbf{f}_i(\mathbf{x}) = \mathbf{x}'} p_i.$$

which in terms of the individual Boolean functions can be expressed as

$$\sum_{i=1}^{s_0} p_i \left[ \prod_{j=1}^{n} \left( 1 - \left| f_{\mathbf{K}_{ij}}^{(j)}(x_1, \ldots, x_n) - x'_j \right| \right) \right],$$

where we treat binary values as real values. This is in fact the transition probability when $p = 0$, as in equation (3.14).

If at least one gene is perturbed, then the transition probability depends on the number of perturbed genes. Given that a perturbation did occur, causing a transition from state $\mathbf{x}$ to state $\mathbf{x}'$, we can conclude that the number of perturbed genes was $\eta(\mathbf{x}, \mathbf{x}')$, which is the Hamming distance between $\mathbf{x}$ and $\mathbf{x}'$. Because $\gamma \in \{0, 1\}^n$ is i.i.d. with $E[\gamma_i] = p$, $i = 1, \ldots, n$, the probability that $\mathbf{x}$ was changed to $\mathbf{x}'$ is equal to $p^{\eta(\mathbf{x}, \mathbf{x}')} \times (1 - p)^{n - \eta(\mathbf{x}, \mathbf{x}')}$. It is clear that the fact that at least one perturbation occurred implies that $\mathbf{x}$ and $\mathbf{x}'$ cannot be equal, and so this expression must be multiplied by $1_{[\mathbf{x} \neq \mathbf{x}']}$.                                                                                  □

If the perturbation vector $\gamma$ is not identically distributed (i.e., some genes are more likely to be flipped), then the above transition probabilities become slightly more complicated, requiring products of individual probabilities $P\{\gamma_i = 1\}$. It can be seen from theorem 3.4 that the transition probability between two different states cannot be zero as long as $p > 0$.

A practical benefit of the randomization afforded by gene perturbation is that it empirically simplifies various computations involving PBNs. For example, consider the computation of the influence $I_k(f_j^{(i)})$ of gene $x_k$ on the predictor function $f_j^{(i)}$. The computation of the influence of a gene on the predictor entails computing the joint distribution $D(\mathbf{x})$ of all the genes used by that predictor in order to compute the expectation of the partial derivatives of the predictors. This distribution, however, should be consistent with the model itself. For example, if we wish to quantify "long-term" influence, we need to obtain the stationary distribution of the Markov chain corresponding to the PBN. Obtaining these long-run probabilities, however, may be problematic from an empirical point of view since the Markov chain may consist of a number of irreducible subchains and these probabilities depend on the initial starting state. In other words, depending on where we start the process, we may end up in different irreducible subchains. Obtaining long-run behavior directly from the state transition matrix $\mathbf{A}$ may also be impractical even for moderate sizes of PBNs, thus requiring simulation-based analysis, which we will discuss later.

The assumption of random gene perturbation, described above, solves this problem by ridding us of the dependence on the initial starting state. Since all the states communicate, according to proposition 3.3, the steady-state distribution is the same as the stationary distribution, and by letting the process run for a sufficiently long time, we can empirically compute the distribution $D(\mathbf{x})$ simply by keeping track of the proportion of time each combination of values of the genes in the domain of the predictor occurs. For instance, if the predictor is a function of some given three variables, then we simply have to tabulate the frequency of appearance of each of the eight combinations of these three variables to obtain the necessary distribution in order to compute the influence on this predictor. Let us illustrate these ideas with an example.

**Example 3.5**  *Consider a PBN consisting of three genes $V = (x_1, x_2, x_3)$ and the function sets $F = (F_1, F_2, F_3)$, where $F_1 = \{f_1^{(1)}, f_2^{(1)}\}$, $F_2 = \{f_1^{(2)}\}$, and $F_3 = \{f_1^{(3)}, f_2^{(3)}\}$. The function truth tables, as well as selection probabilities $c_j^{(i)}$, are given in example 3.1. Let us assume that the initial (starting) distribution of the Markov chain is the uniform distribution; that is, $D(\mathbf{x}) = 1/8$ for all $\mathbf{x} \in \{0, 1\}^3$. Using this distribution, we can compute the influence matrix $\mathbf{\Gamma}$ (Example 3.2). At the next time step, however, the distribution of all the states is no longer uniform. In general, the distribution at a given time step can be obtained simply by multiplying the distribution at the previous time step by the state transition matrix $\mathbf{A}$. Therefore, if we would like to compute the influence matrix at an arbitrary time point, we must have the distribution vector corresponding to that time point. Similarly, if we would like to compute the long-term influence (i.e., the influence after the network has reached equilibrium), we must have the stationary distribution vector. Let us suppose that the perturbation probability is equal to $p = 0.01$ and see how the influence matrix changes over time. That is, for every step in the network, we will recompute the influence matrix. Let us focus on the influence of gene $x_2$ on the other three genes (namely, row 2 of the influence matrix). Figure 3.4 shows the trajectories for these three influences for the first 100 time steps. First, it can be seen that the influences indeed converge to their asymptotic values. Second, it is worthwhile noting that the "transient" influence (e.g., the first 10 time steps in this example) can be very different from the long-term influence. For example, the influence of $x_2$ on $x_3$ is the lowest at the beginning and the highest at the end. The important thing to note here is that the long-term influences are guaranteed to be independent of the initial starting state or distribution because a nonzero gene perturbation probability is used.*

*Remark.*  We restrict ourselves to binary-valued nodes when providing definitions and discussing properties of probabilistic Boolean networks; however, the basic theory and model applications do not depend on binary values but apply to any finite quantization, a particularly important case being ternary quantization, where expression levels take on the values +1 (up-regulated), −1 (down-regulated), and 0 (invariant). Ternary values naturally arise from cDNA microarrays by assigning gene expression values according to a hypothesis test for the ratios (Chen et al., 1997). Nonbinary networks have been referred to as probabilistic gene regulatory networks in the literature. Here, as is commonly done, we do not employ this terminology and continue to refer to nonbinary networks as probabilistic Boolean networks, leaving the context to indicate the quantization.

### 3.2.2 Inference

As already discussed in the context of Boolean networks, a natural way to select a set of predictors for a given gene is to employ the coefficient of determination.

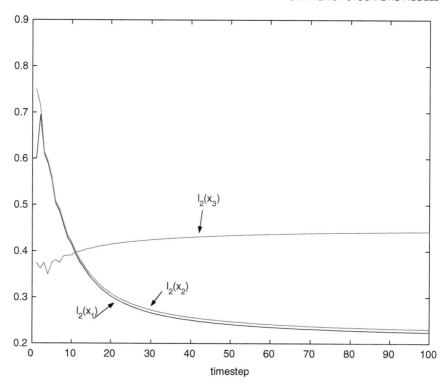

Figure 3.4. The trajectories of the influences $I_2(x_i)$ for $i = 1, 2, 3$ plotted as a function of the time steps taken by the PBN given in Example 3.1. The gene perturbation probability is equal to $p = 0.01$. (Shmulevich et al., 2002b.)

Recall that the CoD measures the degree to which the transcriptional levels of an observed gene set can be used to improve prediction of the transcriptional level of a target gene relative to the best possible prediction in the absence of observations. Let us revisit the CoD in the context of PBNs.

Let $X_i$ be the target gene; $X_1^{(i)}, X_2^{(i)}, \dots, X_{l(i)}^{(i)}$ be sets of genes; and $f_1^{(i)}, f_2^{(i)}, \dots,$ $f_{l(i)}^{(i)}$ be function rules such that $f_1^{(i)}(X_1^{(i)}), \dots, f_{l(i)}^{(i)}(X_{l(i)}^{(i)})$ are optimal predictors of $X_i$ relative to some probabilistic error measure $\varepsilon(X_i, f_k^{(i)}(X_k^{(i)}))$, keeping in mind that $X_i$ and $f_k^{(i)}(X_k^{(i)})$ are random variables. For each $k$, the coefficient of determination for $X_i$ relative to the conditioning set $X_k^{(i)}$ is defined by

$$\theta_k^i = \frac{\varepsilon_i - \varepsilon(X_i, f_k^{(i)}(X_k^{(i)}))}{\varepsilon_i},$$

where $\varepsilon_i$ is the error of the best (constant) estimate of $X_i$ in the absence of any conditional variables. For instance, in the case of minimum mean-square error estimation, $\varepsilon_i$ is the error of the mean of $X_i$, which is the best constant estimate, and $f_k^{(i)}(X_k^{(i)})$ is the conditional expectation of $X_i$ given $X_k^{(i)}$; that is, $f_k^{(i)}(X_k^{(i)}) = E[X_i|X_k^{(i)}]$.

In practice, the CoD must be estimated from training data, with designed approximations being used in place of $f_1^{(i)}, f_2^{(i)}, \ldots, f_{l(i)}^{(i)}$. Consequently, the complexity of the functions $f_1^{(i)}, f_2^{(i)}, \ldots, f_{l(i)}^{(i)}$ and the amount of training data become an issue. For the microarray-based analysis in (Kim et al., 2000), the number of genes in each predictor was kept to a maximum of three. The framework afforded by the PBN model is well suited for dealing with design imprecision due to limited sample size, where the domain of each predictor may need to be constrained because of a lack of training data but several predictors with possibly different domains are collectively employed in a synthesis filter bank fashion.

Let us now assume that a class of gene sets $X_1^{(i)}, X_2^{(i)}, \ldots, X_{l(i)}^{(i)}$ possessing high CoDs has been selected. We can take the designed approximations of the optimal function rules $f_1^{(i)}, f_2^{(i)}, \ldots, f_{l(i)}^{(i)}$ as the rule set for gene $X_i$, with the probability of $f_j^{(i)}$ being selected (equation (3.10)) given by

$$c_k^{(i)} = \frac{\theta_k^i}{\sum_{j=1}^{l(i)} \theta_j^i},$$

where the CoDs are the estimates formed from the training data. According to the above expression, functions corresponding to the highest CoDs will be selected more often in the probabilistic network. The number of chosen predictors, $l(i)$, can be a user-selectable parameter and determines the amount of uncertainty that the model can handle. Another approach to selecting the probabilities $c_k^{(i)}$ is to use the best-fit error size instead of the CoD, followed by normalization such that the probabilities sum to 1.

As an alternative approach, Zhou et al. (2004) have proposed a fully Bayesian approach to constructing PBNs (and their multivalued generalizations) that emphasizes network topology. This method computes the possible parent sets of each gene, the corresponding predictors, and the associated probabilities based on a nonlinear perceptron model, using a reversible jump Markov chain Monte Carlo (MCMC) technique. An MCMC method is employed to search the network configurations to find those with the highest Bayesian scores to construct the network. This method can be naturally applied for context-sensitive PBNs.

### 3.2.3 Dynamics of PBNs

In the last section, we showed that the dynamical behavior of PBNs can be represented as a Markov chain where the state transition matrix $\mathbf{A}$ is completely specified by all of the possible Boolean functions and their selection probabilities. As such, the theory of the limiting behavior of Markov chains is directly applicable to the study of the dynamics of PBNs. For instance, suppose that in example 3.1 the probabilities of the predictors are given by $c_1^{(1)} = 0.6$, $c_2^{(1)} = 0.4$, $c_1^{(2)} = 1$, $c_1^{(3)} = 0.5$, $c_2^{(3)} = 0.5$, as shown in the bottom row of the network truth table. Then the probabilities of the four networks can be computed via equation (3.12). For example, to

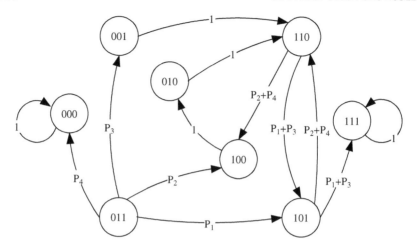

Figure 3.5.  State transition diagram for example 3.1. (Shmulevich et al., 2002a.)

compute $p_2$, we use row 2 in matrix $\mathbf{K}$ and get

$$p_2 = c_1^{(1)} c_1^{(2)} c_2^{(3)} = 0.6 \times 1 \times 0.5 = 0.3.$$

Thus, the probabilities of the four networks are $P_1 = 0.3$, $P_2 = 0.3$, $P_3 = 0.2$, and $P_4 = 0.2$. Substituting these values into the matrix $\mathbf{A}$ and iterating equation (3.8), supposing that the starting joint distribution of all the genes is uniform, that is, $D^0 = [\frac{1}{8}, \ldots, \frac{1}{8}]$, we find that the limiting probabilities are

$$\pi = [0.15, 0, 0, 0, 0, 0, 0, 0.85].$$

This tells us that in the long run, all three genes will either be off (000), with probability 0.15, or all three genes will be on (111), with probability 0.85. These two states are called *absorbing*. This can be easily seen by looking at the state transition diagram corresponding to matrix $\mathbf{A}$ shown in figure 3.5. Once the process moves into state (000) or (111), it never exits. This notion corresponds to the concept of attractors in standard Boolean networks. Similarly, the concept of limit cycle attractors corresponds to the irreducible sets of states in the Markov chain or, in other words, those sets of states such that no state outside them can be reached from any state in them. One can think of the network as being "trapped" in these sets of states. Finally, the transient states in the Markov chain that lead to either absorbing states or irreducible sets of states correspond to the basins of attraction in Boolean networks. Thus, PBNs qualitatively exhibit the same dynamical properties as Boolean networks but are inherently probabilistic and thus can cope with uncertainty.

An important consideration for probabilistic Boolean networks, which is not an issue in Boolean networks, is whether or not there exists a steady-state distribution. For instance, in the above example, the probabilities of ending up in state (000) or (111) would be different if the starting distribution $D^0$ were also different.

To illustrate this, suppose the starting distribution is $D^0 = [1, 0, 0, 0, 0, 0, 0, 0]$; that is, the network begins in state (000) with probability 1. Then it is clear that it will never escape this state, and the limiting probabilities will also be $\pi = D^0$. Therefore, we cannot ask the question about where this PBN will end up in the long run without specifying where it started. Since this is an important issue, let us describe it in more detail.

### The Existence of Steady-State Distributions

When considering the long-run behavior of a Markov chain, it is useful to consider equivalence classes within the set of states. This is especially true for genomic systems in which the state space can be extremely large and may be partitioned according to various subsystems. If an equivalence class is closed, meaning that no state outside the class is accessible from a state within the class, then for long-run analysis this class can be treated as an irreducible Markov chain in its own right: once inside the class, the system cannot leave it. Hence, we will consider long-run dynamics in terms of a single irreducible finite Markov chain.

A key property in the characterization of long-run behavior is periodicity. Since periodicity is a class property, an irreducible Markov chain can be considered to be or not to be aperiodic. A homogeneous Markov chain with finite state space $S = \{1, 2, \ldots, M\}$ is said to possess a *stationary distribution* (or *invariant distribution*) if there exists a probability distribution $\pi = (\pi_1, \pi_2, \ldots, \pi_M)$ such that, for any $j \in S$ and for any number $r$ of time steps,

$$\pi_j = \sum_{i=1}^{M} \pi_i P_{ij}^r,$$

where $P_{ij}^r$ is the $r$-step transition probability. Hence, if the initial distribution is $\pi = (\pi_1, \pi_2, \ldots, \pi_M)$, then the probability of being in state $i$ at time $r$ is equal to $\pi_i$ for all $r$ and the Markov chain is a strictly stationary random process. The Markov chain is said to possess a *steady-state (limiting) distribution* if there exists a probability distribution $\pi = (\pi_1, \pi_2, \ldots, \pi_M)$ such that, for all states $i, j \in S$,

$$\lim_{r \to \infty} P_{ij}^r = \pi_j.$$

If there exists a steady-state distribution, then, regardless of the starting state, the probability of the Markov chain being in state $i$ in the long run is $\pi_i$. In particular, for any initial distribution $D^0 = (D_1^0, D_2^0, \ldots, D_M^0)$, the state probability $D_i^k$ approaches $\pi_i$ as $k \to \infty$. Relative to the probability vector $\pi$, the vector $D^k$ satisfies $\lim_{k \to \infty} D^k = \pi$. Every irreducible, finite-state, homogenous Markov chain possesses a unique probability vector $\pi$, with $0 < \pi_i < 1$, providing the stationary distribution. If the chain is also aperiodic, then $\pi$ also provides the steady-state distribution. Should the chain be only irreducible and not necessarily aperiodic, then it may not possess a steady-state distribution. A more detailed treatment of the above concepts can be found in most textbooks on stochastic processes, such as (Çınlar, 1997).

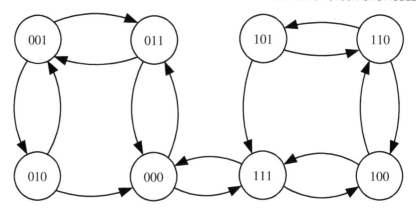

Figure 3.6. State transition diagram of an independent PBN without a steady-state distribution. (Shmulevich et al., 2002a.)

If the chain has a steady-state distribution, we can answer the following question: In the long run, what is the probability that the chain is in state $i$? The answer does not depend on the initial state. Suppose the states are divided into two classes $C_1$ and $C_2$. Then we can answer the following question without concern for the initial state: In the long run, what is the probability that the chain is in class $C_1$ (or $C_2$)? Such a question need not be answerable if there does not exist a steady-state distribution (if the chain is not aperiodic).

To illustrate the lack of a steady-state distribution, let us consider a three-variable, independent, instantaneously random PBN. Since we are not concerned with the probabilities but only with the possible Boolean functions, we can use a simplified notation to list the possible functions. We use a table consisting of eight rows corresponding to the eight states and three columns corresponding to the possible values the Boolean functions can have for the three variables given the state determining the row. The entry * in the table means that the value of the predictor for that gene given the values of the genes in that row can be either 0 or 1. Consider the following function table.

| $x_1 x_2 x_3$ | $f^{(1)}$ | $f^{(2)}$ | $f^{(3)}$ |
|---|---|---|---|
| 000 | * | 1 | 1 |
| 001 | 0 | 1 | * |
| 010 | 0 | 0 | * |
| 011 | 0 | 0 | * |
| 100 | 1 | 1 | * |
| 101 | 1 | 1 | * |
| 110 | 1 | 0 | * |
| 111 | * | 0 | 0 |

For example, there are four possible predictors $f_1^{(1)}$, $f_2^{(1)}$, $f_3^{(1)}$, $f_4^{(1)}$ for the first gene. Similarly, there are 256 possible vector functions (network realizations) of the form $\mathbf{f} = (f^{(1)}, f^{(2)}, f^{(3)})$. The corresponding Markov diagram is given in figure 3.6. Every state has period 2, and therefore the PBN is not aperiodic. Thus,

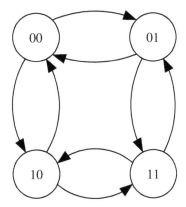

Figure 3.7. Example of a dependent PBN not containing a steady-state distribution. (Shmulevich et al., 2002a.)

there does not exist a steady-state distribution. The requirement for a PBN to possess a steady-state distribution can be imposed so that the associated long-run questions can be posed. If so, this then imposes a constraint on the collections of Boolean functions. Certain sets of Boolean functions, such as the one just considered, are not permissible.

The previous example considered an independent PBN. The steady-state requirement can be even more constraining for dependent PBNs. Consider the following very simple PBN with only two network functions (realizations), $\mathbf{f}_1(x_1, x_2) = (f_1^{(1)}, f_1^{(2)}) = (\bar{x}_1, x_2)$ and $\mathbf{f}_2(x_1, x_2) = (f_2^{(1)}, f_2^{(2)}) = (x_1, \bar{x}_2)$. Since the PBN is dependent, the selection of the predictor for the first gene cannot be viewed independently of the selection of the predictor for the second gene. The above two possible network realizations imply, for instance, that if $f_1^{(1)} = \bar{x}_1$ is selected for the first gene, then $f_2^{(2)} = \bar{x}_2$ cannot be simultaneously selected for the second gene. That is, the probability that the network $\mathbf{f}$ takes on any realization other than the two given above, say, $(f_1^{(1)}, f_2^{(2)})$, is zero. The corresponding Markov diagram is given in figure 3.7. This PBN is not aperiodic and does not possess a steady-state distribution. Note that addition of the network function $\mathbf{f}_3(x_1, x_2) = (f_3^{(1)}, f_3^{(2)}) = (\bar{x}_1, \bar{x}_2)$ makes the PBN aperiodic.

Although the above discussion pertained to instantaneously random PBNs, steady-state distributions for context-sensitive PBNs have been studied by Brun et al. (2005).

### Steady-State Probabilities of Attractors and Basins

Recall that the attractors of a Boolean network characterize its long-run behavior. However, if we incorporate random perturbation, then the network can escape its attractors. In this case, full long-run behavior is characterized by its steady-state distribution, as discussed above. Nonetheless, if the probability of perturbation is very small, the network will lie in its attractor cycles for a large majority of the time, meaning that attractor states will carry most of the steady-state probability

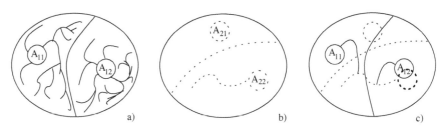

Figure 3.8. The behavior of a context-sensitive PBN. (Brun et al., 2005.)

mass. The amount of time spent in any given attractor depends on its basin. Large basins tend to produce attractors possessing a relatively large steady-state mass.

Let us now consider context-sensitive PBNs. As long as there is no functional switching (i.e., a new realization of the network), the current Boolean network realization of the PBN characterizes the activity of the PBN and it transitions into one of its attractor cycles and remains there until a switch occurs. When a network switch does occur, the present state becomes an initialization for the new realization and the network transitions into the attractor cycle whose basin contains the state. It remains there until another network switch. The attractor family of the PBN is defined as the union of all the attractors in the constituent Boolean networks. Note that the attractors of a PBN need not be disjoint, although those corresponding to each constituent Boolean network must be disjoint.

Figure 3.8 shows an example of the behavior of a context-sensitive PBN relative to its attractors during a change of function. Figure 3.8a shows the attractor cycles $A_{11}$ and $A_{12}$ for a network function $\mathbf{f}_1$, its corresponding basins, and some trajectories. Figure 3.8b shows the attractor cycles $A_{21}$ and $A_{22}$ for a network function $\mathbf{f}_2$ and its corresponding basins. In figure 3.8c, we can see that if the system is using the function $\mathbf{f}_2$ and makes a function change to $\mathbf{f}_1$, then the future of the system depends on which part of the trajectory it is at the moment of the function change. In this example, for the particular trajectory shown with the dotted line toward attractor $A_{22}$, the first part of the trajectory is in the basin corresponding to attractor $A_{11}$, and the end of the trajectory is inside the basin corresponding to attractor $A_{12}$. Therefore, if the change of function occurs before the system crosses the boundary between the basins, it will transition toward attractor $A_{11}$. If the change of function occurs after it crosses the boundary, then it will transition toward attractor $A_{12}$. In particular, we see that attractor $A_{22}$ lies completely inside the basin corresponding to attractor $A_{12}$. In this case, if a change of function occurs when the system is inside attractor $A_{22}$, it will always transition to attractor $A_{12}$.

If one now incorporates perturbation into the PBN model, the stationary distribution characterizes the long-run behavior of the network. If both the switching and perturbation probabilities are very small, then the attractors will still carry most of the steady-state probability mass. This property has been used to validate network inference from data (Kim et al., 2002; Zhou et al., 2004) and to formulate analytic expressions of the probabilities of attractor states (Brun et al., 2005), to which we turn next.

For a context-sensitive PBN, the state $\mathbf{X}_t$ at time $t$ does not constitute a homogeneous Markov chain because the transition probabilities $P[\mathbf{X}_{t+1} = \mathbf{x}|\mathbf{X}_t = \mathbf{y}]$ depend on the function $\mathbf{f}$ selected at time $t$. Instead of representing the states $\mathbf{x}$ as the states of a Markov chain, we can represent the states $(\mathbf{x}, \mathbf{f})$ as states of a homogeneous Markov chain, $(\mathbf{X}_t, \mathbf{F}_t)$, with transition probabilities

$$P_{\mathbf{y},\mathbf{g}}(\mathbf{x}, \mathbf{f}) = P[\mathbf{X}_{t_0+1} = \mathbf{x}, \mathbf{F}_{t_0+1} = \mathbf{f}|\mathbf{X}_{t_0} = \mathbf{y}, \mathbf{F}_{t_0} = \mathbf{g}]$$

for any time $t_0$. The steady-state probabilities $\pi(\mathbf{x})$ are the marginal probabilities of the steady-state probabilities $\pi(\mathbf{x}, \mathbf{f})$. Let

$$P_{\mathbf{y},\mathbf{g}}^{(t)}(\mathbf{x}, \mathbf{f}) = P[\mathbf{X}_{t_0+t} = \mathbf{x}, \mathbf{F}_{t_0+t} = \mathbf{f}|\mathbf{X}_{t_0} = \mathbf{y}, \mathbf{F}_{t_0} = \mathbf{g}],$$

for any time $t_0$, be the probability of transition from $(\mathbf{y}, \mathbf{g})$ to $(\mathbf{x}, \mathbf{f})$ in $t$ steps. The steady-state distribution $\pi$ of the Markov chain is now defined, for any starting state–function pair $(\mathbf{y}, \mathbf{g})$, by $\pi(\mathbf{x}, \mathbf{f}) = \lim_{t \to \infty} P_{\mathbf{y},\mathbf{g}}^{(t)}(\mathbf{x}, \mathbf{f})$. The term *restart* refers to a situation where there occurs a perturbation, a function change, or both at the same time. We define, for each time $t$, the random variables $\tau_p(t)$ and $\tau_c(t)$ that measure the time elapsed between $t$ and the last perturbation and the time elapsed between $t$ and the last function change, respectively. The random variable $\tau(t) = \min(\tau_c(t), \tau_p(t))$ measures the time elapsed between $t$ and the last restart. The events $\tau_p(t) = 0$ and $\tau_c(t) = 0$ indicate the existence of perturbation and function change, respectively, at time $t$. They are determined by the probabilities $p$ and $q$, being $P[\tau_p(t) = 0] = 1 - (1-p)^n$ and $P[\tau_c(t) = 0] = q$, and are independent of time $t$. For each value of $t$, if we assume that there has been at least one perturbation before time $t$, the distribution for $\tau(t)$ is given by

$$P[\tau(t) = j] = \frac{(1-b)b^j}{1-b^t} \mathbf{1}_{[0 \le j \le t-1]}, \tag{3.19}$$

where $b = (1-q)(1-p)^n$.

Since the Markov chain is homogeneous and the probabilities for $\tau_p(t)$ and $\tau_c(t)$ are independent of $t$, for any time $t$ we can split the transition matrix $P_{\mathbf{y},\mathbf{f}_l}(\mathbf{x}, \mathbf{f}_k)$ by conditioning it on the existence of a perturbation and function change:

$$\begin{aligned}
P_{\mathbf{y},\mathbf{f}_l}(\mathbf{x}, \mathbf{f}_k) &= P[\mathbf{X}_t = \mathbf{x}, \mathbf{F}_t = \mathbf{f}_k|\mathbf{X}_{t-1} = \mathbf{y}, \mathbf{F}_{t-1} = \mathbf{f}_l] \\
&= P[\mathbf{X}_t = \mathbf{x}, \mathbf{F}_t = \mathbf{f}_k|\mathbf{X}_{t-1} = \mathbf{y}, \mathbf{F}_{t-1} = \mathbf{f}_l, \tau_p(t) > 0, \tau_c(t) > 0] \\
&\quad \times P[\tau_p(t) > 0, \tau_c(t) > 0] \\
&\quad + P[\mathbf{X}_t = \mathbf{x}, \mathbf{F}_t = \mathbf{f}_k|\mathbf{X}_{t-1} = \mathbf{y}, \mathbf{F}_{t-1} = \mathbf{f}_l, \tau_p(t) > 0, \tau_c(t) = 0] \\
&\quad \times P[\tau_p(t) > 0, \tau_c(t) = 0] \\
&\quad + P[\mathbf{X}_t = \mathbf{x}, \mathbf{F}_t = \mathbf{f}_k|\mathbf{X}_{t-1} = \mathbf{y}, \mathbf{F}_{t-1} = \mathbf{f}_l, \tau_p(t) = 0, \tau_c(t) > 0] \\
&\quad \times P[\tau_p(t) = 0, \tau_c(t) > 0] \\
&\quad + P[\mathbf{X}_t = \mathbf{x}, \mathbf{F}_t = \mathbf{f}_k|\mathbf{X}_{t-1} = \mathbf{y}, \mathbf{F}_{t-1} = \mathbf{f}_l, \tau_p(t) = 0, \tau_c(t) = 0] \\
&\quad \times P[\tau_p(t) = 0, \tau_c(t) = 0].
\end{aligned}$$

Table 3.3  Conditional Transition Probabilities for the Four Cases

| Event $A$ | Probability $P(A)$ | $P[\mathbf{X}_t = \mathbf{x}, \mathbf{F}_t = \mathbf{f}_k \| \mathbf{X}_{t-1} = \mathbf{y},$ $\mathbf{F}_{t-1} = \mathbf{f}_l, A]$ |
|---|---|---|
| $\tau_p(t) > 0, \tau_c(t) > 0$ | $(1-p)^n(1-q)$ | $\mathbf{1}_{[\mathbf{x} = \mathbf{f}_l(\mathbf{y})]} \cdot \mathbf{1}_{[k=l]}$ |
| $\tau_p(t) > 0, \tau_c(t) = 0$ | $(1-p)^n q$ | $P(\mathbf{F} = \mathbf{f}_k) \cdot \mathbf{1}_{[\mathbf{x} = \mathbf{f}_l(\mathbf{y})]}$ |
| $\tau_p(t) = 0, \tau_c(t) > 0$ | $(1 - (1-p)^n)(1-q)$ | $\frac{p^{\eta(\mathbf{x},\mathbf{y})}(1-p)^{n-\eta(\mathbf{x},\mathbf{y})}}{1-(1-p)^n} \cdot \mathbf{1}_{[\mathbf{x} \neq \mathbf{y}]} \cdot \mathbf{1}_{[k=l]}$ |
| $\tau_p(t) = 0, \tau_c(t) = 0$ | $(1 - (1-p)^n)q$ | $\frac{p^{\eta(\mathbf{x},\mathbf{y})}(1-p)^{n-\eta(\mathbf{x},\mathbf{y})}}{1-(1-p)^n} \cdot P(\mathbf{F} = \mathbf{f}_k) \cdot \mathbf{1}_{[\mathbf{x} \neq \mathbf{y}]}$ |

Each condition corresponds to one of four cases listed in Table 3.3, which summarizes the probabilities of the conditional transition matrices for the four terms.

In table 3.3, the last three events correspond to restarts of the system. The restart probability $P^*_{\mathbf{y},\mathbf{f}_l}(\mathbf{x}, \mathbf{f}_k)$, for any $t$, is the probability of being in $(\mathbf{x}, \mathbf{f}_k)$ after a restart, given that the system was in $(\mathbf{y}, \mathbf{f}_l)$ before the restart:

$$P^*_{\mathbf{y},\mathbf{f}_l}(\mathbf{x}, \mathbf{f}_k) = P[\mathbf{X}_t = \mathbf{x}, \mathbf{F}_t = \mathbf{f}_k | \mathbf{X}_{t-1} = \mathbf{y}, \mathbf{F}_{t-1} = \mathbf{f}_l, \tau(t) = 0].$$

A direct way to compute the steady-state probabilities $\pi(A_{ki})$ is as the marginal probabilities of $\pi(A_{ki}, \mathbf{f}_l)$. We will first compute the steady-state probabilities $\pi(B_{ki}, \mathbf{f}_k)$ for the basins, then the steady-state probabilities $\pi(A_{ki}, \mathbf{f}_k)$ for the attractors, and finally the marginal steady-state probabilities $\pi(A_{ki})$.

Computation of the steady-state probabilities $\pi(A_{ki}, \mathbf{f}_k)$ will be split into three stages:

1. computation of the steady-state probabilities $\pi(B_{ki}, \mathbf{f}_k)$ of the basins,
2. computation of the conditional probabilities $\pi(A_{ki}, \mathbf{f}_k | B_{ki}, \mathbf{f}_k)$,
3. approximation to the marginal steady-state probabilities $\pi(A_{ki})$.

BASINS

The next theorem represents the steady-state distribution of the basins as the solution of a system of linear equations in which the coefficients are perturbation-based probabilities. We will need the following definition. For any starting state–function pair $(\mathbf{h}, \mathbf{g})$, let

$$P^*_{B_{lv},\mathbf{f}_l}(B_{ki}, \mathbf{f}_k) = \lim_{t \to \infty} P[\mathbf{X}_t \in B_{ki}, \mathbf{F}_t = \mathbf{f}_k | \mathbf{X}_{t-1} \in B_{lv}, \mathbf{F}_{t-1} = \mathbf{f}_l, \mathbf{X}_0 = \mathbf{h},$$
$$\mathbf{F}_0 = \mathbf{g}, \tau(t) = 0]$$

be the steady-state probability of a restart in basin $B_{ki}$ using the function $\mathbf{f}_k$ after a restart, given that the system was inside basin $B_{lv}$ and using the function $\mathbf{f}_l$ before the restart.

**THEOREM 3.6**

$$\pi(B_{ki}, \mathbf{f}_k) = \sum_{l=1}^{r} \sum_{v=1}^{m_l} P^*_{B_{lv}, \mathbf{f}_l}(B_{ki}, \mathbf{f}_k) \pi(B_{lv}, \mathbf{f}_l)$$

*Proof.* Because we are interested in $\pi(B_{ki}, \mathbf{f}_k) = \lim_{t\to\infty} P^{(t)}_{\mathbf{h}}(B_{ki}, \mathbf{f}_k)$, we can assume that for any $t$ there has been a previous restart and that equation (3.19) applies. We can compute $P^{(t)}_{\mathbf{h}}(B_{ki}, \mathbf{f}_k)$ as a function of the probability $P[\tau(t) = j]$, the probability of reaching basin $B_{ki}$, and the function $\mathbf{f}_k$ at time $t - j$ (after the restart). For any starting state $\mathbf{h}$ and function $\mathbf{g}$,

$$\pi(B_{ki}, \mathbf{f}_k) = \lim_{t\to\infty} P[\mathbf{X}_t \in B_{ki}, \mathbf{F}_t = \mathbf{f}_k | \mathbf{X}_0 = \mathbf{h}, \mathbf{F}_0 = \mathbf{g}]$$

$$= \lim_{t\to\infty} \sum_{j=0}^{\infty} P[\mathbf{X}_t \in B_{ki}, \mathbf{F}_t = \mathbf{f}_k | \mathbf{X}_0 = \mathbf{h}, \mathbf{F}_0 = \mathbf{g}, \tau(t) = j] P[\tau(t) = j]$$

$$= \lim_{t\to\infty} \sum_{j=0}^{\infty} \sum_{l=1}^{r} \sum_{v=1}^{m_l}$$

$$\times P[\mathbf{X}_t \in B_{ki}, \mathbf{F}_t = \mathbf{f}_k | \mathbf{X}_0 = \mathbf{h}, \mathbf{F}_0 = \mathbf{g}, \tau(t) = j, \mathbf{X}_{t-j-1} \in B_{lv}, \mathbf{F}_{t-j-1} = \mathbf{f}_l]$$

$$\times P[\mathbf{X}_{t-j-1} \in B_{lv}, \mathbf{F}_{t-j-1} = \mathbf{f}_l | \mathbf{X}_0 = \mathbf{h}, \mathbf{F}_0 = \mathbf{g}, \tau(t) = j] P[\tau(t) = j]$$

$$= \lim_{t\to\infty} \sum_{j=0}^{\infty} \sum_{l=1}^{r} \sum_{v=1}^{m_l}$$

$$\times P[\mathbf{X}_{t-j} \in B_{ki}, \mathbf{F}_{t-j} = \mathbf{f}_k | \mathbf{X}_0 = \mathbf{h}, \mathbf{F}_0 = \mathbf{g}, \tau(t) = j, \mathbf{X}_{t-j-1} \in B_{lv}, \mathbf{F}_{t-j-1} = \mathbf{f}_l]$$

$$\times P[\mathbf{X}_{t-j-1} \in B_{lv}, \mathbf{F}_{t-j-1} = \mathbf{f}_l | \mathbf{X}_0 = \mathbf{h}, \mathbf{F}_0 = \mathbf{g}, \tau(t) = j] P[\tau(t) = j],$$

where in the fourth equality we have used the fact that, if at time $t - j$ the system reaches basin $B_k$ and is using the function $\mathbf{f}_k$, then it will stay there until time $t$ because the condition $\tau(t) = j$ says that there are no restarts between $t - j$ and $t$. Looking at equation (3.19) for $P[\tau(t) = j]$, we can see that for a $t$ large enough, $1 - b^t > 0.5$. Therefore, for each $j$, the expression inside the sum is bounded by $2 \cdot (1 - b)b^j$. Hence, by dominated convergence, the limit can be brought inside and

$$\pi(B_{ki}, \mathbf{f}_k) = \sum_{l=1}^{r} \sum_{v=1}^{m_l} \sum_{j=0}^{\infty} P^*_{B_{lv}, \mathbf{f}_l}(B_{ki}, \mathbf{f}_k) \pi(B_{lv}, \mathbf{f}_l)(1 - b)b^j$$

$$= \sum_{l=1}^{r} \sum_{v=1}^{m_l} P^*_{B_{lv}, \mathbf{f}_l}(B_{ki}, \mathbf{f}_k) \pi(B_{lv}, \mathbf{f}_l),$$

where in the first equality we have used the fact that the probability of being in the basin at time $t - j - 1$ is independent of the existence of a perturbation at time $t - j$. $\qquad \square$

The expression for the steady-state distribution for the basins depends on the unknown quantities $P^*_{B_{lv}, \mathbf{f}_l}(B_{ki}, \mathbf{f}_k)$. The next step is to find an approximation for these quantities. For any starting state–function pair $(\mathbf{h}, \mathbf{g})$, let

$$\pi^*(\mathbf{x}, \mathbf{f}) = \lim_{t \to \infty} P[\mathbf{X}_{t-1} = \mathbf{x}, \mathbf{F}_{t-1} = \mathbf{f} | \mathbf{X}_0 = \mathbf{h}, \mathbf{F}_0 = \mathbf{g}, \tau(t) = 0]$$

be the steady-state probability for $\mathbf{x}$, conditioned on times previous to perturbation, Then, for any set $B$,

$$\pi^*(\mathbf{x}, \mathbf{f} | B, \mathbf{f}) = \lim_{t \to \infty} P[\mathbf{X}_{t-1} = \mathbf{x}, \mathbf{F}_{t-1} = \mathbf{f} | \mathbf{X}_{t-1} \in B, \mathbf{F}_{t-1} = \mathbf{f}, \mathbf{X}_0 = \mathbf{h}, \tau(t) = 0].$$

**LEMMA 3.7**

$$P^*_{B_{lv}, \mathbf{f}_l}(B_{ki}, \mathbf{f}_k) = \sum_{\mathbf{x} \in B_{ki}} \sum_{\mathbf{y} \in B_{lv}} P^*_{\mathbf{y}, \mathbf{f}_l}(\mathbf{x}, \mathbf{f}_k) \pi^*(\mathbf{y}, \mathbf{f}_l | B_{lv}, \mathbf{f}_l).$$

*Proof.* For any starting state $\mathbf{h}$ and function $\mathbf{g}$,

$$P^*_{B_{lv}, \mathbf{f}_l}(B_{ki}, \mathbf{f}_k) = \lim_{t \to \infty} \sum_{\mathbf{x} \in B_{ki}} P[\mathbf{X}_t = \mathbf{x}, \mathbf{F}_t = \mathbf{f}_k | \mathbf{X}_{t-1} \in B_{lv}, \mathbf{F}_{t-1} = \mathbf{f}_l,$$

$$\mathbf{X}_0 = \mathbf{h}, \mathbf{F}_0 = \mathbf{g}, \tau(t) = 0]$$

$$= \lim_{t \to \infty} \sum_{\mathbf{x} \in B_{ki}} \sum_{\mathbf{y} \in B_{lv}}$$

$$\times P[\mathbf{X}_t = \mathbf{x}, \mathbf{F}_t = \mathbf{f}_k | \mathbf{X}_{t-1} = \mathbf{y}, \mathbf{F}_{t-1} = \mathbf{f}_l, \mathbf{X}_0 = \mathbf{h}, \mathbf{F}_0 = \mathbf{g}, \tau(t) = 0]$$

$$\times P[\mathbf{X}_{t-1} = \mathbf{y} | \mathbf{X}_{t-1} \in B_{lv}, \mathbf{F}_{t-1} = \mathbf{f}_l, \mathbf{X}_0 = \mathbf{h}, \mathbf{F}_0 = \mathbf{g}, \tau(t) = 0]$$

$$= \lim_{t \to \infty} \sum_{\mathbf{x} \in B_{ki}} \sum_{\mathbf{y} \in B_{lv}} P[\mathbf{X}_t = \mathbf{x}, \mathbf{F}_t = \mathbf{f}_k | \mathbf{X}_{t-1} = \mathbf{y}, \mathbf{F}_{t-1} = \mathbf{f}_l, \tau(t) = 0]$$

$$\times P[\mathbf{X}_{t-1} = \mathbf{y} | \mathbf{X}_{t-1} \in B_{lv}, \mathbf{F}_{t-1} = \mathbf{f}_l, \mathbf{X}_0 = \mathbf{h}, \mathbf{F}_0 = \mathbf{g}, \tau(t) = 0]$$

$$= \sum_{\mathbf{x} \in B_{ki}} \sum_{\mathbf{y} \in B_{lv}} P^*_{\mathbf{y}, \mathbf{f}_l}(\mathbf{x}, \mathbf{f}_k) \pi^*(\mathbf{y}, \mathbf{f}_l | B_{lv}, \mathbf{f}_l),$$

where in the third equality we have removed the conditions $\mathbf{X}_0 = \mathbf{h}$ and $\mathbf{F}_0 = \mathbf{g}$ because the transition probability under perturbation is independent of the starting state–function pair.                                                                          □

The only unknown value is now $\pi^*(\mathbf{y}, \mathbf{f}_l | B_{lv}, \mathbf{f}_l)$, the steady-state conditional probability of being in $\mathbf{y}$ given that the system is in $B_{lv}$ using the function $\mathbf{f}_l$ and given that there will be a restart at the next point in time. If the probability of restart is small, then the average time between restarts is large, and for approximation purposes we can assume that the system has already reached an attractor previous to each restart. Therefore, we make the assumption that if there will be a restart at time $t$ and the system is in basin $B_{lv}$ at time $t - 1$, then it is inside attractor $A_{lv}$ corresponding to basin $B_{lv}$ at time $t - 1$. In such a case, for any starting state $\mathbf{h}$ and function **g** and for any time $t$, we have the approximation

$$P[\mathbf{X}_{t-1} = \mathbf{y}|\mathbf{X}_{t-1} \in B_{lv}, \mathbf{F}_{t-1} = \mathbf{f}_l, \mathbf{X}_0 = \mathbf{h}, \mathbf{F}_0 = \mathbf{g}, \tau(t) = 0]$$
$$\approx \frac{1}{|A_{lv}|} 1_{[\mathbf{y} \in A_{lv}]},$$

resulting in

$$\pi^*(\mathbf{y}, \mathbf{f}_l | B_{lv}, \mathbf{f}_l) \approx \frac{1}{|A_{lv}|} 1_{[\mathbf{y} \in A_{lv}]} \tag{3.20}$$

and

$$P^*_{B_{lv}, \mathbf{f}_l}(B_{ki}, \mathbf{f}_k) \approx \frac{1}{|A_{lv}|} \sum_{\mathbf{x} \in B_{ki}} \sum_{\mathbf{y} \in B_{lv}} P^*_{\mathbf{y}, \mathbf{f}_l}(\mathbf{x}, \mathbf{f}_k).$$

If we define the values $C_{ki, lv}$ by

$$C_{ki, lv} = \frac{1}{|A_{lv}|} \sum_{\mathbf{x} \in B_{ki}} \sum_{\mathbf{y} \in B_{lv}} P^*_{\mathbf{y}, \mathbf{f}_l}(\mathbf{x}, \mathbf{f}_k), \tag{3.21}$$

then

$$\pi(B_{ki}, \mathbf{f}_k) \approx \sum_{l=1}^{r} \sum_{v=1}^{m_l} C_{ki, lv} \pi(B_{lv}, \mathbf{f}_l). \tag{3.22}$$

Equations (3.21) and (3.22) give an approximation for the vector of steady-state probabilities $\pi(B_1), \ldots, \pi(B_m)$ as the stationary probability vector of the stochastic matrix $C_{ki, lv}$.

### ATTRACTORS

Once we have obtained the steady-state probabilities $\pi(B_{ki}, \mathbf{f}_k)$, we can compute the steady-state probabilities $\pi(A_{ki}, \mathbf{f}_k)$ given by

$$\pi(A_{ki}, \mathbf{f}_k) = \pi(A_{ki}, \mathbf{f}_k | B_{ki}, \mathbf{f}_k) \pi(B_{ki}, \mathbf{f}_k),$$

with

$$\pi(A_{ki}, \mathbf{f}_k | B_{ki}, \mathbf{f}_k) = \lim_{t \to \infty} P^{(t)}_{\mathbf{h}, \mathbf{g}}(A_{ki}, \mathbf{f}_k | B_{ki}, \mathbf{f}_k).$$

To find the expression for $\pi(A_k)$ we need to prove the following.

**LEMMA 3.8** *For basin $B_{ki}$, starting state* **h**, *starting function* **g**, *fixed value $j \geq 0$, and* $\mathbf{x} \in B_{ki}$,

$$\lim_{t \to \infty} P[\mathbf{X}_{t-j} = \mathbf{x}, \mathbf{F}_{t-j} = \mathbf{f}_k | \mathbf{X}_{t-j} \in B_{ki}, \mathbf{F}_{t-j} = \mathbf{f}_k, \mathbf{X}_0 = \mathbf{h}, \mathbf{X}_0 = \mathbf{g}, \tau(t) = j]$$

$$= \frac{1}{\pi(B_{ki}, \mathbf{f}_k)} \sum_{l=1}^{r} \sum_{v=1}^{m_l} \sum_{\mathbf{y} \in B_{lv}} P^*_{\mathbf{y}, \mathbf{f}_l}(\mathbf{x}, \mathbf{f}_k) \pi^*(\mathbf{y}, \mathbf{f}_l | B_{lv}, \mathbf{f}_l) \pi(B_{lv}, \mathbf{f}_l).$$

*Proof.* Let $L$ be the limit in the lemma. We split the probability over all possible values for $\tau(t)$. For any $\mathbf{x} \in B_{ki}$,

$$L = \lim_{t \to \infty} \frac{P[\mathbf{X}_{t-j} = \mathbf{x}, \mathbf{F}_{t-j} = \mathbf{f}_k | \mathbf{X}_0 = \mathbf{h}, \mathbf{X}_0 = \mathbf{g}, \tau(t) = j]}{P[\mathbf{X}_{t-j} \in B_{ki}, \mathbf{F}_{t-j} = \mathbf{f}_k | \mathbf{X}_0 = \mathbf{h}, \mathbf{X}_0 = \mathbf{g}, \tau(t) = j]}$$

$$= \frac{1}{\pi(B_{ki}, \mathbf{f}_k)} \lim_{t \to \infty} P[\mathbf{X}_{t-j} = \mathbf{x}, \mathbf{F}_{t-j} = \mathbf{f}_k | \mathbf{X}_0 = \mathbf{h}, \mathbf{X}_0 = \mathbf{g}, \tau(t) = j]$$

$$= \frac{1}{\pi(B_{ki}, \mathbf{f}_k)} \lim_{t \to \infty} \sum_{l=1}^{r} \sum_{v=1}^{m_l} \sum_{\mathbf{y} \in B_{lv}}$$

$$\times P[\mathbf{X}_{t-j} = \mathbf{x}, \mathbf{F}_{t-j} = \mathbf{f}_k | \mathbf{X}_{t-j-1} = \mathbf{y}, \mathbf{F}_{t-j-1} = \mathbf{f}_l, \mathbf{X}_0 = \mathbf{h}, \mathbf{X}_0 = \mathbf{g}, \tau(t) = j]$$

$$\times P[\mathbf{X}_{t-j-1} = \mathbf{y}, \mathbf{F}_{t-j-1} = \mathbf{f}_l | \mathbf{X}_{t-j-1} \in B_{lv}, \mathbf{F}_{t-j-1} = \mathbf{f}_l, \mathbf{X}_0 = \mathbf{h}, \mathbf{X}_0 = \mathbf{g}, \tau(t) = j]$$

$$\times P[\mathbf{X}_{t-j-1} \in B_{lv}, \mathbf{F}_{t-j-1} = \mathbf{f}_l | \mathbf{X}_0 = \mathbf{h}, \mathbf{X}_0 = \mathbf{g}, \tau(t) = j]$$

$$= \frac{1}{\pi(B_{ki}, \mathbf{f}_k)} \sum_{l=1}^{r} \sum_{v=1}^{m_l} \sum_{\mathbf{y} \in B_{lv}} P^*_{\mathbf{y}, \mathbf{f}_l}(\mathbf{x}, \mathbf{f}_k) \pi^*(\mathbf{y}, \mathbf{f}_l | B_{lv}, \mathbf{f}_l) \pi(B_{lv}, \mathbf{f}_l),$$

where in the second equality we have used the fact that the probability of the system being in basin $B_{lv}$ is independent of the time since the previous restart or the time to the next restart ($j$ in this case), where the last equality results from calling $t' = t - j$, bringing the limit inside the summation to the products, and again using the fact that the transition probability at the restart is independent of the starting point $(\mathbf{h}, \mathbf{g})$.                                                    $\square$

**LEMMA 3.9** *If $\delta(\mathbf{x}, A_k)$ is the number of iterations of $\mathbf{f}$ needed to reach attractor $A_k$ from state $\mathbf{x}$, then for any $\mathbf{x} \in A_k$, $b < 1$,*

$$\sum_{j=\delta(\mathbf{x}, A_k)}^{\infty} (1 - b)b^j = b^{\delta(\mathbf{x}, A_k)}.$$

*Proof.*

$$\sum_{j=\delta(\mathbf{x}, A_k)}^{\infty} (1 - b)b^j = 1 - \sum_{j=0}^{\delta(\mathbf{x}, A_k)-1} (1 - b)b^j = 1 - (1 - b)\frac{1 - b^{\delta(\mathbf{x}, A_k)}}{1 - b}. \qquad \square$$

**THEOREM 3.10**

$$\pi(A_{ki}, \mathbf{f}_k) = \sum_{l=1}^{r} \sum_{v=1}^{m_l} \left[ \sum_{\mathbf{x} \in B_{ki}} \sum_{\mathbf{y} \in B_{lv}} P^*_{\mathbf{y}, \mathbf{f}_l}(\mathbf{x}, \mathbf{f}_k) \pi^*(\mathbf{y}, \mathbf{f}_l | B_{lv}, \mathbf{f}_l) b^{\delta(\mathbf{x}, A_{ki})} \right] \pi(B_{lv}, \mathbf{f}_l).$$

$$(3.23)$$

*Proof.* We prove the theorem by splitting $P^{(t)}_{\mathbf{h},\mathbf{g}}(A_{ki}, \mathbf{f}_k | B_{ki}, \mathbf{f}_k)$ over the possible values for the random variable $\tau(t)$ and using the fact that the system cannot leave a basin between restarts. In the third equality below, we replace $\mathbf{X}_t \in B_{ki}$ by $\mathbf{X}_{t-j} \in B_{ki}$ because they are the same event: if the system reaches basin $B_{ki}$

at time $t - j$ and is using the function $\mathbf{f}_k$, it will not leave the basin until time $t$ (at least) because there is no restart between $t - j$ and $t$ ($j$ being the time between $t$ and the last restart). Moreover, we split the event $\mathbf{X}_{t-j} \in B_{ki}$ over the possible starting states $\mathbf{x} \in B_{ki}$:

$$\pi(A_{ki}, \mathbf{f}_k) = \pi(B_{ki}, \mathbf{f}_k) \lim_{t \to \infty} P_{\mathbf{h},\mathbf{g}}^{(t)}(A_{ki}, \mathbf{f}_k | B_{ki}, \mathbf{f}_k)$$

$$= \pi(B_{ki}, \mathbf{f}_k) \lim_{t \to \infty} \sum_{j=0}^{\infty} P[\mathbf{X}_t \in A_{ki}, \mathbf{F}_t = \mathbf{f}_k | \mathbf{X}_t \in B_{ki}, \mathbf{F}_t = \mathbf{f}_k, \mathbf{X}_0 = \mathbf{h},$$

$$\mathbf{F}_0 = \mathbf{g}, \tau(t) = j] P[\tau(t) = j]$$

$$= \pi(B_{ki}, \mathbf{f}_k)$$

$$\times \lim_{t \to \infty} \sum_{j=0}^{\infty} \sum_{\mathbf{x} \in B_{ki}} P[\mathbf{X}_t \in A_{ki}, \mathbf{F}_t = \mathbf{f}_k | \mathbf{X}_{t-j} = \mathbf{x}, \mathbf{F}_{t-j} = \mathbf{f}_k, \mathbf{X}_0 = \mathbf{h},$$

$$\mathbf{F}_0 = \mathbf{g}, \tau(t) = j]$$

$$\times P[\mathbf{X}_{t-j} = \mathbf{x} | \mathbf{X}_{t-j} \in B_{ki}, \mathbf{F}_{t-j} = \mathbf{f}_k, \mathbf{X}_0 = \mathbf{h}, \mathbf{F}_0 = \mathbf{g}, \tau(t) = j]$$

$$P[\tau(t) = j].$$

For each state $\mathbf{x}$, the system will reach attractor $A_{ki}$ ($f^j(\mathbf{x}) \in A_{ki}$) if and only if the distance (trajectory length) between $\mathbf{x}$ and $A_{ki}$ is smaller than $j$, that is, $j \geq \delta(\mathbf{x}, A_{ki})$, so that we use the indicator function $\mathbf{1}_{[j \geq \delta(\mathbf{x}, A_{ki})]}$:

$$\pi(A_{ki}, \mathbf{f}_k) = \pi(B_{ki}, \mathbf{f}_k) \sum_{j=0}^{\infty} \sum_{\mathbf{x} \in B_{ki}} \mathbf{1}_{[j \geq \delta(\mathbf{x}, A_{ki})]}$$

$$\times \lim_{t \to \infty} P[\mathbf{X}_{t-j} = \mathbf{x} | \mathbf{X}_{t-j} \in B_{ki}, \mathbf{F}_{t-j} = \mathbf{f}_k, \mathbf{X}_0 = \mathbf{h}, \mathbf{F}_0 = \mathbf{g}, \tau(t) = j]$$

$$\times \lim_{t \to \infty} P[\tau(t) = j].$$

The proof is completed by using dominated convergence because for a large $t$, $P[\tau(t) = j]$ is bounded by $2 \cdot (1 - b)b^j$. We move the limit inside the sum and then apply Lemmas 3.8 and 3.9:

$$\pi(A_{ki}, \mathbf{f}_k) = \pi(B_{ki}, \mathbf{f}_k) \sum_{j=0}^{\infty} \sum_{\mathbf{x} \in B_{ki}} \mathbf{1}_{[j \geq \delta(\mathbf{x}, A_{ki})]}$$

$$\times \left[ \frac{1}{pi(B_{ki}, \mathbf{f}_k)} \sum_{l=1}^{r} \sum_{v=1}^{m_l} \sum_{\mathbf{y} \in B_{lv}} P_{\mathbf{y}, \mathbf{f}_l}^*(\mathbf{x}, \mathbf{f}_k) \pi^*(\mathbf{y}, \mathbf{f}_l | B_{lv}, \mathbf{f}_l) \pi(B_{lv}, \mathbf{f}_l) \right] (1 - b)b^j$$

$$= \sum_{\mathbf{x} \in B_{ki}} \left[ \sum_{l=1}^{r} \sum_{v=1}^{m_l} \sum_{\mathbf{y} \in B_{lv}} P_{\mathbf{y}, \mathbf{f}_l}^*(\mathbf{x}, \mathbf{f}_k) \pi^*(\mathbf{y}, \mathbf{f}_l | B_{lv}, \mathbf{f}_l) \pi(B_{lv}, \mathbf{f}_l) \right] \sum_{j \geq \delta(\mathbf{x}, A_{ki})}^{\infty} (1 - b)b^j$$

$$= \sum_{l=1}^{r} \sum_{v=1}^{m_l} \left[ \sum_{\mathbf{x} \in B_{ki}} \sum_{\mathbf{y} \in B_{lv}} P_{\mathbf{y}, \mathbf{f}_l}^*(\mathbf{x}, \mathbf{f}_k) \pi^*(\mathbf{y}, \mathbf{f}_l | B_{lv}, \mathbf{f}_l) b^{\delta(\mathbf{x}, A_{ki})} \right] \pi(B_{lv}, \mathbf{f}_l) \qquad \square$$

Finally, as $\pi^*(\mathbf{y}, \mathbf{f}_l | B_{lv}, \mathbf{f}_l)$ is unknown, if the probability $q(1 - (1 - p)^n)$ of a restart is small, we can approximate $\pi(A_{ki}, \mathbf{f}_k)$ by

$$\pi(A_{ki}, \mathbf{f}_k) \approx \sum_{l=1}^{r} \sum_{v=1}^{m_l} \frac{1}{|A_{lv}|} \left[ \sum_{\mathbf{x} \in B_{ki}} \sum_{\mathbf{y} \in B_{lv}} P^*_{\mathbf{y},\mathbf{f}_l}(\mathbf{x}, \mathbf{f}_k) b^{\delta(\mathbf{x}, A_{ki})} \right] \pi(B_{lv}, \mathbf{f}_l). \qquad (3.24)$$

Once we have the steady-state probabilities for the pairs $(A_{ki}, \mathbf{f}_k)$, we are interested in the steady-state probabilities $\pi(A_{ki})$ of being in attractor $A_{ki}$ of the Boolean function $\mathbf{f}_k$. Because we can be inside this attractor $A_{ki}$ even using a different function $\mathbf{f}_l$, with $l \neq k$, this probability is

$$\pi(A_{ki}) = \sum_{l=1}^{r} \pi(A_{ki}, \mathbf{f}_l).$$

Equations (3.23) and (3.24) give the steady-state probability $\pi(A_{ki}, \mathbf{f}_l)$ of being in attractor $A_{ki}$ and using the Boolean function $\mathbf{f}_l$ only when $l = k$. We need to compute $\pi(A_{ki}, \mathbf{f}_l)$ for $l \neq k$. We will approximate these probabilities assuming again that the system is almost always inside some attractor; that is, $A_{li}, i = 1, 2, \ldots, m_l$ is a partition of $X$. Using the approximation

$$\pi(A_{ki}, \mathbf{f}_l | A_{lj}, \mathbf{f}_l) \approx \frac{|A_{ki} \cap A_{lj}|}{|A_{lj}|}, \qquad (3.25)$$

we can approximate $\pi(A_{ki}, \mathbf{f}_l)$ by

$$\begin{aligned}
\pi(A_{ki}, \mathbf{f}_l) &\approx \sum_{j=1}^{m_l} \pi(A_{ki}, \mathbf{f}_l | A_{lj}, \mathbf{f}_l) \pi(A_{lj}, \mathbf{f}_l) \\
&\approx \sum_{j=1}^{m_l} \frac{|A_{ki} \cap A_{lj}|}{|A_{lj}|} \pi(A_{lj}, \mathbf{f}_l).
\end{aligned} \qquad (3.26)$$

Equation (3.26) is still valid for $l = k$ because in this case $A_{ki} \cap A_{kj} = \emptyset$ for $i \neq j$. Therefore, the marginal steady-state probability for the attractors can be approximated by

$$\pi(A_{ik}) \approx \sum_{l=1}^{r} \sum_{j=1}^{m_l} \frac{|A_{ki} \cap A_{lj}|}{|A_{lj}|} \pi(A_{lj}, \mathbf{f}_l). \qquad (3.27)$$

UNION OF ATTRACTORS

Equation (3.27) provides representation of the steady-state probabilities for attractors. In the Boolean model, this at once provides representation for unions of attractors because of their disjointness. For PBNs, they need not be disjoint. Hence, we wish to apply the probability addition theorem, but first we need to obtain the steady-state probability of an intersection of attractors. Choosing one of the attractors to condition on, we have

$$
\pi\left(\bigcap_{u=1}^{n} A_{k_u,i_u}\right) = \pi\left(\bigcap_{u=1}^{n-1} A_{k_u,i_u} \mid A_{k_n,i_n}\right) \pi(A_{k_n,i_n}).
$$

We obtain the approximation

$$
\pi\left(\bigcap_{u=1}^{n} A_{k_u,i_u}\right) \approx \frac{\left|\bigcap_{u=1}^{n} A_{k_u,i_u}\right|}{|A_{k_n,i_n}|} \pi(A_{k_n,i_n}). \tag{3.28}
$$

The intersection is null if any two $k_u$ are equal, and therefore the probability in this case is 0. Since conditioning in equation (3.28) can be with respect to any of the attractors in the intersection and since the accuracy of the approximation depends on approximation of the conditional probability, to avoid the unfortunate consequence of choosing a conditioning attractor for which the approximation is particularly bad, we take the average over all the attractors in the intersection for our final approximation:

$$
\pi\left(\bigcap_{u=1}^{n} A_{k_u,i_u}\right) \approx \frac{1}{n}\sum_{j=1}^{n} \frac{\left|\bigcap_{u=1}^{n} A_{k_u,i_u}\right|}{|A_{k_j,i_j}|} \pi(A_{k_j,i_j}),
$$

where the attractor probabilities are defined in equation (3.27). The probability addition theorem yields

$$
\pi\left(\bigcup_{u=1}^{n} A_{k_u,i_u}\right) = \sum_{u=1}^{n}\pi(A_{k_u,i_u}) + \sum_{v=2}^{n}(-1)^{v+1}\sum_{1\leq u_1<u_2<u_3<\cdots<u_v\leq n} \pi\left(\bigcap_{w=1}^{v} A_{k_{u_w},i_{u_w}}\right)
$$

$$
\approx \sum_{u=1}^{n}\pi(A_{k_u,i_u}) + \sum_{v=2}^{n}(-1)^{v+1}\frac{1}{n}\sum_{1\leq u_1<u_2<u_3<\cdots<u_v\leq n}\sum_{j=1}^{v} \frac{\left|\bigcap_{w=1}^{v} A_{k_{u_w},i_{u_w}}\right|}{|A_{k_{u_j},i_{u_j}}|} \pi(A_{k_{u_j},i_{u_j}}).
$$

### 3.2.4 Steady-State Analysis of Instantaneously Random PBNs

As discussed above, a key aspect of the analysis of the dynamics of PBNs is the determination of their steady-state (long-run) behavior. This is a crucial task in many contexts. For instance, as already shown in example 3.5, the steady-state distribution is necessary in order to compute the (long-term) influences. As another example, suppose we are interested in the long-term joint behavior of several selected genes. That is, we would like to obtain their limiting joint distribution. This information can supply answers to these types of questions: What is the probability that gene A will be expressed in the long run? or What is the probability that gene B and gene C will *both* be expressed in the long-run? Steady-state analysis is necessary for answering such questions.

One useful approach in determining such steady-state probabilities is to use Monte Carlo simulation methods. An important issue is the convergence rate of the Markov chain corresponding to the PBN, as it is crucial to ensure that the chain reaches stationarity before collecting information of interest.

Most approaches to steady-state analysis use the state transition matrix in some form or another. In the case of PBNs, this consists of constructing the state transition matrix $\mathbf{A}$ (theorem 3.5) and then applying numerical methods. A variety of approaches using iterative, projection, decompositional, and other methods could potentially be used (Stewart, 1994). Unfortunately, however, in the case of PBNs, the size of the state space grows exponentially in the number of genes and becomes prohibitive for a matrix-based numerical analysis of large networks.

On a more positive note, it should be recognized that even larger state spaces are commonly encountered in Markov chain Monte Carlo methods for many applications, including Markov random field modeling in image processing (Winkler, 1995), where efficient simulation and estimation are routinely performed. Thus, Monte Carlo methods represent a viable alternative to numerical matrix-based methods for obtaining steady-state distributions. Informally speaking, this consists of running the Markov chain for a sufficiently long time until convergence to the stationary distribution is reached and observing the proportion of time the process spends in the parts of the state space that represent the information of interest, such as the joint stationary distribution of several specific genes. A key factor is convergence, which to a large extent depends on the perturbation probability $p$. In general, a larger $p$ results in quicker convergence, but making $p$ too large is not biologically meaningful.

In order for us to perform long-term analysis of the Markov chain corresponding to a PBN using Monte Carlo methods, we need to be able to estimate the convergence rate of the process. Only after we are sufficiently sure that the chain has reached its stationary distribution can we begin to collect information of interest. Typical approaches for assessing convergence are based on the second-largest eigenvalue of the transition probability matrix $\mathbf{A}$. Unfortunately, as mentioned above, for even a moderate number of genes, obtaining the eigenvalues of the transition matrix may be impractical. Thus, it is advantageous to be able to determine the number of iterations necessary until satisfactory convergence is reached. One approach for obtaining a priori bounds on the number of iterations is based on the *minorization condition* for Markov chains (Rosenthal, 1995). This approach was discussed by Shmulevich et al. (2003) in the context of PBNs. However, their analysis indicated that even if one makes assumptions about the relative magnitudes of the perturbation and transition probabilities, one is not likely to obtain a useful bound on convergence via this approach. We now turn to diagnosing convergence to the steady-state distribution.

*Diagnosing Convergence*

In a practical situation, it is important to be able to empirically determine when to stop the chain and produce our estimates. For this purpose, there are a number of monitoring methods available (Cowles and Carlin, 1996; Robert, 1995). Consider, for example, the Kolmogorov-Smirnov test—a nonparametric test of stationarity that can be used to assess convergence.

### KOLMOGOROV-SMIRNOV TEST

When the chain is stationary, distributions $\pi^{(k_1)}$ and $\pi^{(k_2)}$ are the same for arbitrary times $k_1$ and $k_2$. Thus, given a sample $\mathbf{x}^{(1)}, \ldots, \mathbf{x}^{(T)}$, we can compare the two halves: $\mathbf{x}^{(1)}, \ldots, \mathbf{x}^{(T/2)}$ and $\mathbf{x}^{(T/2+1)}, \ldots, \mathbf{x}^{(T)}$. In order to correct for non-i.i.d. (correlated) samples, we introduce a "batch size" $G$ leading to the construction of two (quasi-) independent samples (Robert and Casella, 1999). We thus select subsamples $\mathbf{x}_1^{(G)}, \mathbf{x}_1^{(2G)}, \ldots$ and $\mathbf{x}_2^{(G)}, \mathbf{x}_2^{(2G)}, \ldots$ and use the Kolmogorov-Smirnov statistic with lexicographical ordering to define the indicator:

$$
K = \frac{1}{M} \max_{\eta} \left| \sum_{g=1}^{M} 1_{[0\ldots0,\eta)}(\mathbf{x}_1^{(gG)}) - \sum_{g=1}^{M} 1_{[0\ldots0,\eta)}(\mathbf{x}_2^{(gG)}) \right|, \tag{3.29}
$$

where the maximum is over the state space, the vertices of an $n$-dimensional Boolean hypercube. As $\sqrt{M}K$ has the cumulative distribution function $R(x) = 1 - \sum_{k=1}^{\infty}(-1)^{k-1}e^{-2k^2x^2}$ (Robert and Casella, 1999), the corresponding $p$ value can be computed for each $T$ until it reaches a desired level.

This can also be used to assess convergence for a selected group of genes $j_1, \ldots, j_m$ by replacing state vectors $\mathbf{x}_1^{(gG)}$ and $\mathbf{x}_2^{(gG)}$ in equation (3.29) with just the vectors of their $j_1$th, $\ldots, j_m$th coordinates and modifying the domain of $\eta$ into the hypercube over these coordinates only. For example, if the distribution for only the first gene is of interest, namely, $\pi(\mathbf{x}(1) = 0)$, the maximum in equation (3.29) degenerates into just an absolute difference in the numbers of zeros in the first coordinate between the two samples.

### Two-State Markov Chain Approach

Another approach originally proposed by Raftery and Lewis (1992) can be useful in the context of PBNs (Shmulevich et al., 2003). This method reduces the study of the convergence of the chain to the study of the convergence of a two-state Markov chain. Suppose that we are interested in knowing the steady-state probability of the event {gene A is on and gene B is off}. Then, we can partition the state space into two disjoint subsets such that one subset contains all the states on which the event occurs and the other subset contains the rest of the states. Consider the two "meta-states" corresponding to these two subsets. Although the sequence of these meta-states does not form a Markov chain in itself, it can be approximated by a first-order Markov chain if every $k$ states from the original Markov chain are discarded (i.e., the chain is subsampled) if one assumes that the dependence in the sequence of meta-states falls off fairly rapidly with lag. It turns out in practice that $k$ is usually set to 1, meaning that nothing is discarded and the sequence of meta-states is treated as a homogeneous Markov chain (see (Raftery and Lewis, 1992) for details) with transition probabilities $\alpha$ and $\beta$ between the two meta-states. This is illustrated in figure 3.9.

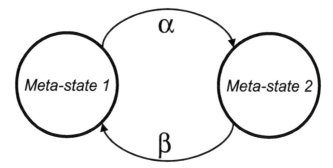

Figure 3.9. An illustration of the two meta-states in the approach of Raftery and Lewis (1992), with transition probabilities $\alpha$ and $\beta$.

Using standard results for two-state Markov chains, it can be shown that the "burn-in" period (the number of iterations necessary to achieve stationarity) $m_0$ satisfies

$$m_0 \geq \log\left(\frac{\varepsilon(\alpha+\beta)}{\max(\alpha,\beta)}\right) \bigg/ \log(1-\alpha-\beta). \qquad (3.30)$$

For the examples that we give below, we set $\varepsilon = 0.001$. In addition, it can be shown that the minimum total number of iterations $\nu$ necessary to achieve a desired accuracy $r$ (we use $r = 0.01$ in our examples below) is

$$\nu = \frac{\alpha\beta(2-\alpha-\beta)}{(\alpha+\beta)^3}\left(\frac{r}{\Phi(\frac{1}{2}(1+s))}\right)^{-2}, \qquad (3.31)$$

where $\Phi(\cdot)$ is the standard normal cumulative distribution function and $s$ is a parameter that we set to 0.95 in our examples. For detailed explanations of the "precision" parameters $\varepsilon$, $r$, and $s$, see (Raftery and Lewis, 1992). The question becomes how to estimate the transition probabilities $\alpha$ and $\beta$, as these are unknown. The solution is to perform a test run from which $\alpha$ and $\beta$ can be estimated and from which $m_0$ and $\nu$ can be computed. Then, another run with the computed burn-in period $m_0$ and the total number of iterations $N$ is performed and parameters $\alpha$ and $\beta$ are reestimated, from which $m_0$ and $\nu$ are recomputed. This can be done several times in an iterative manner until the estimates of $m_0$ and $\nu$ are smaller than the number of iterations already achieved. The above method has been used to determine the steady-state probabilities of some genes of interest from microarray gene expression data (Shmulevich et al., 2003).

*Examples of Analysis*

Using the human glioma gene expression data set described in (Fuller et al., 1999), Shmulevich et al. (2003) constructed a small subnetwork consisting of 15 genes.

Table 3.4 Steady-state Analysis of Several Pairs of Genes

| Tie-2 | NF$\kappa$B | Percent | Tie-2 | TGFB3 | Percent | TGFB3 | NF$\kappa$B | Percent |
|-------|------|---------|-------|-------|---------|-------|------|---------|
| Off | Off | 15.68 | Off | Off | 14.75 | Off | Off | 10.25 |
| Off | On | 41.58 | Off | On | 42.50 | Off | On | 12.47 |
| On | Off | 9.21 | On | Off | 7.96 | On | Off | 14.64 |
| On | On | 31.53 | On | On | 32.78 | On | On | 60.65 |

The entire 597-gene network was inferred using the coefficient of determination (Dougherty et al., 2000), as described in (Shmulevich et al., 2002a). The algorithm for building the subnetwork starting from "seed" genes, which uses influences of genes and ensures that the subnetwork functions fairly autonomously from the rest of the genes, was described by Hashimoto et al. (2004) and will be discussed later in this chapter.

The joint steady-state probabilities of several combinations of two genes were analyzed: Tie-2 and NF$\kappa$B; Tie-2 and TGFB3; and TGFB3 and NF$\kappa$B. For example, for Tie-2 and NF$\kappa$B, the two-state Markov chain method described above was applied to an initial run of 10,000 iterations and produced a burn-in period of $m_0 = 87$ and a total number of iterations of $v = 48,268$. The transition probabilities $\alpha$ and $\beta$ were both approximately equal to 0.03. The perturbation probability $p$ was set to 0.001. Running the network for another 38,268 steps produced recomputed values of $m_0$ and $v$ of 91 and 50,782, respectively. Running the network for yet another 3000 iterations was sufficient for the given accuracy, and the steady-state probabilities of these two genes could be determined. The steady-state probabilities for all pairs of considered genes are shown in table 3.4 as percentages. Figure 3.10 shows the joint steady-state probabilities for all three of these genes using a bar graph.

Tie-2 is a receptor tyrosine kinase expressed on endothelial cells. Its two ligands, angiopoietin 1 and 2, bind Tie-2 and regulate vasculogenesis (Sato et al., 1993), an important process in embryonic development and tumor development. Other related regulators for vasculogenesis are vascular endothelial growth factor (VEGF) and VEGF receptors, which are often overexpressed in the advanced stage of gliomas (Cheng et al., 1996). Although no experimental evidence supports a direct transcriptional regulation of these regulators by the transcriptional factor NF$\kappa$B, which is also frequently activated in glioma progression (Hayashi et al., 2001) as predicted in this analysis, the results show that NF$\kappa$B, at least indirectly, influences the expression of Tie-2. Thus, it may not be surprising that when NF$\kappa$B is on, Tie-2 is on about $31.53/(41.58 + 31.53) = 43$ percent of the time. Because Tie-2 is only one of the regulators for the important vasculogenesis in glioma progression, it is consistent that this analysis of long-term (steady-state) gene expression activities shows that about 40 percent of the time Tie-2 is on. In contrast, NF$\kappa$B is on 73 percent of the time, implying that fewer redundancies exist for NF$\kappa$B activity.

Interestingly, a similar relationship exists between Tie-2 and TGFB3, as can be seen by comparing the percentages in columns 3 and 6 in table 3.4. This suggests

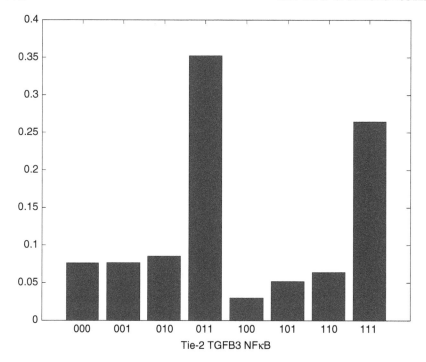

Figure 3.10. Steady-state analysis of all three genes: Tie-2, TGFB3, and NFκB. The gene
combinations are coded in binary. For example, 010 means that Tie-2 is off,
TGFB3 is on, and NFκB is off. The bars show the joint steady-state probabilities. (Shmulevich et al., 2003.)

that TGFB3 and NFκB are more directly linked, which is also shown in the last
three columns of the table (60 percent of the time, they are both on). This relationship is very likely because TGFB1, a homologue of TGFB3, was shown to have
a direct regulatory relationship with NFκB (Arsura et al., 1996). Furthermore, a
recent study by Strauch et al. (2003) showed that NFκB indeed regulates TGFB3
expression (and cell migration).

*Other Remarks*

Steady-state analysis of PBNs is an important problem. For even $n = 20$ genes,
working with $2^{20} \times 2^{20}$ matrices becomes cumbersome and quickly prohibitive for
larger $n$. However, Monte Carlo techniques can be successfully used as long as we
are sufficiently confident that the Markov chain corresponding to the PBN has con-
verged to its equilibrium distribution. Moreover, Monte Carlo techniques exhibit
favorable scaling behavior with respect to the number of genes. Despite the fact
that the size of the state space grows exponentially with $n$, efficient steady-state
analysis can still be carried out. We again note that MCMC methods on images
containing $512 \times 512 = 2.6214 \times 10^5$ pixels, resulting in state spaces on the

order of $10^{78,912}$ for binary models, can nonetheless be effectively performed. The actual inference step, however, is more challenging. When selecting the best predictor genes (inputs) for one target gene (output) using some measure such as the coefficient of determination, we face a combinatorial explosion: for 600 genes, there are $\binom{600}{3} = 3.582 \times 10^7$ possible 3-gene predictors to be tested for each of the 600 genes. Although massively parallel supercomputers are useful in carrying out such an inference, other suboptimal efficient methods are available (Hashimoto et al., 2003).

### 3.2.5 Relationships of PBNs to Bayesian Networks

We discussed Bayesian networks in section 3.1. Let us take a closer look at the basic building blocks of probabilistic Boolean networks and illustrate the relationships between these two models. Recall that in Bayesian networks we express the joint probability as a product of conditional probabilities, as in equation (3.1). Let us consider these conditional probability building blocks of Bayesian networks in the context of probabilistic Boolean networks.

Recall that for a node (gene) $x_i$ there is a corresponding set $F_i$ of possible predictor functions along with their probabilities $c_1^{(i)}, \ldots, c_{l(i)}^{(i)}$. We are interested in computing the conditional probability of gene $X_i$ given its "parents," namely, the set of all genes involved in predicting it. To make this more formal, let the set $X_j^{(i)} \subseteq \{x_1, \ldots, x_n\}$, $j = 1, \ldots, l(i)$, denote the set of essential variables used by predictor $f_j^{(i)}$ for gene $x_i$. Then, the set

$$\mathrm{Pa}(x_i) = \bigcup_{j=1}^{l(i)} X_j^{(i)}$$

is the set of all variables used to predict gene $x_i$, or simply, the *parents* of $x_i$. For convenience of notation, let us expand the domains of all the predictors by adding fictitious variables so that all are functions of the variables in $\mathrm{Pa}(x_i)$. Computationally, this step is not necessary. Then, the conditional probability that $X_i = 1$, given the fact that predictor $f_j^{(i)}$ is used, is equal to

$$P\{X_i = 1 | f_j^{(i)} \text{ is used}\} = \sum_{\mathbf{x} \in \{0,1\}^{|\mathrm{Pa}(x_i)|}} D_i(\mathbf{x}) f_j^{(i)}(\mathbf{x}), \qquad (3.32)$$

where $D_i(\mathbf{x})$ is the joint distribution over the variables in $\mathrm{Pa}(x_i)$. Because we are in the binary setting, equation (3.32) can also be interpreted as the conditional expectation of $X_i$ given the predictor $f_j^{(i)}$. Computationally, we need to consider only the essential variables of $f_j^{(i)}$ and their joint distribution, which can be obtained from the entire joint distribution $D(\mathbf{x})$ by "integrating out" the fictitious variables. Moreover, equation (3.32) can be computed as a vector multiplication of the joint distribution and the truth table of the predictor.

Ultimately, we are interested in obtaining $P\{X_i = 1\}$.  Recall that $c_j^{(i)} = P\{f_j^{(i)} \text{ is used}\}$. Then,

$$P\{X_i = 1\} = \sum_{j=1}^{l(i)} P\{X_i = 1 | f_j^{(i)} \text{ is used}\} \cdot c_j^{(i)}$$

$$= \sum_{j=1}^{l(i)} c_j^{(i)} \cdot \sum_{\mathbf{x} \in \{0,1\}^{|Pa(x_i)|}} D_i(\mathbf{x}) f_j^{(i)}(\mathbf{x}). \tag{3.33}$$

Equation (3.33) provides the expression of the probability of the target gene $X_i$ in terms of the probabilities of the predictors and the joint distribution of the target's parent genes.  To establish the relationships further, we need to consider DBNs since PBNs are inherently temporal. As shown by Lähdesmäki et al. (2006), PBNs and a certain subclass of DBNs can represent the same joint probability distribution over their common variables.

Now suppose that we are interested in the joint probability distribution of the variables expanded over a finite number of updating steps, say, $T$ steps, that is, $\{\mathbf{X}(i) : 0 \le i \le T\}$. Given the initial state $\mathbf{x}(t-1)$, recall from equation (3.10) that the probability of moving to some state $\mathbf{x}(t)$ after one step in the network is

$$A(\mathbf{x}(t-1), \mathbf{x}(t)) = \sum_{j: \mathbf{f}_j(\mathbf{x}(t-1)) = \mathbf{x}(t)} P\{\mathbf{F} = \mathbf{f}_j\}. \tag{3.34}$$

In an instantaneously random PBN, the network realizations are selected independently for each time instant, and thus the joint probability distribution over all possible time series of length $T + 1$ can be expressed as

$$P\{\mathbf{X}(0) = \mathbf{x}(0), \dots, \mathbf{X}(T) = \mathbf{x}(T)\} \tag{3.35}$$

$$= P\{\mathbf{X}(0) = \mathbf{x}(0)\} \prod_{t=1}^{T} A(\mathbf{x}(t-1), \mathbf{x}(t)),$$

where $P\{\mathbf{X}(0) = \mathbf{x}(0)\}$ denotes the distribution of the first state.

Let us concentrate on independent PBNs and rewrite the state transition probabilities $A(\mathbf{x}(t-1), \mathbf{x}(t))$. The dependent PBN case is treated in detail by Lähdesmäki et al. (2006). Let $(\mathbf{x}(t))_i$, $1 \le i \le n$, denote the $i$th element of $\mathbf{x}(t)$ and let $A(\mathbf{x}(t-1), (\mathbf{x}(t))_i)$ denote the probability that the $i$th element of $\mathbf{x}(t)$ will be $(\mathbf{x}(t))_i$ after one step in the network given that the current state is $\mathbf{x}(t-1)$. Since all the nodes are assumed to be independent, each node is updated independently, and equation (3.34) can be written as

$$A(\mathbf{x}(t-1), \mathbf{x}(t)) = \prod_{i=1}^{n} A(\mathbf{x}(t-1), (\mathbf{x}(t))_i). \tag{3.36}$$

So, given equations (3.35) and (3.36), the joint probability distribution over all possible time series of length $T + 1$ can be expressed as

$$P\{\mathbf{X}(0) = \mathbf{x}(0), \dots, \mathbf{X}(T) = \mathbf{x}(T)\} \tag{3.37}$$

$$= P\{\mathbf{X}(0) = \mathbf{x}(0)\} \prod_{t=1}^{T} \prod_{i=1}^{n} A(\mathbf{x}(t-1), (\mathbf{x}(t))_i).$$

*Relationships Between Independent PBNs and DBNs*

In order to establish the relationship between PBNs and DBNs, we will add an extra feature to the definition of a PBN. We will assume that the initial state $\mathbf{X}(0)$ of a PBN can have any joint probability distribution, as in equations (3.35) and (3.37). In particular, we assume $\mathbf{X}(0)$ to have the same distribution as defined by $B_0$ for the first slice of a DBN. For instance, in the context of genetic regulatory network modeling this initial distribution can be set equal to the stationary distribution of the corresponding Markov chain, as described in the section 3.1. Also note that this discussion pertains only to the class of binary-valued DBNs. This requirement can be relaxed, as demonstrated by Lähdesmäki et al. (2006).

We first illustrate a way of expressing an independent PBN $G(V, F)$ as a DBN $(B_0, B_1)$. Let an independent PBN $G(V, F)$ be fixed. The nodes in graphs $G_0$ and $G_1$ must clearly correspond to the nodes in the PBN. In order to distinguish between the nodes from different models, the nodes in the PBN are denoted by $X_i$, $1 \le i \le n$, as above, and the corresponding nodes in the DBN are denoted by $\hat{X}_i$ (similarly for the vector-valued random variables $\mathbf{X}$ and $\hat{\mathbf{X}}$). We refer to nodes $X_i$ and $\mathbf{X}$ as the PBN counterparts of $\hat{X}_i$ and $\hat{\mathbf{X}}$, respectively, and vice versa. Note that in order for us to establish these relationships, we must treat the nodes (genes) in the PBN as random variables, denoted by capital letters.

Because an initial Bayesian network $B_0$ is capable of representing any joint probability distribution over its nodes, the distribution of the initial state of the PBN, $P\{\mathbf{X}(0)\}$, can be represented by $B_0$. It is easy to see that both PBNs and DBNs obey the first-order Markovian property. Thus, we need to consider only one-step transition probabilities $P\{\mathbf{X}(t)|\mathbf{X}(t-1)\}$ when expressing the PBN as a DBN. Further, we can see that for PBNs and DBNs, the joint probability can also be decomposed over their nodes in the same way. Thus, when constructing a transition Bayesian network $B_1$, we can further concentrate on only a single node, say $X_i$, with the other nodes being handled similarly. Let $\mathbf{X}_j^{(i)}(t-1) \subseteq \mathbf{X}(t-1)$ denote the set of essential variables (nodes) used by predictor function $f_j^{(i)}$ for gene $X_i$ at time $t$. The set

$$\mathrm{Pa}(X_i(t)) = \bigcup_{j=1}^{l(i)} \mathbf{X}_j^{(i)}(t-1) \tag{3.38}$$

denotes the set of all variables used to predict the value of gene $X_i$. Let us expand the domain of all the predictor functions in $F_i$ by adding fictitious variables so that they are functions of variables in $\mathrm{Pa}(X_i(t))$. Thus, we can define $B_1 = (G_1, \Theta_1)$ as follows. Graph $G_1$ has directed edges from $\hat{\mathbf{X}}(t-1)$ to $\hat{\mathbf{X}}(t)$ such that the parents of $\hat{X}_i(t)$ are equal to the DBN counterparts of the nodes shown in equation (3.38). Given the distribution $D_i$ over the predictor variables of node $X_i$ (denoted as $\mathrm{Pa}(X_i) \sim D_i$), equation (3.33) can be interpreted as $P\{X_i(t) = 1|\mathrm{Pa}(X_i(t)) \sim D_i\}$. Thus, by specifying $D_i$ to be "deterministic" such that it corresponds to a particular predictor node configuration $\mathrm{Pa}(X_i(t)) = \mathbf{z}$, that is, $D_i(\mathbf{x}) = 1$ if $\mathbf{x} = \mathbf{z}$ and $D_i(\mathbf{x}) = 0$ otherwise, we have

$$P\{X_i(t) = 1|\mathrm{Pa}(X_i(t)) = \mathbf{z}\} = \sum_{j=1}^{l(i)} f_j^{(i)}(\mathbf{z}) c_j^{(i)}. \tag{3.39}$$

Then the set $\Theta_1$ of local conditional probability distributions (or conditional probability tables, CPTs, in the discrete case) induced by the graph structure $G_1$ in the DBN has exactly the same entries as shown in equation (3.39) for each node. Thus, any PBN can be expressed as a binary DBN.

To establish the converse of the above result, let us see how a binary-valued DBN $(B_0, B_1)$ can be represented as an independent PBN $G(V, F)$. Let a DBN $(B_0, B_1)$ be given. The set of nodes $V$ must clearly correspond to the nodes in graphs $G_0$ and $G_1$, and the distribution of the first state $\mathbf{X}(0)$ in the PBN must be the same as for the DBN.

Following the same reasoning as above, we can again conclude that when constructing a PBN, one needs to consider only the predictor functions for only a single node $X_i$ between consecutive time instants $t - 1$ and $t$. Let us assume that the local conditional probability distributions in $\Theta_1$ are given in the form of CPTs and that the number of parents of the $i$th node is denoted as $q = |\text{Pa}(\hat{X}_i(t))|$. Let us enumerate the entries in the CPTs assigned to the $i$th node as triplets $(z_l, \mathbf{y}_l, p_l)$, where $z_l \in \{0, 1\}$, $\mathbf{y}_l \in \{0, 1\}^q$,

$$p_l = P\{\hat{X}_i(t) = z_l | \text{Pa}(\hat{X}_i(t)) = \mathbf{y}_l\},$$

and $1 \leq l \leq 2^{q+1}$. Let us also suppose that the triplets are enumerated such that the first $r = 2^q$ triplets have $z_l = 1$ and that they are sorted in increasing order; that is, $1 \leq k < l \leq r \Rightarrow p_k \leq p_l$. Let $\mathbf{y}_l = y_{l_1} y_{l_2} \ldots y_{l_q}$ and interpret a sequence of symbols $\mathbf{x}_l = x_{l_1}^{y_{l_1}} x_{l_2}^{y_{l_2}} \ldots x_{l_q}^{y_{l_q}}$ as a conjunction, where $x_{l_i}^{y_{l_i}} = x_{l_i}$ if $y_{l_i} = 1$ and $x_{l_i}^{y_{l_i}} = \bar{x}_{l_i}$ if $y_{l_i} = 0$. The variables in the conjunction correspond to a set of specific variables $\{X_{l_1}, \ldots, X_{l_q}\}$ in the PBN; that is, they are the PBN counterparts of $\text{Pa}(\hat{X}_i(t))$.

Then, for the $i$th node in the PBN, the set of functions $F_i$ can be generated as follows: $F_i = \{f_1^{(i)}, \ldots, f_r^{(i)}, f_{r+1}^{(i)}\}$, where

$$f_l^{(i)} = \mathbf{x}_l \vee \mathbf{x}_{l+1} \vee \cdots \vee \mathbf{x}_r, \quad \text{for } 1 \leq l \leq r, \tag{3.40}$$

is a disjunction of conjunctions, $f_{r+1}^{(i)}$ is a zero function, that is, $f_{r+1}^{(i)}(\mathbf{x}) \equiv 0$ for all $\mathbf{x} \in \{0, 1\}^q$, and $r = 2^q$. Note that $f_1^{(i)}$ is essentially the unity function. Also, some of the Boolean expressions in equation (3.40) can possibly be expressed in more compact form. The corresponding selection probabilities are set to $c_1^{(i)} = p_1$, $c_l^{(i)} = p_l - p_{l-1}$ for $2 \leq l \leq r$, and $c_{r+1}^{(i)} = 1 - p_r$. If some of the $c_j^{(i)}$ values happen to be zeros, the corresponding functions $f_j^{(i)}$ can be removed from $F_i$. By applying equation (3.39), we can verify that the above construction indeed gives a PBN equivalent to the given DBN. Let us first compute the one-step prediction probabilities for cases where $X_i(t) = 1$ and the parent nodes have values $\mathbf{y}_l$. We then have

$$P\{X_i(t) = 1 | \text{Pa}(X_i(t)) = \mathbf{y}_l\} = \sum_{j=1}^{r+1} f_j^{(i)}(\mathbf{y}_l) c_j^{(i)}$$

$$= \sum_{j=1}^{l} f_j^{(i)}(\mathbf{y}_l) c_j^{(i)} = p_1 + \sum_{j=2}^{l} (p_j - p_{j-1}) = p_l,$$

where the second equality follows from the fact that only the first $l$ functions $f_1^{(i)}, \ldots, f_l^{(i)}$ contain the conjunction $\mathbf{x}_l$ (see equation (3.40)); that is, $f_j^{(i)}(\mathbf{y}_l) = 1$ only for $1 \leq j \leq l$. The probability of the corresponding complement event $P\{X_i(t) = 0 | \mathrm{Pa}(X_i(t)) = \mathbf{y}_l\}$ is clearly $1 - p_l$. So, a binary-valued DBN can be represented as an independent PBN. Thus, we can state the following theorem.

**THEOREM 3.11** *Independent PBNs $G(V, F)$ and binary-valued DBNs $(B_0, B_1)$ whose initial and transition Bayesian networks $B_0$ and $B_1$ are assumed to have only within and between consecutive slice connections, respectively, can represent the same joint distribution over their common variables.*

It is important to note that the mapping from a binary DBN to an independent PBN is not unique. Instead, there are many PBNs that have the same probabilistic structure. This is best illustrated by an example. Assume that gene $X_1$ is regulated by gene $X_2$ and that all the values in the CPT for gene $X_1$ are equal to 0.5. Then the following two function sets have the same conditional probabilities: $F_1 = \{f_1^{(1)}, f_2^{(1)}\}$, with $c_1^{(1)} = c_2^{(1)} = 0.5$ and where the functions $f_1^{(1)} = 0$ and $f_2^{(1)} = 1$ are constant zero and unity functions, respectively, and $F_{1'} = \{f_{1'}^{(1)}, f_{2'}^{(1)}\}$, with $c_{1'}^{(1)} = c_{2'}^{(1)} = 0.5$ and where the functions $f_{1'}^{(1)} = x_2$ and $f_{2'}^{(1)} = \overline{x}_2$ are the identity and the negation functions, respectively. In other words, two fundamentally different function sets can have the same probabilistic structure. In the case of many parent variables, this issue becomes more complicated. In practice, one may want to construct the predictor functions for each node in the PBN such that the predictor functions have as few variables as possible or such that the number of predictor functions is minimized.

Let us first consider minimizing the number of variables in the predictor functions. The above construction method produces predictor functions with the maximal number of variables. In some cases, when the CPTs are "separable," one can construct predictor functions with fewer variables but which at the same time are consistent with the original conditional probabilities. For example, consider a case where node $X_1$ is regulated by a set of nodes $X_2, \ldots, X_n$. Assume that the first parent node $X_2$ has a canalizing effect on the target node $X_1$ such that all the values in the CPT for which $X_1 = 1$ and $X_2 = 1$ are equal to 0.9. Assume further, for simplicity, that all the other values in the CPT for which $X_1 = 1$ (namely, $X_1 = 1$ and $X_2 = 0$) are smaller than 0.9, the largest of these being, say, 0.8. Instead of blindly using the aforementioned algorithm for generating the predictor functions for $X_1$, it can be useful to take the special form of the CPT into account. Following the above algorithm, the first functions can be constructed as already explained. However, the effect of the canalizing variable $X_2$ can be accounted for using only a single one-variable predictor function $f^{(1)} = x_2$ with $c^{(1)} = 0.9 - 0.8 = 0.1$. Alternatively, if no single-variable predictor functions can be constructed as explained above, then combinations of variables, starting with two variables, three variables, and so on, can be considered. For example, if all the values in the CPT for which $X_1 = 1$, $X_2 = 1$, and $X_3 = 1$ are equal, then a single two-variable function can be defined as $f^{(1)} = x_2 x_3$.

The above search for predictor variables is explained in terms of the original conditional probabilities (i.e., CPTs). Alternatively, one can also try to optimize the obtained predictor functions. That is, each function in $F_i$ should be expressed in some optimal form. The first thing to be considered is the removal of fictitious variables from the functions. This step can potentially result in functions having far fewer input variables. One can further optimize the actual expressions of the Boolean functions using methods such as the well-known Quine-McCluskey algorithm, which minimizes the number of terms in the disjunctive normal form. The above discussion, however, does not provide an optimal method for predictor function construction (apart from optimal representation of the functions). An optimal method can be described in terms of the number of predictor functions, which is discussed next.

In the worst case, the minimum number of predictor functions is determined by the number of different values in the CPT. The above construction method automatically selects this number of predictors (plus a possible constant zero function). In some special cases, the number of predictor functions can be reduced. Consider again the same triplets as above, $(z_l, \mathbf{y}_l, p_l)$, for a single node $X_i$ and again focus on only those triplets for which $z_l = 1$. This leaves us with a set of probability values $p_1, \ldots, p_r, r = 2^q$. The general optimality criterion can be stated as follows. Find the smallest set of selection probabilities $\{c_1^{(i)}, \ldots, c_m^{(i)}\}$, where each $c_j^{(i)} \in [0, 1]$ and $\sum_{j=1}^{m} c_j^{(i)} = 1$, such that each $p_l$, $1 \leq l \leq r$, can be written as

$$p_l = \sum_{j \in I_l} c_j^{(i)}, \tag{3.41}$$

where $I_l \subseteq \{1, \ldots, m\}$. Let $J_j = \{l \mid j \in I_l\} \subseteq \{1, \ldots, 2^q\}$ be the set of indices of those $p_l$ for which $c_j^{(i)}$ is used in the sum in equation (3.41). Then the function set $F_i = \{f_1^{(i)}, \ldots, f_m^{(i)}\}$ is defined as

$$f_j^{(i)} = \bigvee_{l \in J_j} \mathbf{x}_l,$$

where $\mathbf{x}_l$ is as above. We can verify the correctness of the predictor functions and the corresponding selection probabilities

$$P\{X_i(t) = 1 \mid \mathrm{Pa}(X_i(t)) = \mathbf{y}_l\} = \sum_{j=1}^{m} f_j^{(i)}(\mathbf{y}_l) c_j^{(i)}$$

$$= \sum_{j: f_j^{(i)}(\mathbf{y}_l)=1} c_j^{(i)} = \sum_{j: l \in J_j} c_j^{(i)} = \sum_{j \in I_l} c_j^{(i)} = p_l$$

because (on the second line) $f_j^{(i)}(\mathbf{y}_l) = 1$ only if the function $f_j^{(i)}$ contains the conjunction $\mathbf{x}_l$, that is, $l \in J_j$. The second-to-last equality follows from the fact that $l \in J_j \Rightarrow j \in I_l$, and the last equality from equation (3.41). Note that the optimal function set can still be nonunique.

The obtained functions in $F_i$ can be modified further, as explained above, by removing the possible fictitious variables and by applying the Quine-McCluskey optimization algorithm. A computationally efficient algorithm for the search for the optimal function sets remains to be developed. Fortunately, the search problem is usually limited in the sense that each gene contains only a few parent variables.

Theorem 3.11 says that the two model classes can represent the same probabilistic behavior. However, there are many statistically equivalent PBNs that correspond to a DBN. This means that the PBN formalism is redundant from the probabilistic point of view. On the other hand, the PBN formalism is richer from the functional point of view because it can explain the regulatory roles of different gene sets in more detail than the conditional probabilities in DBNs. A detailed discussion of these issues, including dependent PBNs and discrete-valued DBNs, is contained in (Lähdesmäki et al., 2006).

*Remark.* At this point it might be good to comment on the issue of model complexity. One can argue endlessly about whether a particular model is sufficiently complex relative to the variables it contains, the relations between these variables, and the valuation space. Referring back to the preface, ultimately it is the ability of a model to predict future outcomes that matters. Added complexity can help or hurt, depending on whether it is needed to adequately describe the variables of interest, how the variables are to be transformed, and the availability of data and computational resources. In many instances, the disputes are essentially meaningless. For instance, it may be that a stochastic differential equation provides an elegant and complete description but that it can be satisfactorily approximated by a discrete model (Ivanov and Dougherty, 2006). Which is better depends on the circumstances. A similar statement can be made with respect to the level of quantization—as is well known in signal processing. As we will see when we discuss classification, a complex model can be deleterious and the overcomplexity can be quantified. All these matters relate to the epistemology of science, in particular to the epistemology of computational biology (Dougherty and Braga-Neto, 2006).

### 3.2.6 Growing Subnetworks from Seed Genes

It is likely that genetic regulatory networks function in what might be called a multiscale manner. One of the basic principles in multiscale modeling is that meaningful and useful information about a system or object exists on several different "levels" simultaneously. In the context of genetic networks, this implies that genes form small groups (or clusters) wherein genes have close interactions. Some of these clusters are functionally linked, forming larger "meta-clusters," and these meta-clusters have interactions as well. This process can continue on several different scales.

An important goal is to discover relatively small subnetworks, out of the larger overall network, that function more or less independently of the rest of the network. Such a small subnetwork would require little or sometimes even no

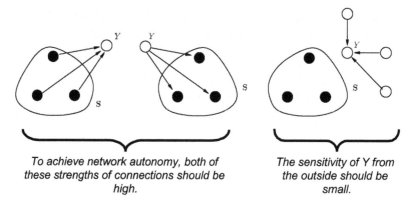

Figure 3.11. In order to maintain the self-autonomy of subnetwork **S**, the collective strength of the genes in **S** on gene $Y$—a candidate for inclusion in the subnetwork— should be high. The strength of $Y$ on **S** should be high as well, thus maintaining high interdependency between the genes in the subnetwork. At the same time, the strength from genes outside the subnetwork on gene $Y$ should be low. (Hashimoto et al., 2004.)

information from the "outside." We can proceed by starting with a seed consisting of one or more genes that are believed to participate in such a subnetwork. Then, we iteratively adjoin new genes to this subnetwork such that we maintain the aforementioned "autonomy" of the subnetwork as much as possible, using the notions of gene influence or the coefficient of determination. Such an algorithm for growing subnetworks from seed genes has been developed by Hashimoto et al. (2004).

Subnetwork construction proceeds in a way that enhances a strong collective strength of connections among the genes within the subnetwork and also limits the collective strength of the connections from outside the subnetwork. Consider figure 3.11. Suppose we have a subnetwork **S** and are considering the candidate gene $Y$ for inclusion in this subnetwork. We would like the collective strength (to be defined in a moment) of the genes in **S** on the candidate gene $Y$, as well as the strength of gene $Y$ on the genes in **S**, to be high. In other words, the genes in **S** and $Y$ should be tightly interdependent. At the same time, other genes from outside the subnetwork should have little impact on $Y$ if we are to maintain the subnetwork's autonomy or "self-determinacy." Thus, their collective strength on $Y$ should be low. At each step, the subnetwork grows by one new gene so as to ensure maximal autonomy. An overall measure of subnetwork autonomy, which serves as an objective function in the subnetwork growing algorithm, is a combination of the three types of strength just described (Hashimoto et al., 2004). Finally, the strength itself can be naturally captured either by the coefficient of determination or by the influence. Examples of subnetworks with insulin-like growth factor binding protein 2 (IGFBP2) or VEGF as seeds are shown in figures 3.12 and 3.13. In both glioma subnetworks, the influence was used as the strength of connection. The numbers above the arrows represent influences.

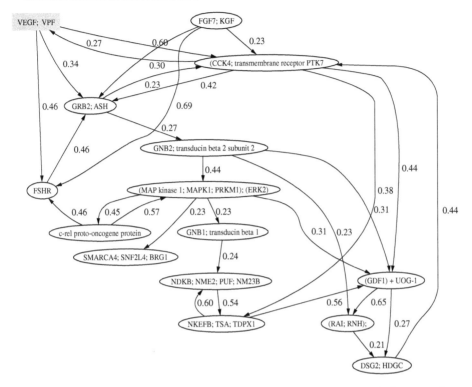

Figure 3.12. A subnetwork grown from the vascular endothelial growth factor seed. (Hashimoto et al., 2004.)

*Interpretation and Validation of Subnetworks with Prior Biological Knowledge and Experimental Results*

Having constructed subnetworks in figures 3.12 and 3.13 from expression data via the seed-based growing algorithm, we would like to interpret and validate (in so far as possible) the constructions using prior knowledge and independent experimental results. IGFBP2 and VEGF are genes that have been extensively studied and are well characterized in the literature. It is known that IGFBP2 and VEGF are overexpressed in high-grade gliomas, glioblastoma multiforme (GBM)—the most advanced stage of tumor (Kleihues and Cavenee, 2000)—as compared to other types of glioma (Fuller et al., 1999). This finding was confirmed by two independent studies (Sallinen et al., 2000; Elmlinger et al., 2001).

Studies that were completely independent of the PBN modeling work showed that NFκB activity is activated in cells stably overexpressing IGFBP2. This was done by using a luciferase reporter gene linked to a promoter element that contains an NFκB-binding site. An analysis of the IGFBP2 promoter sequence showed that there are several NFκB-binding sites, suggesting that NFκB transcriptionally regulates IGFBP2. A review of the literature revealed that Cazals et al. (1999) indeed demonstrated that NFκB activated the IGFBP2 promoter in lung alveolar epithelial cells. Interestingly, in the IGFBP2 network (figure 3.13), we see an

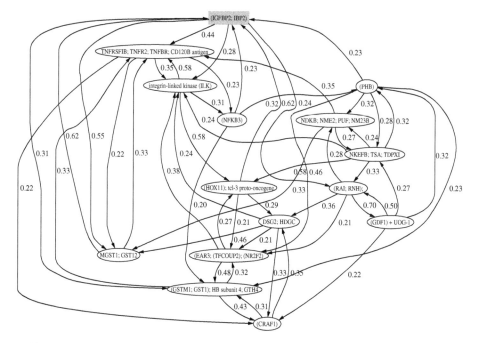

Figure 3.13.  A subnetwork grown from the insulin-like growth factor binding protein 2 seed.

arrow linking NFκB3 to IGFBP2, and we see a two-step link from IGFBP2 to
NFκB through TNF receptor 2 (TNFR2) and integrin-linked kinase (ILK). This
parallels what was observed by Wei Zhang, a close collaborator of the authors and
director of the Cancer Genomics Laboratory of the M.D. Anderson Cancer Center
in Houston, Texas.  The presence of NFκB-binding sites on the IGFBP2 promoter
implies a direct influence of NFκB on IGFBP2. Although higher NFκB activity
in cells overexpressing IGFBP2 was found, a transient transfection of IGFBP2-
expressing vector together with an NFκB promoter reporter gene construct did
not lead to increased NFκB activity, suggesting an indirect effect of IGFBP2 on
NFκB that requires time to take place.  In fact, because of this indirect effect,
this observation was not pursued for a year until the PBN-based subnetwork was
linked to the laboratory experiments. IGFBP2 also contains an RGD domain,
implying its interaction with integrin molecules.  Integrin-linked kinase is in the
integrin signal transduction pathway.  The second subnetwork starting with
VEGF (figure 3.12) offers even more compelling insight and supporting
evidence.

    Gliomas, like other cancers, are highly angiogenic, reflecting the need of cancer
tissues for nutrients.  To satisfy this need, expression of the vascular endothelial
growth factor gene is often elevated.  VEGF protein is secreted outside the cells
and then binds to its receptor on endothelial cells to promote their growth (Folkman, 1995).  Blockage of the VEGF pathway has been an intensive research area
in cancer therapeutics (Bikfalvi and Bicknell, 2002).  A scrutiny of the VEGF network (figure 3.12) reveals some very interesting insight, which is highly consistent

with prior biological knowledge derived from biochemical and molecular biology experiments. Let us elaborate. From the graph, VEGF, FGF7, FSHR, and PTK7 all influence Grb2. FGF7 is a member of the fibroblast growth factor family (Rubin et al., 1989). FSHR is a follicle-stimulating hormone receptor. PTK7 is another tyrosine kinase receptor (Banga et al., 1997). The protein products of all four genes are part of signal transduction pathways that involve surface tyrosine kinase receptors. These receptors, when activated, recruit a number of adaptor proteins to relay the signal to downstream molecules. Grb2 is one of the most crucial adaptors that has been identified (Stoletov et al., 2002). We should note that Grb2 is a target for cancer intervention (Wei et al., 2003) because of its link to multiple growth factor signal transduction pathways including VEGF, EGF, FGF, and PDGF. Thus, the gene transcript relationships among the above five genes in the VEGF subnetwork appear to reflect their known or likely functional and physical relationship in cells. Molecular studies reported in the literature have further demonstrated that activation of the protein tyrosine kinase receptor–Grb-2 complex in turn activates the ras–MAP kinase–NF$\kappa$B pathway to complete the signal relay from outside the cells to the nuclei (Bancroft et al., 2001). Although ras is not present in our VEGF network, a ras family member, GNB2, or transducin beta 2, is directly influenced by Grb2. GNB2 then influences MAP kinase 1 or ERK2, which in turn influences the NF$\kappa$B component c-rel (Pearson et al., 2001).

In the VEGF subnetwork shown in figure 3.12, we also observe some potential feedback loop relationships. For example, c-rel influences FSHR, which influences Grb2-GNB2-MAPK1, and then c-rel itself. This may be a feedback regulation, a crucial feature of biological regulatory systems in cells needed to maintain homeostasis. Other feedback regulation may also exist. RAI, or rel-A (another NF$\kappa$B component) associated inhibitor (Yang et al., 1999), influences GNB2, which is two steps away from c-rel. RAI is further linked to PTK7 through GDF1, reflecting potentially another feedback regulatory mechanism. Whether these relationships are true negative feedback control mechanisms will need to be validated experimentally in the future. In this regard, the networks built from these models provide valuable theoretical guidance for experiments.

## 3.3 INTERVENTION

One objective of PBN modeling is to use the PBN to design different approaches for affecting evolution of the gene activity profile of the network. Such strategies can be useful in the identification of potential drug targets, for example, in cancer therapy. To date, such intervention studies have used three different approaches:

1. Resetting the state of the PBN, as necessary, to a more desirable initial state and letting the network evolve from there (Shmulevich et al., 2002b),
2. Changing the steady-state (long-run) probability distribution of the network by minimally altering its rule-based structure (Shmulevich et al., 2002c),
3. Manipulating external (control) variables that affect the transition probabilities of the network and can, therefore, be used to desirably affect its dynamical evolution (Datta et al., 2003, 2004).

### 3.3.1 Gene Intervention

A random gene perturbation may cause the real regulatory network to transition into an undesirable cellular state, which itself will be stable under most subsequent gene perturbations. In the context of (deterministic) Boolean networks, this corresponds to a transition from one attractor to another. One is then faced with the challenge of determining which genes are good potential candidates for intervention in order to reverse the effects or force the regulatory network to transition to another, desirable stable state. Thus, it is important not only to study the effects of gene perturbation, especially on long-run network behavior, but also to develop tools for discovering intervention targets. While we distinguish between random gene perturbation and intentional gene intervention, the PBN model class allows us to take a unified viewpoint. We will also make a distinction between *transient* and *permanent* perturbation or intervention. The former type can be reversed by the network itself, while the latter is unchangeable or fixed. Although for the most part we focus on transient perturbation or intervention, we will discuss this distinction a bit later.

*Framework for Gene Intervention*

Let us consider the effects of deliberately affecting a particular gene by means of intervention. One of the key goals of PBN modeling is the determination of possible intervention targets (genes) such that the network can be "persuaded," if not forced, to transition into a desired state or set of states. Whereas in Boolean networks attractors are hypothesized to correspond to functional cellular states (Huang, 1999), in instantaneously random PBNs, this role is played by irreducible subchains. When the probability of perturbation, $p$, is equal to zero, a PBN is unable to escape from an irreducible subchain, implying that the cellular state cannot be altered. When $p$ becomes positive, there is a chance that the current cellular state may switch to another cellular state by means of a random gene perturbation. Clearly, perturbation of certain genes is more likely to achieve the desired result than perturbation of some other genes. Our goal, then, is to discover which genes are the best potential "lever points," borrowing the terminology from Huang (1999), in the sense of having the greatest possible impact on desired network behavior so that we can intervene with them by changing their values (1 or 0) as needed. In addition, we wish to be able to intervene with as few genes as possible in order to achieve our goals. To motivate the discussion, let us illustrate the idea with an example.

Consider figure 3.5. For the predictor probabilities given in example 3.1, the probabilities of the four possible network realizations are $p_1 = 0.3$, $p_2 = 0.3$, $p_3 = 0.2$, and $p_4 = 0.2$. Suppose that we are currently in state (111) and wish to eventually transition to state (000). Finally, let us assume, for the moment, that the probability of random perturbation is zero ($p = 0$). The question is, With which of the three genes—$x_1$, $x_2$, or $x_3$—should we intervene so that the probability is greatest that we will end up in (000)? By direct inspection of figure 3.5, we can see that if we make $x_1 = 0$, then with probability $p_4 = 0.2$ we will transition into (000), whereas if we make $x_2 = 0$ or $x_3 = 0$, it will be impossible for us to end up in (000), and with probability 1, we will eventually come back to (111), which is where we started. In other words, the network will be resistant to perturbations of the second or third gene and will eventually maintain the same state. Thus, the answer to our question in this rather simple example is that only by intervening with gene $x_1$ do we have a chance of achieving our goal. In order for us to answer such questions in general, we need to develop several tools.

When $p > 0$, by proposition 3.3, the entire Markov chain is ergodic, and thus every state will eventually be visited. Thus, the question of intervention should be posed in the sense of *reaching a desired state as early as possible*. For instance, in the example considered above, if $p$ is very small and we are in state (111), then it will be a long time until we reach (000), and setting $x_1 = 0$ is much more likely to get us there faster. We are, therefore, interested in the probability $F_k(\mathbf{x}, \mathbf{y})$ that, starting in state $\mathbf{x}$, the first time the PBN reaches some given state $\mathbf{y}$ will be at time $k$. This is often referred to as the *first-passage time* from state $\mathbf{x}$ to state $\mathbf{y}$. A related measure of interest is the mean first-passage time from state $\mathbf{x}$ to state $\mathbf{y}$, defined as

$$M(\mathbf{x}, \mathbf{y}) = \sum_k k F_k(\mathbf{x}, \mathbf{y}). \tag{3.42}$$

This measure tells us how long, on average, it will take to get from state $\mathbf{x}$ to state $\mathbf{y}$.

It is easy to see that for $k = 1$, $F_k(\mathbf{x}, \mathbf{y}) = A(\mathbf{x}, \mathbf{y})$, which is just the transition probability from $\mathbf{x}$ to $\mathbf{y}$. For $k \geq 2$, it is also straightforward to show (e.g., Çınlar, 1997) that $F_k(\mathbf{x}, \mathbf{y})$ satisfies

$$F_k(\mathbf{x}, \mathbf{y}) = \sum_{\mathbf{z} \in \{0,1\}^n - \{\mathbf{y}\}} A(\mathbf{x}, \mathbf{z}) F_{k-1}(\mathbf{z}, \mathbf{y}). \tag{3.43}$$

Every required entry in matrix $\mathbf{A}$ can be computed directly using theorem 3.4. Let us illustrate this computation with the same example given above.

Suppose, as before, that $p = 0.01$. Then, the steady-state distribution equals [0.0752 0.0028 0.0371 0.0076 0.0367 0.0424 0.0672 0.7310], where the leftmost element corresponds to (000) and the rightmost to (111). As expected, the PBN spends much more time in state (111) than in any other state. In fact, more than 70 percent of the time is spent in this state. Let our starting state $\mathbf{x}$ be (111) and the destination state $\mathbf{y}$ be (000), as before. The question with which we concern ourselves is whether we should intervene with gene $x_1$, $x_2$, or $x_3$. In other words, we would like to compute $F_k((011), (000))$, $F_k((101), (000))$, and $F_k((110), (000))$, where the states are written in their binary representations. We can then assess our results by plotting

$$H_{k_0}(\mathbf{x}, \mathbf{y}) = \sum_{k=1}^{k_0} F_k(\mathbf{x}, \mathbf{y})$$

for the states $\mathbf{x}$ of interest and for a sufficiently large $k_0$. The intuition behind this approach is the following. Since the events {the first-passage time from $\mathbf{x}$ to $\mathbf{y}$ will

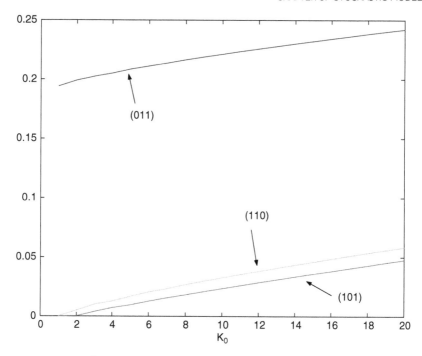

Figure 3.14. $H_{k_0}(\mathbf{x}^{(i)}, \mathbf{y})$ for $k_0 = 1, \ldots, 20$ for starting states (011), (101), and (110), corresponding to perturbations of first, second, and third genes, respectively. (Shmulevich et al., 2002b.)

be at time $k$} are disjoint for different values of $k$, the sum of their probabilities for $k = 1, \ldots, k_0$ is equal to the probability that the network, starting in state $\mathbf{x}$, will visit state $\mathbf{y}$ before time $k_0$. As a special case, when $k_0 = \infty$, this is equal to the probability that the chain will *ever* visit state $\mathbf{y}$, starting at state $\mathbf{x}$, which of course is equal to 1 since our chains are ergodic if $p > 0$. Figure 3.14 shows plots of $H_{k_0}(\mathbf{x}, \mathbf{y})$ for $k_0 = 1, \ldots, 20$ and for the three states of interest, namely, (011), (101), and (110).

The plots indicate that if we start with state (011), we are much more likely to enter state (000) sooner than if we start with state (110) or (101). For example, during the first 20 steps, we have an almost 25 percent chance of entering (000) if we start with (011), whereas if we start with (110) or (101), we have only about a 5 percent chance. This in turn indicates that we should intervene with gene $x_1$ rather than with gene $x_2$ or $x_3$. Of course, in this rather simple example, we could have discerned this by visual inspection of figure 3.5, but for larger networks this method provides a tool for answering these kinds of questions.

In biology, there are numerous examples where the (in)activation of one gene or protein can lead much quicker (or with a higher probability) to a certain cellular functional state or phenotype than the (in)activation of another gene or protein. For instance, let us use a stable cancer cell line as an example. Without any intervention, the cells keep proliferating. Let us assume that the goal of the intervention is to push the cell into programmed cell death (apoptosis). Let us further assume that we will achieve this intervention with two gene candidates: p53 and telomerase. The p53 gene is the most well-known tumor suppressor gene, encoding a protein that regulates the expression of several genes such as Bax and Fas/APO1, which function to promote apoptosis (Miyashita and Reed, 1995; Owen-Schaub et al., 1995), and p21/WAF1, which functions to inhibit cell growth (El-Deiry et al., 1993). The telomerase gene encodes telomerase, which maintains the integrity of the ends of chromosomes (telomeres) in our germ cells, which are responsible for propagating our complete genetic material to the following generation, as well as progenitor cells, which are responsible for replenishing our cells during normal cell turnover (homeostasis). In somatic cells, the telomerase gene is turned off, resulting in telomeres shortening each time the cell divides—a key reason for the limited life span of normal cells (Harley, 1991). In the majority of tumor cells, telomerase is activated, which is believed to contribute to the prolonged life-spans of these cells (Kim et al., 1994) and worsened prognoses for cancer patients (Hiyama et al., 1995; Zhang et al., 1996). Extensive experimental results indicate that when p53 is activated in cells, for example, in response to radiation, the cells undergo rapid growth inhibition and apoptosis in as short a time as a few hours (Lowe, 1993; Kobayashi et al., 1998). In contrast, inhibition of the telomerase gene also leads to cell growth inhibition, differentiation, and cell death, but only after cells go through a number of cell divisions (allowing telomere shortening), which takes a longer time to occur than via p53.

Another valuable computational tool is the mean first-passage time given in equation (3.42). Intuitively, the best candidate gene for intervention should be the one that results in the smallest mean first-passage time to the destination state. For the same example as above, the three mean first-passage times corresponding to the perturbation of genes $x_1$, $x_2$, and $x_3$ are equal to 337.51, 424.14, and 419.20, respectively. Since the first one is the smallest, this again supports the idea that gene $x_1$ is the best candidate for intervention.

To summarize, we simply generate different states $\mathbf{x}^{(i)} = \mathbf{x} \oplus \mathbf{e}_i$, $i = 1, \ldots, n$, where $\mathbf{e}_i$ is the unit binary vector with a 1 in the $i$th coordinate, by perturbing each of the $n$ genes and compute $H_{k_0}(\mathbf{x}^{(i)}, \mathbf{y})$ for some desired destination state $\mathbf{y}$ and constant $k_0$. Then, the best gene for intervention is the one for which $H_{k_0}(\mathbf{x}^{(i)}, \mathbf{y})$ is maximum. That is, given a fixed $k_0$, the optimal gene $x_{i_{opt}}$ satisfies

$$i_{opt} = \arg\max_i H_{k_0}(\mathbf{x}^{(i)}, \mathbf{y}). \tag{3.44}$$

Alternatively, by minimizing the mean first-passage times, the optimal gene satisfies

$$i_{opt} = \arg\min_i M(\mathbf{x}^{(i)}, \mathbf{y}). \tag{3.45}$$

Another approach related to the one in equation (3.44) might be to first fix a probability $h_0$ and wait until one of the $H_{k_0}(\mathbf{x}^{(i)}, \mathbf{y})$ reaches it first. Note that because of ergodicity, for every state $\mathbf{x}^{(i)}$ there will always be a $k_0^{(i)}$ large enough such that $H_{k_0^{(i)}}(\mathbf{x}^{(i)}, \mathbf{y}) > h_0$. In that sense, the optimal gene for intervention, $x_{i_{\text{opt}}}$, is one for which

$$i_{\text{opt}} = \arg\min_i \min_{k_0^{(i)}} \{k_0^{(i)} : H_{k_0^{(i)}}(\mathbf{x}^{(i)}, \mathbf{y}) > h_0\}. \tag{3.46}$$

At first glance, it might appear as if both approaches, equations (3.44) and (3.46), will yield the same answer since figure 3.14 seems to suggest that the plots do not intersect and that if one of them is maximum for a given $k_0$, it will be the first to reach any fixed $h_0$ thereafter. While it is true that for sufficiently large $k_0$ the plots will not intersect, this is not in general true for smaller values of $k_0$.

The criteria embedded in equations (3.44) and (3.46) have underlying different interpretations. The first aims to *maximize the probability* of reaching a particular state before a certain fixed time, while the second aims to *minimize the time* needed to reach a certain state with a given fixed probability. These two approaches are complementary and can be used in conjunction. Finally, the approach in eqaution (3.45) based on minimizing mean first-passage times is another simple alternative. We will come back to mean first-passage times when we discuss sensitivity analysis of PBNs.

*Sets of States, Avoidance of States, and Permanent Intervention*

So far we have discussed the notion of intervention in terms of a single starting state and a single destination state. However, we may often be more interested in the same types of questions but concerning *sets* of states. For example, two different sets of states can correspond to different functional cellular states, such as proliferation and quiescence, much in the same way that attractors play this role in standard Boolean networks (Huang, 1999). In PBNs, this role is typically played by irreducible subchains when no perturbations can occur ($p = 0$). In other words, once the network enters an irreducible subchain (cf. attractor), it cannot escape. When the perturbation probability is positive, there are no longer any irreducible subchains (see proposition 3.3), but the sets of states that correspond to these irreducible subchains when $p = 0$ still represent the functional states of the organism that is being modeled—there is now simply a probability of escaping due to random perturbations. The sets of states that correspond to irreducible subchains when $p = 0$ can be referred to as *implicitly irreducible* subchains. They are essentially "islands" of states, and the probability of perturbation controls the number of "bridges" between these islands. When $p = 0$, there are no bridges, and when $p$ becomes larger, it becomes easier to "travel" between the islands.

Going back to the question of intervention, we may be interested in posing it as follows. Given that we are in a set of states $X$, what gene is the best candidate for intervention if we want to end up in the set of states $Y$ ? The question can be posed in the sense of eqaution (3.44), (3.45), or (3.46). Fortunately, the mathematical framework does not really change when we talk about sets of states. For example, if $X = \{\mathbf{x}\}$ consists of just one state but $Y$ is a set composed of many states, then the first-passage probabilities $F_k(\mathbf{x}, \mathbf{y})$ can simply be summed over all states $\mathbf{y} \in Y$ and we can define $F_k(\mathbf{x}, Y) = \sum_{\mathbf{y} \in Y} F_k(\mathbf{x}, \mathbf{y})$. Then, the same approaches as discussed above can be used to find the best gene for intervention.

The situation where $X$ is composed of a number of states is conceptually a bit more complicated since now the starting set of states $X$, rather than just one starting state $\mathbf{x}$, represents a type of uncertainty in our knowledge of the current state of the network. That is, we may not know exactly what state the network is in at a particular time, but we may know that it is in a certain set of states. This may be relevant not only from an experimental perspective, as it may be difficult to determine precisely the current state at a given time, but perhaps more importantly, we may not be interested in restricting ourselves to just one state but rather wish to consider a whole set of states $X$ that is believed to correspond to the current functional cellular state.

Consequently, a gene that may be the best candidate for intervention for one of the starting states in $X$ may not be the best for another state in $X$. Therefore, the best we can do in such a case is to combine the individual results for all states $\mathbf{x} \in X$ but weigh them by their respective probabilities of occurrence. The latter are furnished by the steady-state probabilities $\pi_{\mathbf{x}}$. In other words, we can define

$$F_k(X, Y) = \frac{\sum_{\mathbf{x} \in X} \sum_{\mathbf{y} \in Y} F_k(\mathbf{x}, \mathbf{y}) \cdot \pi_{\mathbf{x}}}{\sum_{\mathbf{x} \in X} \pi_{\mathbf{x}}} \qquad (3.47)$$

to be the first-passage probability from set $X$ to set $Y$.

In addition to reaching a desired state or set of states, we may also be interested in *avoiding* a state or set of states. This is quite natural in terms of inducing a network not to enter into some sets of states corresponding to unwanted functional cellular states (e.g., proliferation). This goal is in a sense complementary to what has been described above in terms of reaching a desired state either as soon as possible with a given probability or with as high a probability as possible before a given time. For example, in equation (3.45), our goal was to minimize the mean first-passage time to a destination state. In order to avoid a destination state, we simply have to maximize the mean first-passage time to that state. So, the underlying mechanism is quite the same, and we will not give a separate example illustrating the avoidance of states. We would like to point out, however, that it may be possible that performing no intervention whatsoever is the best option regardless of whether we want to reach or to avoid a state or set of states. In other words, depending on the network, as well as on the starting and destination states or sets of states, it may be the case that not intervening with any gene is optimal in terms of the criteria given in equation (3.44), (3.45), or (3.46).

In our model, the interventions and perturbations we have considered up to this point are in the gene's expression state, which is generally a transient phenomenon. Thus, it can be termed *transient* intervention or perturbation. That is, the effect on a gene, whether by random perturbation or forced intervention, is applied at only one time point, and the network itself is responsible for determining the values of that gene thereafter. It could be said that the effect has the potential to be reversed by the network itself. For example, in figure 3.5, if we are in state (111) and the second gene changes value, resulting in (101), at the next time step, regardless of where the network transitions, (110) or (111), the second gene will always be changed back to 1. Since in this example (111) is an absorbing state, the network will eventually return to it, and the perturbation or intervention—whatever the means was of changing the second gene—will have been "compensated" by the network itself. This inherent resistance to perturbations is a key factor for stability and robustness in PBNs just as it is in Boolean networks.

We can also consider a *permanent* intervention or perturbation. In this scenario, a gene changes value and remains at that value forever. From a genetic perspective, permanent intervention is achieved through removing a gene or "transplanting" a gene, as is done in gene therapy. From a network perspective, the permanent intervention (or perturbation) of a gene essentially reduces the state space by half since all the states in which the gene is not equal to the fixed value cannot appear. The rest of the genes are predicted as usual, via Boolean functions, and their selection probabilities $c_j^{(i)}$ remain unaltered. The Boolean function corresponding to the fixed gene is the identity function (0 or 1) with selection probability 1.

Permanent intervention by gene manipulation is used by both nature and humans. It is an efficient way to generate mutations and also hoped to be an efficient way to correct mutations (therapy). Perhaps the best example of the first scenario is viral infection. Let us use simian virus 40 (SV40) as an example. SV40 virus was discovered in the 1950s during the development of a vaccine for poliovirus (Carbone et al., 1997). It was found that SV40 could transform monkey kidney cells and develop tumors when injected into rodents (Abrahams and van der Eb, 1975). SV40 was not believed to cause tumors in human cells, however, SV40 DNA has been found in some human brain tumors in recent years (Kouhata et al., 2001), suggesting that SV40 may have a tumorigenic effect in humans too, although with a long latent period.

Extensive research has been carried out to elucidate how SV40 causes cancer in mouse cells. Though SV40 does not have a big genome, one of the most important proteins encoded by SV40 is large T-antigen. Large T-antigen interacts with host cell molecules and triggers a series of events that are beneficial for viral replication and bad for the host cells. For example, T-antigen inactivates the functions of p53 (Zhu et al., 1991; Bargonetti et al., 1992), which may be the key mechanism for the tumorigenic effect of SV40 T-antigen. We should point out that SV40 T-antigen also interacts with other molecules such as retinoblastoma (DeCaprio et al., 1988)—an important protein whose activation inhibits DNA synthesis. From a network perspective, the permanent mutation caused by SV40 T-antigen may permanently alter the dynamics of the network, causing it to shift into a set of states associated with tumorigenesis. To further prove that T-antigen itself is sufficient to cause this effect, T-antigen was "transplanted" into mouse brain using a tissue-specific transgenic mouse model (second scenario, man-made event). As expected, brain tumors were found in many of the transgenic mice (Brinster et al., 1984). Since SV40 DNA has been detected in some human brain tumors, one cannot help but speculate that SV40 may be causing human brain tumors too.

From the point of view of human intervention, it may be that permanent rather than transient intervention is the only way to reach a desired set of states. That is, it may be the case that the network is so resistant to transient intervention with any gene that it is extremely unlikely that it will ever reach (or avoid) the desired states. Permanent intervention, though less desirable, as it introduces permanent changes to the network, may be the only alternative for reaching a set of states with a sufficiently high probability. As before, the question is, What genes are the most likely "lever points" for controlling the global behavior of the network?

For example, based on what is known, p53 is one such gene. This is clearly demonstrated by the fact that p53 gene deletion or mutation (permanent perturbation) is one of the most frequent genetic changes in cancers (Hollstein et al., 1991). Removing p53 genes from mice through embryonic stem cell gene knockout technology, researchers generated p53-null mice. A mouse can be born normally and develop into an adult normally, but most of them develop cancers at 4.5 months (Donehower et al., 1992). So p53 may be an important lever gene for the regulation of homeostasis—a delicate balance between cell growth and cell death. Thus, it may not be surprising that p53 is often selected as a therapeutic target for permanent intervention. In cultured cells, the introduction of p53 back into p53-null cells leads to cell growth inhibition or cell death (El-Deiry et al., 1993). Thus, one properly chosen lever gene has the potential to lead the network into a specific implicitly irreducible subchain (cf. attractor in standard Boolean networks). The p53 gene is also being used in gene therapy, where the target gene (p53 in this case) is cloned into a viral vector (an adenovirus vector is a common one). The modified virus serves as a vehicle to transport the p53 gene into tumor cells to generate permanent intervention (Swisher et al., 1999; Bouvet et al., 1998).

### Sensitivity of Stationary Distributions to Gene Perturbations

Let us now address the question of the sensitivity of stationary distributions to random gene perturbations. This is an important issue because it characterizes the effect of perturbations on long-term network behavior. It is clear that whatever is meant by sensitivity, it no doubt depends on the probability $p$ of random perturbation. The general question is, If we perturb the transition probabilities, how much will the stationary distributions or, equivalently, the limiting probabilities change? This question has generally been addressed in the area known as perturbation theory of stochastic matrices and dates back to (Schweitzer, 1968). If $A$ and $\tilde{A} = A - E$ are the original and the "perturbed" Markov matrices, where $E$ represents the perturbation and $\pi$ and $\tilde{\pi}$ are their respective stationary distributions, then most results are of the form

$$\|\tilde{\pi} - \pi\| \leq \kappa \|E\| \qquad \text{or} \qquad \left| \frac{\pi_j - \tilde{\pi}_j}{\pi_j} \right| \leq \kappa_j \|E\|,$$

for some matrix norm $\| \cdot \|$, and $\kappa$, $\kappa_j$ are called condition numbers and are used as measures of sensitivity. Recently, a new approach used to measure the sensitivity of the Markov chain to perturbations, in terms of mean first-passage times, has been proposed by Cho and Meyer (2000). This approach has the advantage that it does not require computing or estimating the condition numbers.

**THEOREM 3.12 (Cho and Meyer, 2000)** *Let* $\mathbf{A}$ *and* $\tilde{\mathbf{A}} = \mathbf{A} - \mathbf{E}$ *be transition probability matrices for two irreducible Markov chains with respective stationary distributions* $\pi$ *and* $\tilde{\pi}$. *Denote by* $\|\mathbf{E}\|_\infty$ *the infinity norm of* $\mathbf{E}$, *which is the maximum over the row sums* $\sum_j |\mathbf{E}(i, j)|$. *Let* $M(\mathbf{x}, \mathbf{y}) = \sum_k k F_k(\mathbf{x}, \mathbf{y})$ *denote the mean first-passage time from state* $\mathbf{x}$ *to state* $\mathbf{y}$ *in the chain corresponding to* $\mathbf{A}$. *Then, the relative change in the limiting probability for state* $\mathbf{y}$ *is*

$$\frac{|\pi_{\mathbf{y}} - \tilde{\pi}_{\mathbf{y}}|}{\pi_{\mathbf{y}}} \le \frac{1}{2} \|\mathbf{E}\|_\infty \max_{\mathbf{x} \ne \mathbf{y}} M(\mathbf{x}, \mathbf{y}).$$

Cho and Meyer (2000) also showed that their bound is tight in the sense that there always exists a perturbation $\mathbf{E}$ that attains the bound. Let us now consider this result in the context of random gene perturbations.

**THEOREM 3.13** *Given a PBN* $G(V, F)$ *with an existing steady-state distribution, let* $\pi_{\mathbf{y}}$ *be a limiting probability of state* $\mathbf{y}$ *when* $p = 0$ *(no perturbations) and let* $\tilde{\pi}_{\mathbf{y}}$ *be the limiting probability of the same state when* $0 < p < 1/2$. *Then,*

$$\frac{|\pi_{\mathbf{y}} - \tilde{\pi}_{\mathbf{y}}|}{\pi_{\mathbf{y}}} \le (1 - (1 - p)^n) \max_{\mathbf{x} \ne \mathbf{y}} M(\mathbf{x}, \mathbf{y}).$$

*Proof.* The perturbation matrix $\mathbf{E}$ from theorem 3.12 can be expressed directly from theorem 3.4 as follows. Let $\mathbf{E}(\mathbf{x}, \mathbf{x}')$ be the entry in $\mathbf{E}$ corresponding to the transition probability from $\mathbf{x}$ to $\mathbf{x}'$ for $\mathbf{x}, \mathbf{x}' \in \{0, 1\}^n$. Recall that

$$A(\mathbf{x}, \mathbf{x}') = \sum_{i=1}^{s_0} p_i \left[ \prod_{j=1}^{n} (1 - |f_{\mathbf{K}_{ij}}^{(j)}(x_1, \ldots, x_n) - x_j'|) \right]$$

denotes the transition probability when $p = 0$ and $\tilde{A}(\mathbf{x}, \mathbf{x}')$ denotes the transition probability given in theorem 3.4, where a nonzero perturbation probability is assumed. In other words,

$$\tilde{A}(\mathbf{x}, \mathbf{x}') = A(\mathbf{x}, \mathbf{x}') \times (1 - p)^n + p^{\eta(\mathbf{x}, \mathbf{x}')} \times (1 - p)^{n - \eta(\mathbf{x}, \mathbf{x}')} \times 1_{[\mathbf{x} \ne \mathbf{x}']}. \tag{3.48}$$

Then, if we consider the matrices constructed from these probabilities, $\mathbf{E}(\mathbf{x}, \mathbf{x}') = A(\mathbf{x}, \mathbf{x}') - \tilde{A}(\mathbf{x}, \mathbf{x}')$ and for each row of $\mathbf{E}$ we have

$$\sum_{\mathbf{x}'} |\mathbf{E}(\mathbf{x}, \mathbf{x}')|$$

$$= \sum_{\mathbf{x}'} |A(\mathbf{x}, \mathbf{x}') \times (1 - (1 - p)^n) - p^{\eta(\mathbf{x}, \mathbf{x}')} \times (1 - p)^{n - \eta(\mathbf{x}, \mathbf{x}')} \times 1_{[\mathbf{x} \ne \mathbf{x}']}|$$

$$\le \sum_{\mathbf{x}'} (|A(\mathbf{x}, \mathbf{x}') \times (1 - (1 - p)^n)| + |p^{\eta(\mathbf{x}, \mathbf{x}')} \times (1 - p)^{n - \eta(\mathbf{x}, \mathbf{x}')} \times 1_{[\mathbf{x} \ne \mathbf{x}']}|).$$

$$\tag{3.49}$$

First, we observe that since $\sum_{\mathbf{x}'} \mathbf{A}(\mathbf{x}, \mathbf{x}') = 1$, the first term of the summation in equation (3.49) is simply equal to $(1 - (1 - p)^n)$. Next, we have

$$\sum_{\mathbf{x}'} |p^{\eta(\mathbf{x},\mathbf{x}')} \times (1 - p)^{n - \eta(\mathbf{x},\mathbf{x}')} \times 1_{[\mathbf{x} \neq \mathbf{x}']}| = \sum_{\mathbf{x}' \neq \mathbf{x}} p^{\eta(\mathbf{x},\mathbf{x}')} \times (1 - p)^{n - \eta(\mathbf{x},\mathbf{x}')}, \quad (3.50)$$

where we can remove the absolute value symbols since each summand is positive. Since the summation in equation (3.50) is taken over all possible values of $\mathbf{x}'$ except $\mathbf{x}' = \mathbf{x}$, the Hamming distance $\eta(\mathbf{x}, \mathbf{x}')$ ranges from 1 to $n$. As there are $\binom{n}{k}$ states $\mathbf{x}'$ that are Hamming distance $k$ from $\mathbf{x}$ (i.e., $|\{\mathbf{x}' : \eta(\mathbf{x}, \mathbf{x}') = k\}| = \binom{n}{k}$), equation (3.50) can be rewritten as

$$\sum_{\mathbf{x}' \neq \mathbf{x}} p^{\eta(\mathbf{x},\mathbf{x}')} \times (1 - p)^{n - \eta(\mathbf{x},\mathbf{x}')} = \sum_{k=1}^{n} \binom{n}{k} p^k (1 - p)^{n-k} \quad (3.51)$$

$$= 1 - (1 - p)^n.$$

Thus, every row of $\mathbf{E}$ satisfies

$$\sum_{\mathbf{x}'} |\mathbf{E}(\mathbf{x}, \mathbf{x}')| \leq 2(1 - (1 - p)^n),$$

and so

$$\|\mathbf{E}\|_\infty \leq 2(1 - (1 - p)^n) \quad (3.52)$$

as well.

Using equation (3.52) together with theorem 3.12 gives the desired result. □

Theorem 3.13 allows us to bound the sensitivity of the limiting probabilities of any state of the PBN relative to the probability of random gene perturbation. The mean first-passage times $M(\mathbf{x}, \mathbf{y})$ can be computed in a straightforward way by using the recursive formula in equation (3.43). The same type of analysis as above can be performed between two PBNs with different perturbation probabilities $p_1 < p_2$, and the relative sensitivity of the limiting probabilities can be expressed in terms of $p_1$, $p_2$, and the mean first-passage times. One important implication of theorem 3.13 is that if a particular state of a PBN can be "easily reached" from other states, meaning that the mean first-passage times are small, then its steady-state probability will be relatively unaffected by perturbations. Such sets of states, if we hypothesize that they correspond to some functional cellular states, are thus relatively insensitive to random gene perturbations.

*Other Remarks*

The complex interplay of the elements in a genetic regulatory network implies that any individual element or group of elements exerts an effect on the entire network. The extent of this effect depends on the nature of the relationships between the elements as well as on the state of the network. We have addressed two related questions. Given the possibility of a random gene perturbation with a certain prob-ability, to what extent do such perturbations affect the long-term behavior of the entire network? Alternately, given a desire to elicit certain behavior from the net-work, What genes would make the best candidates for intervention so as to increase the likelihood of this behavior?

The first question can be addressed by constructing an explicit formulation of the state transition probabilities in terms of the Boolean functions and the probability of perturbation and then deriving a bound on the steady-state probabilities given in theorem 3.13. In concordance with intuition, an interesting implication of this theorem is that the steady-state probabilities of the states of the network to which it is easy to transition from other states, in terms of mean first-passage times, are more resilient to random gene perturbations. The first-passage times provide a conceptual link with the second question in that they furnish the means by which we develop the tools for finding the best candidate genes for intervention.

### 3.3.2  Structural Intervention

The transient intervention we described above—one that allows one to transiently intervene with the *value* of a gene—can be useful in modulating the dynamics of the network, but it is not able to alter its underlying structure. Accordingly, the sta-tionary distribution remains unchanged. However, an imbalance between certain sets of states, which is characteristic of neoplasia in view of gene regulatory net-works, can be caused by mutations in the wiring of certain genes, thus permanently altering the state transition structure and, consequently, the long-run behavior of the network.

Therefore, it is prudent to develop a methodology for altering the steady-state probabilities of certain states or sets of states with minimal modifications to the rule-based structure. The motivation is that these states may represent different phenotypes or cellular functional states, such as cell invasion and quiescence, and that we would like to decrease the probability that the whole network will end up in an undesirable set of states and increase the probability that it will end up in a desirable set of states. The mechanism by which we accomplish this consists of altering some Boolean functions (predictors) in the PBN. An additional goal is to alter as few functions as possible. Such alterations to the rules of regulation may be possible by the introduction of a factor or drug that alters the extant behavior. Let us give an example.

We know that women can age much faster after menopause. In developed countries, estrogen is often taken by women after menopause to alter this trend. However, the dose of estrogen is important because an overdose may increase the probabilities of developing breast and ovarian cancers. Although the mechanism is not clear yet, it is conceivable that this phenomenon has a gene regulation basis. Estrogen binds its receptors, the complex being transported into the nucleus to bind the enhancer element on the target genes, and functions like transcriptional factors affecting genes such as the preproenkephalin (PENK) gene (Vasudevan et al., 2001). Interestingly, there are several different estrogen receptors that com-pete with each other for binding estrogen, as well as for the coactivator, which is also required for efficient transcriptional regulation by estrogen (Zhang and Teng, 2001). It can be envisioned that estrogen binds one receptor better than another and that these complexes bind DNA and the coactivator with opposite efficiency. That is, complex C1 binds DNA better than complex C2, but complex C2 binds the coactivator better than complex C1. Thus, under low-estrogen conditions, when there is not much competition for DNA binding, there is sufficient binding of C2 to DNA so as to turn on the downstream target gene. However, when estrogen is present at a high concentration, both complexes exist at very high levels and complex C2, taking up most of the coactivator away from C1, has little chance to bind to DNA. Consequently, the better DNA-binding complex (C1) does not have the necessary coactivator to activate the target gene. If the target gene plays a role in tumor suppression, for instance, this could explain why high levels of estrogen have a tumorigenic effect. Thus, by changing the concentration of estrogen, one is able to alter the rule determining the value of a gene (e.g., PENK) in terms of the levels of estrogen receptor complexes C1 and C2.

For example, under a low-estrogen condition, assuming Boolean values for all genes, PENK can be expressed as PENK $= C1 \vee C2$ (recall that $\vee$ stands for logical OR). That is, the presence of at least one complex (C1 or C2) would be sufficient to turn on PENK. However, under a high-estrogen condition, in view of the above squelching effect, PENK $= C1 \oplus C2$ ($\oplus$ stands for exclusive OR). That is, when either C1 or C2 is present individually with no competition from the other, PENK is turned on, but when both C1 and C2 are present together, PENK is turned off. The ability to alter such rules would provide a means of at least partially controlling the steady-state behavior of the network. Formal methods and algorithms for addressing such problems were developed by Shmulevich et al. (2002c).

*Problem Setting*

Consider two sets of states $A, B \subseteq \{0, 1\}^n$. As mentioned above, each state $\mathbf{x} \in \{0, 1\}^n$ has a positive stationary probability $\pi(\mathbf{x})$. Thus, we can define $\pi(A) = \sum_{\mathbf{x} \in A} \pi(\mathbf{x})$ and $\pi(B)$ similarly. Suppose we are interested in altering the stationary probabilities of these two sets of states in such a way that the stationary probability of $A$ is decreased and the stationary probability of $B$ is increased by $0 < \lambda < 1$. We could have also stated the problem in a more general manner, where instead of decreasing the stationary probability of $A$ by $\lambda$ and increasing that of $B$ by $\lambda$, we would have specified the desired stationary probabilities of $A$ and $B$ separately, under the obvious constraint that their sum be less than or equal to 1. Indeed, this is what we will do when we discuss the computational solution to this problem. As already mentioned above, these two sets of states may represent two different cellular functional states or phenotypes.

To achieve our goal, suppose we alter function $f_{j_0}^{(i_0)}$ by replacing it with a new function $g_{j_0}^{(i_0)}$. The probability $c_{j_0}^{(i_0)}$ corresponding to $g_{j_0}^{(i_0)}$ must remain the same as for $f_{j_0}^{(i_0)}$ since $\sum_{j=1}^{l(i)} c_j^{(i)} = 1$. Thus, we have a new PBN whose stationary distribution we can denote by $\mu$. Let $\mu(A)$ and $\mu(B)$ be the stationary probabilities of sets $A$ and $B$ under the altered PBN model. We can then pose the following optimization problem: given sets $A$ and $B$, predictor functions $f_j^{(i)}$ together with their selection probabilities $c_j^{(i)}$, $i = 1, \ldots, n$, $j = 1, \ldots, l(i)$, and $\lambda \in (0, 1)$, select an $i_0$, a $j_0$, and a function $g_{j_0}^{(i_0)}$ to replace $f_{j_0}^{(i_0)}$ such that

$$\varepsilon(\pi(A) - \lambda, \mu(A)) \tag{3.53}$$

and

$$\varepsilon(\pi(B) + \lambda, \mu(B)) \tag{3.54}$$

are minimum among all $i$, $j$, $g_j^{(i)}$. Since we want both equations (3.53) and (3.54) to be minimum, we can construct a single objective function that incorporates both. For example, the sum of the two functions is an obvious choice. Here $\varepsilon(a, b)$ represents some chosen error function such as the absolute error (i.e., $\varepsilon(a, b) = |a - b|$). An additional constraint can be that $g_{j_0}^{(i_0)}$ has no more essential variables than $f_{j_0}^{(i_0)}$. In this scenario, we are allowing the alteration of only one predictor function. More generally, we can preselect a number of predictor functions that we are willing to alter.

Let us illustrate these ideas with a small, concrete example. We will use the same example as example 3.1. Recall that the state transition diagram, assuming no perturbations ($p = 0$), is shown in figure 3.5. As can be seen from this figure, two states, namely, (000) and (111), are absorbing states. Let us hypothesize, for the sake of this example, that (111) corresponds to cell invasion (and rapid proliferation) and state (000) corresponds to quiescence. Let us fix a probability of perturbation, $p = 0.01$. Then a simple analysis based on the probability transition matrix reveals that the stationary probabilities of states (000) and (111) are 0.0752 and 0.7310, respectively. Thus, in the long run, the network will be in quiescence only 7 percent of the time and will be in proliferation 73 percent of the time. Now, suppose we wish to alter this imbalance and require the two stationary probabilities to be approximately 0.4 for both (000) and (111). The other 6 states will then be visited only 20 percent of the time. In the framework of the above optimization problem, $A = \{(111)\}$, $B = \{(000)\}$, $\pi(A) = 0.7310$, $\pi(B) = 0.0752$, $\mu(A) = \mu(B) = 0.4$, and $\lambda = 0.3279$. Finally, suppose we are allowed to change only one predictor function. This corresponds to changing only one of the truth tables while keeping the selection probabilities $c_j^{(i)}$ unchanged. Thus, there are 5 possible columns (predictors) and 256 possibilities for each.

For the purposes of this example, we have generated each of the $5 \times 256 = 1280$ possible alterations. For each, we have computed and plotted the stationary probabilities $\mu(000)$ and $\mu(111)$, as shown in figure 3.15. The optimal values of $\mu(000)$ and $\mu(111)$ for the error function $\varepsilon(a, b) = |a - b|$ are indicated by an arrow. The objective function to be minimized is

$$|\mu(000) - 0.4| + |\mu(111) - 0.4|, \tag{3.55}$$

which corresponds to the sum of the two objective functions in equations (3.53) and (3.54). The marker types represent which predictor is altered. For example, the squares denote that predictor $f_1^{(1)}$ is altered.

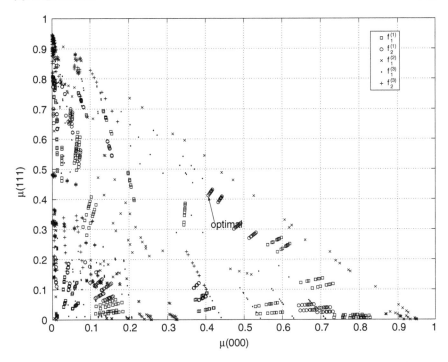

Figure 3.15. Each plotted point represents one of the 1280 possible alterations to the predictors. The $x$-axis is $\mu(000)$, and the $y$-axis is $\mu(111)$. The optimal choice is shown with an arrow, as it comes closest to 0.4 for both stationary probabilities. The marker types of the points represent the predictor that is altered (see key). (Shmulevich et al., 2002c.)

In this example, the optimal predictor is the one that alters $f_2^{(1)}$ for gene 1 (column 2 in the truth tables in example 3.1), and the truth table for the optimal predictor is $(00010101)^T$. This predictor achieves the stationary probabilities $\mu(000) = 0.4068$ and $\mu(111) = 0.4128$, which are quite close to the desired probabilities. The structure of the plot in figure 3.15 reveals an interesting phenomenon: the two stationary probabilities exhibit regularities, forming clusters of points arranged in a linear fashion, with different directions. In fact, this phenomenon has been observed in numerous examples, also in higher dimensions.

Such regularities seem to suggest that a brute-force search for the optimal predictor alteration may very well be avoided. That is, a number of search directions should be followed simultaneously, and the more promising ones should be explored further. Such a strategy is the hallmark of genetic algorithms, which have been successfully used in many optimization problems (Goldberg, 1989). We turn to this subject next.

*Solution via Genetic Algorithms*

Genetic algorithms (GAs) are abstractions of biological evolution. In a GA, a population of chromosomes, which are represented as binary vectors (cf. the truth tables for our gene predictors), moves to a new population of chromosomes using "natural selection" strategies such as crossover and mutation. Some chromosomes are selected and allowed to reproduce, and the more fit chromosomes produce more offspring. Recombination is achieved via the crossover operator, which exchanges parts of two chromosomes. Mutation can randomly change some of the locations or alleles (genes) in a chromosome and is very much like the perturbation in our probabilistic boolean networks.

Chromosomes are used to encode possible solutions to an optimization problem. At each stage in a search for a solution, the choice of the possible chromosomes depends on the results of the previous stage of the search. That is, if the parents represent promising solutions in different regions of the search space, then their offspring, produced by recombination, are likely to be promising solutions as well. The quality of a solution (chromosome) is represented by its fitness, which in the case of an optimization problem is the value of the objective function, such as in equation (3.55), at this proposed solution.

The general operation of a GA can be described as follows. First, an initial population of chromosomes (encoded as binary vectors of a certain length) is generated, and the fitness of each chromosome is calculated. Then, a pair of parent chromosomes is selected such that more fit chromosomes are more likely to be selected. With a certain probability, the two parents are recombined via the crossover operator at a random point (locus) in the chromosomes, producing two offspring. If the parents are not recombined, then their offspring are identical copies of the parents. Finally, the offspring are mutated at each locus with a certain mutation probability. This process is repeated until the number of offspring is equal to the number of parents and the old population is replaced by the new population, forming a new "generation." GAs typically make no assumptions about the characteristics of the problem, such as continuity or differentiability, and are well suited for multimodal function optimization as they are much less likely to become stuck in local optima.

Let us return to the example we considered above. A candidate solution essentially consists of two parts: the choice of the predictor to be altered (here we have a total of $s_0 = 5$ choices) and the truth table of the new, altered predictor. The former can be coded with $\lceil \log_2 s_0 \rceil$ bits, and the latter with $2^k$ bits, where $k$ is the maximum number of input variables over all the predictors. In the above example, $s_0 = 5$ and $k = 3$, so each candidate solution (chromosome) is encoded with $\lceil \log_2 5 \rceil + 2^3 = 11$ bits. One minor inconvenience is the fact that 3 bits is more than we need to code the choice of the predictor (5 possibilities). Thus, the mapping between the length-3 binary strings and the integers $1, \ldots, 5$ should be as uniform as possible so as to avoid certain predictors being chosen more often than others. This means that several different length-3 binary strings can code for the same integer. Of course, when $s_0$ is a power of 2, this problem does not exist.

By applying a GA to this example, using the above encoding scheme and a fitness function given in equation (3.55), the correct result ($f_2^{(1)} = (00010101)^T$) can typically be obtained after about 300 fitness function evaluations (Shmulevich et al., 2002c). Thus, only $300/1280 \approx 23$ percent of the work is performed, compared to the brute-force approach, even for such a small example. This is no small gain, considering that every fitness function evaluation entails recomputing the entire state transition matrix and then finding its left eigenvector to obtain the stationary probabilities. When the numbers of genes and predictors per gene grow large, the advantages of GAs should become dramatically more apparent. This was illustrated by Shmulevich et al. (2002c) with computer experiments where the genetic algorithm was able to locate the optimal solution (structural alteration) in only 200 steps (evaluations of the fitness function) out of a total of 21 billion possibilities, which is the number of steps a brute-force approach would require. The reason for such high efficiency of the genetic algorithm is that the embedded structure in the PBN can be exploited. It should be mentioned that genetic algorithms have also been used for modifying the structure of Boolean networks in the context of evolution (Stern, 1999; Bornholdt and Sneppen, 2000).

*Other Remarks*

The structural intervention formalism described above is quite generic in the sense that any objective function using any number of states and any number of possible predictors to be modified can be specified. In view of genetic regulatory networks, it is advisable to minimize this type of structural manipulation as much as possible, which translates into limiting the number of gene predictors that can be simultaneously modified or altered. Clearly, this seems to work against the goals of the optimization—the fewer choices we have (at any one particular time), the less power we have to alter the long-run behavior of the network. Thus, it may be the case that, given the constraints on the number of predictors that can be simultaneously altered, some solutions may not be feasible. However, in practice, we are interested in only approximate solutions. For example, if we wish to make the stationary probability of a certain (undesirable) state zero, we would be content if it were simply close to zero. Thus, many "nearly optimal" solutions may exist. Genetic algorithms indeed do not always guarantee a global optimum but may find many good suboptimal solutions as they explore a number of promising regions of the search space.

In addition, we may wish to allow only the use of predictors belonging to a certain class of Boolean functions, such as canalizing functions or functions with a limited number of input variables. We can even insist that the new, altered predictor have the same input variables as the predictor it is replacing. It is quite straightforward to incorporate such constraints during the GA optimization stage.

### 3.3.3 External Control

The aforementioned intervention methods, namely, gene intervention and structural intervention, do not provide effective "knobs" that can be used to externally guide the time evolution of the network toward more desirable states. By considering possible external intervention control inputs, ideas from optimal control theory can be applied to develop a general optimal intervention theory for Markovian gene regulatory networks, in particular, for PBNs. This strategy can make use of dynamical programming. The costs and benefits of using interventions are incorporated into a single performance index, which also penalizes the state where the network ends up following the intervention. The use of auxiliary variables makes sense from a biological perspective. For instance, in the case of diseases like cancer, auxiliary treatment inputs such as radiation, chemotherapy, and so on, can be employed to move the state probability distribution vector away from one that is associated with uncontrolled cell proliferation or markedly reduced apoptosis. The auxiliary variables can include genes that serve as external master regulators for all the genes in the network. To be consistent with the binary nature of the expression status of individual genes in the PBN, we will assume that the auxiliary variables (*control inputs*) can take on only the binary values zero and one. The values of the individual control inputs can be changed from one time step to the other in an effort to make the network behave in a desirable fashion. Interventions using a finite treatment horizon and full information (Datta et al., 2003) and partial information (Datta et al., 2004) have been considered for instantaneously random PBNs for which the states of the Markov chain are the states of the PBN. Following (Datta et al., 2003), we summarize the full-information case here.

### The Optimal Control Problem

To develop the control strategy, let $\mathbf{x}(k) = [x_1(k), x_2(k), \ldots, x_n(k)]$ denote the state vector (gene activity profile) at step $k$ for the $n$ genes in the network. The state vector $\mathbf{x}(k)$ at any time step $k$ is essentially an $n$-digit binary number whose decimal equivalent is given by

$$z(k) = 1 + \sum_{j=1}^{n} 2^{n-1} x_j(k).$$

As $\mathbf{x}(k)$ ranges from $000\ldots0$ to $111\ldots1$, $z(k)$ takes on all values from 1 to $2^n$. The map from $\mathbf{x}(k)$ to $z(k)$ is one-to-one, onto, and hence invertible. Instead of the binary representation $\mathbf{x}(k)$ for the state vector, we can equivalently work with the decimal representation $z(k)$.

Suppose that the PBN has $m$ control inputs $u_1, u_2, \ldots, u_m$. Then, at any given time step $k$, the row vector $\mathbf{u}(k) = [u_1(k), u_2(k), \ldots, u_m(k)]$ describes the complete status of all the control inputs. Clearly, $\mathbf{u}(k)$ can take on all binary values from $000\ldots0$ to $111\ldots1$. An equivalent decimal representation of the control input is given by

$$v(k) = 1 + \sum_{i=1}^{m} 2^{m-1} u_i(k).$$

As $\mathbf{u}(k)$ takes on binary values from $000\ldots0$ to $111\ldots1$, $v(k)$ takes on all values from 1 to $2^m$. We can equivalently use $v(k)$ as an indicator of the complete control input status of the PBN at time step $k$.

As shown by Datta et al. (2003), the one-step evolution of the probability distribution vector in the case of such a PBN with control inputs takes place according to the equation

$$\mathbf{w}(k+1) = \mathbf{w}(k)\mathbf{A}(v(k)), \tag{3.56}$$

where $\mathbf{w}(k)$ is the $2^n$-dimensional state probability distribution vector and $\mathbf{A}(v(k))$ is the $2^n \times 2^n$ matrix of control-dependent transition probabilities. Since the transition probability matrix is a function of the control inputs, the evolution of the probability distribution vector of the PBN with control depends not only on the initial distribution vector but also on the values of the control inputs at different time steps.

In the control literature, equation (3.56) is referred to as a *controlled Markov chain* (Bertsekas, 1976). Given a controlled Markov chain, the objective is to come up with a sequence of control inputs, usually referred to as a *control strategy*, such that an appropriate cost function is minimized over the entire class of allowable control strategies. To arrive at a meaningful solution, the cost function must capture the costs and benefits of using any control. The design of a good cost function is application-dependent and likely to require considerable expert knowledge. In the case of diseases like cancer, treatment is often applied over a finite time horizon. For instance, in the case of radiation treatment, the patient may be treated with radiation over a fixed interval of time, following which the treatment is suspended for some time as the effects are evaluated. After that, the treatment may be applied again, but the important point to note is that the treatment window at each stage is often finite. Thus, we are interested in a finite horizon problem where the control is applied over only a finite number of steps.

Suppose that the number of steps over which the control input is to be applied is $m_0$ and we are interested in controlling the behavior of the PBN over the interval $k = 0, 1, 2, \ldots, m_0 - 1$. We can define a cost $C_k(z(k), v(k))$ as being the cost of applying the control input $v(k)$ when the state is $z(k)$. The expected cost of control over the entire treatment horizon is

$$E\left[\sum_{k=0}^{M-1} C_k(z(k), v(k))|z(0)\right]. \qquad (3.57)$$

Even if the network starts from a given (deterministic) initial state $z(0)$, the subsequent states will be random because of the stochastic nature of the evolution in equation (3.56). Consequently, the cost in equation (3.57) must be defined using an expectation. Equation (3.57) gives us one component of the finite horizon cost, namely, the cost of control.

Regarding the second component of the cost, the net result of the control actions $v(0), v(1), \ldots, v(m_0 - 1)$ is that the state of the PBN will transition according to equation (3.56) and will end up in some state $z(m_0)$. Owing to the stochastic nature of the evolution, the terminal state $z(m_0)$ is a random variable that can potentially take on any of the values $1, 2, \ldots, 2^n$. We assign a penalty, or terminal cost, $C_{m_0}(z(m_0))$ to each possible state. To do this, divide the states into different categories depending on their desirability and assign higher terminal costs to the undesirable states. For instance, a state associated with rapid cell proliferation leading to cancer should be associated with a high terminal penalty, while a state associated with normal behavior should be assigned a low terminal penalty. For our purposes here, we assume that the assignment of terminal penalties has been carried out, and we have a terminal penalty $C_{m_0}(z(m_0))$ that is a function of the terminal state. This is the second component of our cost function. $C_{m_0}(z(m_0))$ is a random variable, and so we must take its expectation when defining the cost function to be minimized. In view of equation (3.57), the finite horizon cost to be minimized is given by

$$E\left[\sum_{k=0}^{m_0-1} C_k(z(k), v(k)) + C_{m_0}(z(m_0))|z(0)\right].$$

To proceed further, let us assume that at time $k$ the control input $v(k)$ is a function of the current state $z(k)$, namely, $v(k) = \mu_k(z(k))$. The optimal control problem can now be stated: given an initial state $z(0)$, find a control law $\pi = [\mu_0, \mu_1, \ldots, \mu_{m_0-1}]$ that minimizes the cost functional

$$J_\pi(z(0)) = E\left[\sum_{k=0}^{m_0-1} C_k(z(k), \mu_k(z(k))) + C_{m_0}(z(m_0))|z(0)\right]$$

subject to the probability constraint

$$P[z(k+1) = j|z(k) = i] = a_{ij}(v(k)),$$

where $a_{ij}(v(k))$ is the $i$th row, $j$th column entry in the matrix $\mathbf{A}(v(k))$. Optimal control problems of the type described by the preceding two equations can be solved by using *dynamic programming*, a technique pioneered by Bellman in the 1960s. We will not pursue the solution here but instead refer the reader to (Datta et al., 2003) for the complete solution. We will, however, follow (Datta et al., 2003) in providing an application.

### Control of WNT5A in Metastatic Melanoma

In expression profiling studies concerning metastatic melanoma, the abundance of messenger RNA for the gene WNT5A was found to be a highly discriminating difference between cells with properties typically associated with high metastatic competence versus those with low metastatic competence (Bittner et al., 2000; Weeraratna et al., 2002). In this study, experimentally increasing the levels of the Wnt5a protein secreted by a melanoma cell line via genetic engineering methods directly altered the metastatic competence of the cell as measured by the standard in vitro assays for metastasis. A further finding of interest was that an intervention that blocked the Wnt5a protein from activating its receptor, the use of an antibody that binds Wnt5a protein, could substantially reduce its ability to induce a metastatic phenotype. This suggests a study of control based on interventions that alter the contribution of the WNT5A gene's action to biological regulation since the available data suggest that disruption of this influence could reduce the chance of a melanoma metastasizing, a desirable outcome.

The methods for choosing the 10 genes involved in a small local network that includes the activity of the WNT5A gene and the rules of interaction have been

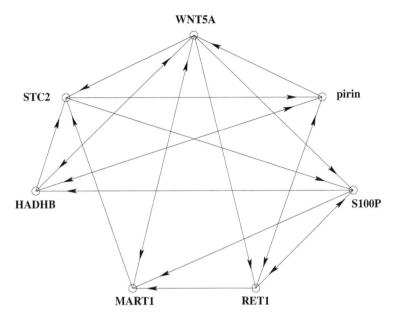

Figure 3.16. The 7-gene WNT5A network. (Datta et al., 2003.)

described in (Kim et al., 2002). The expression status of each gene was quantized to one of three possible levels: −1 (down-regulated), 0 (unchanged), and 1
(up-regulated). Although the network is ternary-valued instead of binary-valued,
the control theory extends directly. Indeed, to apply the control algorithm in (Datta
et al., 2003), it is not necessary to actually construct a PBN; all that is required are
the transition probabilities between the different states under the different controls.
For this study, the number of genes was reduced from 10 to 7 by using coefficient
of determination analysis. The resulting genes along with their multivariate relationships are shown in figure 3.16.

The control objective for this 7-gene network is to externally down-regulate the
WNT5A gene. The reason is that it is biologically known that WNT5A ceasing to
be down-regulated is strongly predictive of the onset of metastasis. For each gene
in this network, the two best 2-gene predictors were determined along with their
corresponding CoDs. The CoD information was used to determine the $7 \times 7$ matrix
of transition probabilities for the Markov chain corresponding to the dynamical
evolution of the 7-gene network.

The optimal control problem can now be completely specified by choosing (1)
the treatment/intervention window, (2) the terminal penalty, and (3) the types of
controls and the costs associated with them. For the treatment window, a window
of length 5 was arbitrarily chosen; that is, control inputs were applied only at time
steps 0, 1, 2, 3, and 4. The terminal penalty at time step 5 was chosen as follows.
Since the objective is to ensure that WNT5A is down-regulated, a penalty of 0 was
assigned to all states for which WNT5A equals −1, a penalty of 3 to all states
for which WNT5A equals 0, and a penalty of 6 to all states for which WNT5A

equals 1. Here the choice of the numbers 3 and 6 is arbitrary, but they do reflect our attempt to capture the intuitive notion that states where WNT5A equals 1 are less desirable than those where WNT5A equals 0. Two types of possible controls are considered in (Datta et al., 2003); here only one of them is considered, where WNT5A is controlled via pirin.

The control objective is to keep WNT5A down-regulated. The control action consists of either forcing pirin to $-1$ or letting it remain wherever it is. A control cost of 1 is incurred if and only if pirin has to be forcibly reset to $-1$ at that time step. Using the resulting optimal controls, the evolution of the state probability distribution vectors has been studied with and without control. For every possible initial state, the resulting simulations have indicated that, at the final state, the probability of WNT5A being equal to $-1$ is higher with control than that without control; however, the probability of WNT5A being equal to $-1$ at the final time point is not, in general, equal to 1. This is not surprising given that one is trying to control the expression status of WNT5A using another gene and that the control horizon of length 5 simply may not be adequate for achieving the desired objective with such a high probability. Nevertheless, even in this case, if the network starts from the state corresponding to STC2 $= -1$, HADHB $= 0$, MART1 $= 0$, RET1 $= 0$, S100P $= -1$, pirin $= 1$, WNT5A $= 1$ and evolves under optimal control, then the probability of WNT5A $= -1$ at the final time point is 0.673521. This is quite good in view of the fact that the same probability would have been equal to 0 in the absence of any control action.

*Other Remarks*

The control problem discussed above was posed in terms of minimizing a certain performance index over a finite number of time steps. The problem formulation was based on the assumption that the state of the PBN was available for observation. Such an assumption may not hold in many real-world situations. For instance, in a cancer treatment application, we may be able to track the expression status of only a limited number of genes and not necessarily all the ones appearing in the PBN of interest. This may be necessitated by cost, accessibility, or other considerations. Clearly, in such a case, the control algorithm cannot be implemented.

Datta et al. (2004) presented a control strategy that can be used when perfect information about the state of the Markov chain is not available. The control strategy in this case is based on the feedback of available measurements which, though different from the exact internal states, are probabilistically related to the latter. Finally, intervention via external control variables in context-sensitive PBNs was considered by Pal et al. (2005). We have restricted ourselves to finite horizon control strategies. Recently, infinite horizon strategies have been studied with the goal of affecting the steady-state distribution (Pal et al., 2006).

# Bibliography

Abrahams PJ, Van der Eb AJ. (1975) In vitro transformation of rat and mouse cells by DNA from simian virus 40. *J Virol* 16(1):206–9.

Albert R, Othmer HG. (2003) The topology of the regulatory interactions predicts the expression pattern of the segment polarity genes in *Drosophila melanogaster*. *J Theor Biol* 223(1):1–18.

Arsura M, Wu M, Sonenshein GE. (1996) TGF beta 1 inhibits NF-kappa B/Rel activity inducing apoptosis of B cells: transcriptional activation of I kappa B alpha. *Immunity* 5(1):31–40.

Bancroft CC, Chen Z, Dong G, Sunwoo JB, Yeh N, Park C, Van Waes C. (2001) Coexpression of proangiogenic factors IL-8 and VEGF by human head and neck squamous cell carcinoma involves coactivation by MEK-MAPK and IKK-NF-kappaB signal pathways. *Clin Cancer Res.* 7(2):435–42.

Banga SS, Ozer HL, Park SK, Lee ST. (1997) Assignment of PTK7 encoding a receptor protein tyrosine kinase-like molecule to human chromosome 6p21.1→p12.2 by fluorescence in situ hybridization. *Cytogenet Cell Genet* 76(1–2):43–4.

Bargonetti J, Reynisdottir I, Friedman PN, Prives C. (1992) Site-specific binding of wild-type p53 to cellular DNA is inhibited by SV40 T antigen and mutant p53. *Genes Dev* 6(10):1886–98.

Bernard A, Hartemink AJ. (2005) Informative structure priors: joint learning of dynamic regulatory networks from multiple types of data. *Pac Symp Biocomput* 2005:459–70.

Bertsekas DP. (1976) *Dynamic Programming and Stochastic Control*, Academic Press, New York.

Bikfalvi A, Bicknell R. (2002) Recent advances in angiogenesis, antiangiogenesis and vascular targeting. *Trends Pharmacol Sci* 23(12):576–82.

Bittner M, Meltzer P, Chen Y, Jiang Y, Seftor E, Hendrix M, Radmacher M, Simon R, Yakhini Z, Ben-Dor A, Sampas N, Dougherty E, Wang E, Marincola F, Gooden C, Lueders J, Glatfelter A, Pollock P, Carpten J, Gillanders E, Leja D, Dietrich K, Beaudry C, Berens M, Alberts D, Sondak V. (2000) Molecular classification of cutaneous malignant melanoma by gene expression profiling. *Nature* 406(6795):536–40.

Bornholdt S, Sneppen K. (2000) Robustness as an evolutionary principle. *Proc Biol Sci* 267(1459):2281–6.

Bouvet M, Bold RJ, Lee J, Evans DB, Abbruzzese JL, Chiao PJ, McConkey DJ, Chandra J, Chada S, Fang B, Roth JA. (1998) Adenovirus-mediated wild-type p53 tumor suppressor gene therapy induces apoptosis and suppresses growth of human pancreatic cancer. *Ann Surg Oncol* 5(8):681–8.

Brinster RL, Chen HY, Messing A, van Dyke T, Levine AJ, Palmiter RD. (1984) Transgenic mice harboring SV40 T-antigen genes develop characteristic brain tumors. *Cell* 37(2):367–79.

Brun M, Dougherty ER, Shmulevich I. (2005) Steady-state probabilities for attractors in probabilistic boolean networks. *Signal Process* 85(4):1993–2013.

Carbone M, Rizzo P, Pass HI. (1997) Simian virus 40, poliovaccines and human tumors: a review of recent developments. *Oncogene* 15(16):1877–8.

Cazals V, Nabeyrat E, Corroyer S, de Keyzer Y, Clement A. (1999) Role for NF-kappa B in mediating the effects of hyperoxia on IGF-binding protein 2 promoter activity in lung alveolar epithelial cells. *Biochim Biophys Acta* 1448(3):349–62.

Chen Y, Dougherty ER, Bittner M. (1997) Ratio-based decisions and the quantitative analysis of cDNA microarray images. *J Biomed Opt* 2(4):364–74.

Cheng SY, Huang HJ, Nagane M, Ji XD, Wang D, Shih CC, Arap W, Huang CM, Cavenee WK. (1996) Suppression of glioblastoma angiogenicity and tumorigenicity by inhibition of endogenous expression of vascular endothelial growth factor. *Proc Natl Acad Sci USA* 93(16):8502–7.

Chickering DM. (1996) Learning Bayesian networks is NP-complete. In *Learning from Data: Artificial Intelligence and Statistics V*, Fisher D, Lenz HJ, eds., Springer-Verlag, New York.

Cho GE, Meyer CD. (2000) Markov chain sensitivity measured by mean first passage times. *Linear Algebra Appl* 316:21–8.

Çınlar E. (1997) *Introduction to Stochastic Processes*, Prentice Hall, Englewood Cliffs, NJ.

Cowles MK, Carlin BP. (1996) Markov chain Monte Carlo convergence diagnostics: a comparative study. *J Am Stat Assoc* 91:883–904.

Datta A, Choudhary A, Bittner M, Dougherty ER. (2003) External control in Markovian genetic regulatory networks. *Mach Learning* 52:169–81.

Datta A, Choudhary A, Bittner ML, Dougherty ER. (2004) External control in Markovian genetic regulatory networks: the imperfect information case. *Bioinformatics* 20(6):924–30.

DeCaprio JA, Ludlow JW, Figge J, Shew JY, Huang CM, Lee WH, Marsilio E, Paucha E, Livingston DM. (1988) SV40 large tumor antigen forms a specific complex with the product of the retinoblastoma susceptibility gene. *Cell* 54(2):275–83.

Donehower LA, Harvey M, Slagle BL, McArthur MJ, Montgomery CA Jr, Butel JS, Bradley A. (1992) Mice deficient for p53 are developmentally normal but susceptible to spontaneous tumours. *Nature* 356(6366):215–21.

Dougherty ER, Braga-Neto U. (2006) Epistemology of computational biology: mathematical models and experimental prediction as the basis of their validity. *J Biol Syst* 14(1):65–90.

Efron B, Tibshirani R. (1993) *An Introduction to the Bootstrap*. Chapman & Hall, New York.

Elmlinger MW, Deininger MH, Schuett BS, Meyermann R, Duffner F, Grote EH, Ranke MB. (2001) In vivo expression of insulin-like growth factor-binding protein-2 in human gliomas increases with the tumor grade. *Endocrinology* 142(4):1652–8.

el-Deiry WS, Tokino T, Velculescu VE, Levy DB, Parsons R, Trent JM, Lin D, Mercer WE, Kinzler KW, Vogelstein B. (1993) WAF1, a potential mediator of p53 tumor suppression. *Cell* 75(4):817–25.

Folkman J. (1995) Angiogenesis in cancer, vascular, rheumatoid and other disease. *Nat Med* 1(1):27–31.

Friedman N, Murphy K, Russell S. (1998) Learning the structure of dynamic probabilistic networks. *Proceedings of the 14th Conference on Uncertainty in Artificial Intelligence*, Madison, WI, pp. 139–147.

Friedman N, Nachman I, Pe'er D. (1999) Learning Bayesian network structure from massive datasets: the "sparse candidate" algorithm. *Proceedings of the 15th Conference on Uncertainty in Artificial Intelligence*, Stockholm, Sweden, pp. 196–205.

Friedman N, Linial M, Nachman I, Pe'er D. (2000) Using a Bayesian network to analyze expression data. *J Comput Biol* 7(3–4):601–20.

Friedman N, Koller D. (2000) Being Bayesian about network structure. *Proceedings of the 16th Conference on Uncertainty in Artificial Intelligence*, Stanford, CA, pp. 201–10.

Fuller GN, Rhee CH, Hess KR, Caskey LS, Wang R, Bruner JM, Yung WK, Zhang W. (1999) Reactivation of insulin-like growth factor binding protein 2 expression in glioblastoma multiforme: a revelation by parallel gene expression profiling. *Cancer Res* 59(17):4228–32.

Gartel AL, Tyner AL. (1999) Transcriptional regulation of the p21((WAF1/CIP1)) gene. *Exp Cell Res* 246(2):280–9.

Goldberg D. (1989) *Genetic Algorithms in Search, Optimization, and Machine Learning*, Addison-Wesley, Reading, MA.

Goutsias J, Kim S. (2004) A nonlinear discrete dynamical model for transcriptional regulation: construction and properties. *Biophys J* 86(4):1922–45.

Harley CB. (1991) Telomere loss: mitotic clock or genetic time bomb? *Mutat Res* 256(2–6):271–82.

Hartemink AJ, Gifford DK, Jaakkola TS, Young RA. (2002) Combining location and expression data for principled discovery of genetic regulatory network models. *Pac Symp Biocomput* 2002:437–49.

Hashimoto RF, Dougherty E, Brun M, Zhou ZZ, Bittner ML, Trent JM. (2003) Efficient selection of feature sets possessing high coefficients of determination based on incremental determinations. *Signal Process* 83(4):695–712.

Hashimoto RF, Kim S, Shmulevich I, Zhang W, Bittner ML, Dougherty ER. (2004) Growing genetic regulatory networks from seed genes. *Bioinformatics* 20(8): 1241–7.

Hayashi S, Yamamoto M, Ueno Y, Ikeda K, Ohshima K, Soma G, Fukushima T. (2001) Expression of nuclear factor-kappa B, tumor necrosis factor receptor type 1, and c-Myc in human astrocytomas. *Neurol Med Chir (Tokyo)* 41(4):187–95.

Heckerman D. (1998) A tutorial on learning with Bayesian networks. In *Learning in Graphical Models*, Jordan MI, ed., Kluwer, Dordrecht, Netherlands.

Hiyama E, Hiyama K, Yokoyama T, Matsuura Y, Piatyszek MA, Shay JW. (1995) Correlating telomerase activity levels with human neuroblastoma outcomes. *Nat Med* 1(3):249–55.

Hollstein M, Sidransky D, Vogelstein B, Harris CC. (1991) p53 mutations in human cancers. *Science* 253(5015):49–53.

Huang S. (1999) Gene expression profiling, genetic networks, and cellular states: an integrating concept for tumorigenesis and drug discovery. *J Mol Med* 77(6): 469–80.

Huang S. (2001) Genomics, complexity and drug discovery: insights from Boolean network models of cellular regulation. *Pharmacogenomics* 2(3):203–22.

Huang S, Ingber DE. (2000) Shape-dependent control of cell growth, differentiation, and apoptosis: switching between attractors in cell regulatory networks. *Exp Cell Res* 261(1):91–103.

Imoto S, Higuchi T, Goto T, Tashiro K, Kuhara S, Miyano S. (2004) Combining microarrays and biological knowledge for estimating gene networks via Bayesian networks. *J Bioinform Comput Biol* 2(1):77–98.

Ito T, Chiba T, Ozawa R, Yoshida M, Hattori M, Sakaki Y. (2001) A comprehensive two-hybrid analysis to explore the yeast protein interactome. *Proc Natl Acad Sci USA* 98(8):4569–74.

Ivanov I, Dougherty ER. (2006) Modeling genetic regulatory networks: continuous or discrete? *J Biol Syst* 14(2):1–11.

Jensen FV. (2001) *Bayesian Networks and Decision Graphs.* Springer-Verlag, New York.

Kahn J, Kalai G, Linial N. (1988) The influence of variables on Boolean functions. *29th Annual Symposium on Foundations of Computer Science*, IEEE Washington, DC, pp. 68–80.

Kauffman SA, Levin S. (1987) Towards a general theory of adaptive walks on rugged landscapes. *J Theor Biol* 128:11–45.

Kauffman SA. (1993) *The Origins of Order: Self-organization and Selection in Evolution*, Oxford University Press, New York.

Kim NW, Piatyszek MA, Prowse KR, Harley CB, West MD, Ho PL, Coviello GM, Wright WE, Weinrich SL, Shay JW. (1994) Specific association of human telomerase activity with immortal cells and cancer. *Science* 266(5193):2011–5.

Kim S, Dougherty ER, Chen Y, Sivakumar K, Meltzer P, Trent JM, Bittner M. (2000) Multivariate measurement of gene expression relationships. *Genomics* 67(2):201–9.

Kim S, Li H, Dougherty ER, Cao N, Chen Y, Bittner M, Suh EB. (2002) Can Markov chain models mimic biological regulation? *J. Biol. Syst* 10(4):431–45.

Kim SY, Imoto S, Miyano S. (2003) Inferring gene networks from time series microarray data using dynamic Bayesian networks. *Brief Bioinform* 4(3):228–35.

Kleihues P, Cavenee WK, eds. (2000) *Pathology and Genetics—Tumours of the Nervous System.* Oxford University Press, Oxford, UK.

Kobayashi T, Ruan S, Jabbur JR, Consoli U, Clodi K, Shiku H, Owen-Schaub LB, Andreeff M, Reed JC, Zhang W. (1998) Differential p53 phosphorylation and activation of apoptosis-promoting genes Bax and Fas/APO-1 by irradiation and ara-C treatment. *Cell Death Differ* 5(7):584-91.

Kouhata T, Fukuyama K, Hagihara N, Tabuchi K. (2001) Detection of simian virus 40 DNA sequence in human primary glioblastomas multiforme. *J Neurosurg* 95(1):96–101.

Lähdesmäki H, Hautaniemi S, Shmulevich I, Yli-Harja O. (2005) Relationships between probabilistic Boolean networks and dynamic Bayesian networks as models of gene regulatory networks. *Signal Process* 86(4):814–34.

Lauritzen S. (1996) *Graphical Models.* Oxford University Press, Oxford, UK.

Lowe SW, Schmitt EM, Smith SW, Osborne BA, Jacks T. (1993) p53 is required for radiation-induced apoptosis in mouse thymocytes. *Nature* 362(6423):847–9.

Miyashita T, Reed JC. (1995) Tumor suppressor p53 is a direct transcriptional activator of the human bax gene. *Cell* 80(2):293–9.

Murphy K, Mian S. (1999) *Modelling Gene Expression Data Using Dynamic Bayesian Networks*, Technical Report, University of California, Berkeley.

Owen-Schaub LB, Zhang W, Cusack JC, Angelo LS, Santee SM, Fujiwara T, Roth JA, Deisseroth AB, Zhang WW, Kruzel E, Radinsky R. (1995) Wild-type human p53 and a temperature-sensitive mutant induce Fas/APO-1 expression. *Mol Cell Biol* 15(6):3032–40.

Pal R, Datta A, Bittner ML, Dougherty ER. (2005) Intervention in context-sensitive probabilistic Boolean networks. *Bioinformatics* 21(7):1211–8.

Pal R, Datta A, Dougherty ER. (2006) Optimal infinite horizon control for probabilistic Boolean networks. *IEEE Trans Sig Proc* 54(6): 2375–87.

Pearl J. (1988) *Probabilistic Reasoning in Intelligent Systems: Networks of Plausible Inference*. Morgan Kaufmann, San Mateo, CA.

Pearl J. (2000) *Casuality: Models, Reasoning, and Inference*, Cambridge University Press, Cambridge, UK.

Pearson G, English JM, White MA, Cobb MH. (2001) ERK5 and ERK2 cooperate to regulate NF-kappaB and cell transformation. *J Biol Chem* 276(11):7927–31.

Pe'er D, Regev A, Tanay A. (2002) Minreg: inferring an active regulator set. *Bioinformatics* 18 Suppl 1:S258–67.

Perrin BE, Ralaivola L, Mazurie A, Bottani S, Mallet J, D'Alche-Buc F. (2003) Gene networks inference using dynamic Bayesian networks. *Bioinformatics* 19 Suppl 2:II138–48.

Pournara I, Wernisch L. (2004) Reconstruction of gene networks using Bayesian learning and manipulation experiments. *Bioinformatics* 20(17):2934–42.

Raftery AE, Lewis S. (1992) How many iterations in the Gibbs sampler? In *Bayesian Statistics 4*, Berger JO. et al., eds., Oxford University Press, New York, pp. 763–73.

Robert CP. (1995) Convergence control techniques for Markov chain Monte Carlo algorithms. *Stat. Sci* 10(3):231–53.

Robert CP, Casella G. (1999) *Monte Carlo Statistical Methods*, Springer, New York.

Rosenthal JS. (1995) Minorization conditions and convergence rates for Markov chain Monte Carlo. *J. Am. Stat. Assoc* 90(430):558–66.

Rubin JS, Osada H, Finch PW, Taylor WG, Rudikoff S, Aaronson SA. (1989) Purification and characterization of a newly identified growth factor specific for epithelial cells. *Proc Natl Acad Sci USA* 86(3):802–6.

Sallinen SL, Sallinen PK, Haapasalo HK, Helin HJ, Helen PT, Schraml P, Kallion-iemi OP, Kononen J. (2000) Identification of differentially expressed genes in human gliomas by DNA microarray and tissue chip techniques. *Cancer Res* 60(23):6617–22.

Sato TN, Qin Y, Kozak CA, Audus KL. (1993) Tie-1 and tie-2 define another class of putative receptor tyrosine kinase genes expressed in early embryonic vascular system. *Proc Natl Acad Sci USA* 90(20):9355–8.

Schweitzer PJ. (1968) Perturbation theory and finite Markov chains. *J Appl Probab* 5:401–13.

Segal E, Shapira M, Regev A, Pe'er D, Botstein D, Koller D, Friedman N. (2003) Module networks: identifying regulatory modules and their condition-specific regulators from gene expression data. *Nat Genet* 34(2):166–76.

Shmulevich I, Dougherty ER, Kim S, Zhang W. (2002a) Probabilistic Boolean networks: a rule-based uncertainty model for gene regulatory networks. *Bioinformatics* 18(2):261–74.

Shmulevich I, Dougherty ER, Zhang W. (2002b) Gene perturbation and intervention in probabilistic Boolean networks. *Bioinformatics* 18(10):1319–31.

Shmulevich I, Dougherty ER, Zhang W. (2002c) Control of stationary behavior in probabilistic Boolean networks by means of structural intervention. *J Biol Syst* 10(4):431–45.

Shmulevich, I, Dougherty ER, Zhang W. (2002d) From Boolean to probabilistic Boolean networks as models of genetic regulatory networks. *Proc IEEE* 90(11):1778–92.

Shmulevich I, Gluhovsky I, Hashimoto R, Dougherty ER, Zhang W. (2003) Steady-state analysis of probabilistic Boolean networks. *Comp Funct Genomics* 4(6): 601–8.

Somogyi R, Sniegoski C. (1996) Modeling the complexity of gene networks: understanding multigenic and pleiotropic regulation. *Complexity* 1:45–63.

Spellman PT, Sherlock G, Zhang MQ, Iyer VR, Anders K, Eisen MB, Brown PO, Botstein D, Futcher B. (1998) Comprehensive identification of cell cycle-regulated genes of the yeast *Saccharomyces cerevisiae* by microarray hybridization. *Mol Biol Cell* 9(12):3273–97.

Spirtes P, Glymour C, Scheines R. (2001) *Causation, Prediction, and Search*, 2nd ed., MIT Press, Cambridge, MA.

Stern MD. (1999) Emergence of homeostasis and "noise imprinting" in an evolution model. *Proc Natl Acad Sci USA* 96(19):10746–51.

Stewart WJ. (1994) *Introduction to the Numerical Solution of Markov Chains*, Princeton University Press, Princeton, NJ.

Stoletov KV, Ratcliffe KE, Terman BI. (2002) Fibroblast growth factor receptor substrate 2 participates in vascular endothelial growth factor-induced signaling. *FASEB J* 16(10):1283–5.

Strauch ED, Yamaguchi J, Bass BL, Wang JY. (2003) Bile salts regulate intestinal epithelial cell migration by nuclear factor–kappa B-induced expression of transforming growth factor–beta. *J Am Coll Surg* 197(6):974–84.

Swisher SG, Roth JA, Nemunaitis J, Lawrence DD, Kemp BL, Carrasco CH, Connors DG, El-Naggar AK, Fossella F, Glisson BS, Hong WK, Khuri FR, Kurie JM, Lee JJ, Lee JS, Mack M, Merritt JA, Nguyen DM, Nesbitt JC, Perez-Soler R, Pisters KM, Putnam JB Jr, Richli WR, Savin M, Schrump DS, Shin DM, Shulkin A, Walsh GL, Wait J, Weill D, Waugh MK. (1999) Adenovirus-mediated p53 gene transfer in advanced non-small-cell lung cancer. *J Natl Cancer Inst* 91(9):763–71.

Uetz P, Giot L, Cagney G, Mansfield TA, Judson RS, Knight JR, Lockshon D, Narayan V, Srinivasan M, Pochart P, Qureshi-Emili A, Li Y, Godwin B, Conover D, Kalbfleisch T, Vijayadamodar G, Yang M, Johnston M, Fields S, Rothberg JM. (2000) A comprehensive analysis of protein-protein interactions in *Saccharomyces cerevisiae*. *Nature* 403(6770):623–7.

Vasudevan N, Zhu YS, Daniel S, Koibuchi N, Chin WW, Pfaff D. (2001) Crosstalk between oestrogen receptors and thyroid hormone receptor isoforms results in differential regulation of the preproenkephalin gene. *J Neuroendocrinol* 13(9):779–90.

von Dassow G, Meir E, Munro EM, Odell GM. (2000) The segment polarity network is a robust developmental module. *Nature* 406(6792):188–92.

von Mering C, Krause R, Snel B, Cornell M, Oliver SG, Fields S, Bork P. (2002) Comparative assessment of large-scale data sets of protein-protein interactions. *Nature* 417(6887):399–403.

Weeraratna AT, Jiang Y, Hostetter G, Rosenblatt K, Duray P, Bittner M, Trent JM. (2002) Wnt5a signaling directly affects cell motility and invasion of metastatic melanoma. *Cancer Cell* 1(3):279–88.

Wei CQ, Gao Y, Lee K, Guo R, Li B, Zhang M, Yang D, Burke TR Jr. (2003) Macrocyclization in the design of Grb2 SH2 domain-binding ligands exhibiting high potency in whole-cell systems. *J Med Chem* 46(2):244–54.

Winkler G. (1995) *Image Analysis, Random Fields and Dynamic Monte Carlo Methods: A Mathematical Introduction*, Springer Verlag, New York.

Yang JP, Hori M, Sanda T, Okamoto T. (1999) Identification of a novel inhibitor of nuclear factor-kappaB, RelA-associated inhibitor. *J Biol Chem* 274(22):15662–70.

Zhang W, Piatyszek MA, Kobayashi T, Estey E, Andreeff M, Deisseroth AB, Wright WE, Shay JW. (1996) Telomerase activity in human acute myelogenous leukemia: inhibition of telomerase activity by differentiation-inducing agents. *Clin Cancer Res* 2(5):799–803.

Zhang Z, Teng CT. (2001) Estrogen receptor alpha and estrogen receptor-related receptor alpha-1 compete for binding and coactivator. *Mol Cell Endocrinol* 172(1–2):223–33.

Zhou X, Wang X, Dougherty ER. (2003) Construction of genomic networks using mutual-information clustering and reversible-jump Markov-chain-Monte-Carlo predictor design. *Signal Process* 83(4):745–61.

Zhou X, Wang X, Pal R, Ivanov I, Bittner M, Dougherty ER. (2004) A Bayesian connectivity-based approach to constructing probabilistic gene regulatory networks. *Bioinformatics* 20(17):2918–27.

Zhu JY, Abate M, Rice PW, Cole CN. (1991) The ability of simian virus 40 large T antigen to immortalize primary mouse embryo fibroblasts cosegregates with its ability to bind p53. *J Virol* 65(12):6872–80.

Zou M, Conzen SD. (2005) A new dynamic Bayesian network (DBN) approach for identifying gene regulatory networks from time course microarray data. *Bioinformatics* 21(1):71–9.

# Chapter Four

## Classification

Pattern classification plays an important role in genomic signal analysis. For instance, cDNA microarrays can provide expression measurements for thousands of genes at once, and a key goal is to perform classification via different expression patterns. This requires designing a classifier (decision function) that takes a vector of gene expression levels as input and outputs a class label that predicts the class containing the input vector. Classification can be between different kinds of cancer, different stages of tumor development, or a host of such differences. Classifiers are designed from a sample of expression vectors. This requires assessing expression levels from RNA obtained from the different tissues with microarrays, determining genes whose expression levels can be used as classifier variables, and then applying some rule to design the classifier from the sample microarray data. Expression values have randomness arising from both biological and experimental variability. Design, performance evaluation, and application of classifiers must take this randomness into account. Three critical issues arise. First, given a set of variables, how does one design a classifier from the sample data that provides good classification over the general population? Second, how does one estimate the error of a designed classifier when data are limited? Third, given a large set of potential variables, such as the large number of expression levels provided by each microarray, how does one select a set of variables as the input vector to the classifier? The problem of error estimation impacts variable selection in a devilish way. An error estimator may be unbiased but have a large variance and therefore often be low (or high). This can produce a large number of gene (variable) sets and classifiers with low (or high) error estimates.

The typical setting for microarray-based classification is a large set of potential variables (gene expressions) and a small sample size (number of microarrays). This makes expression-based classification difficult because small samples tend to result in imprecise classifier design, poor error estimation, and poor feature selection (Dougherty, 2001). This chapter covers basic theory for classification. There is no effort to be encyclopedic. To the contrary, our intent is to elucidate fundamental issues relating to classifier complexity, design precision, sample size, and error estimation, with emphasis placed on small-sample settings.

## 4.1 BAYES CLASSIFIER

Pattern classification involves a *feature vector* $\mathbf{X} = (X_1, X_2, \ldots, X_d)$ on $d$-dimensional Euclidean space $\mathfrak{R}^d$ composed of random variables (*features*), a binary random variable $Y$, and a *classifier* $\psi$: $\Re^d \to \{0, 1\}$ to serve as a predictor of $Y$, which means that $Y$ is to be predicted by $\psi(\mathbf{X})$. The values of $Y$, 0 or 1, are treated as class *labels*. The error $\varepsilon[\psi]$ of $\psi$ is the probability of erroneous classification, namely, $\varepsilon[\psi] = P(\psi(\mathbf{X}) \neq Y)$. It equals the expected (mean) absolute difference $E[|Y - \psi(\mathbf{X})|]$ between the label and the classification. Because $\psi(\mathbf{X})$ and $Y$ are binary, it also equals the mean square error (MSE). $X_1, X_2, \ldots, X_d$ can be discrete or real-valued. An optimal classifier $\psi_d$ is one having minimal error $\varepsilon_d$ among all binary functions on $\Re^d$, so that it is the optimal MSE predictor of $Y$. Here $\psi_d$ and $\varepsilon_d$ are called the *Bayes classifier* and the *Bayes error*, respectively. Classification accuracy depends on the probability distribution of the feature-label pair $(\mathbf{X}, Y)$.

The posterior distribution for $\mathbf{X}$ is defined by $\eta(\mathbf{x}) = f_{\mathbf{X},Y}(\mathbf{x}, 1)/f_{\mathbf{X}}(\mathbf{x})$, where $f_{\mathbf{X},Y}(\mathbf{x}, y)$ and $f_{\mathbf{X}}(\mathbf{x})$ are the densities for $(\mathbf{X}, Y)$ and $\mathbf{X}$, respectively (and can be generalized functions). $\eta(\mathbf{x})$ gives the probability that $Y = 1$ given $\mathbf{X} = \mathbf{x}$, and $\eta(\mathbf{x}) = E[Y|\mathbf{x}]$ is the conditional expectation of $Y$ given $\mathbf{x}$. The error of an arbitrary classifier can be expressed as

$$\varepsilon[\psi] = \int_{\{\mathbf{x}:\psi(\mathbf{x})=0\}} \eta(\mathbf{x})\, f_{\mathbf{X}}(\mathbf{x})\, d\mathbf{x} + \int_{\{\mathbf{x}:\psi(\mathbf{x})=1\}} (1 - \eta(\mathbf{x}))\, f_{\mathbf{X}}(\mathbf{x})\, d\mathbf{x}. \qquad (4.1)$$

Indeed,

$$P(\psi(\mathbf{X}) \neq Y) = P(Y = 1, \psi(\mathbf{X}) = 0) + P(Y = 0, \psi(\mathbf{X}) = 1)$$

$$= \int_{\{(\mathbf{x},y):\psi(\mathbf{x})=0, y=1\}} f_{\mathbf{X},Y}(\mathbf{x}, y)\, d\mathbf{x}\, dy + \int_{\{(\mathbf{x},y):\psi(\mathbf{x})=1, y=0\}} f_{\mathbf{X},Y}(\mathbf{x}, y)\, d\mathbf{x}\, dy$$

$$= \int_{\{\mathbf{x}:\psi(\mathbf{x})=0\}} f_{\mathbf{X},Y}(\mathbf{x}, 1)\, d\mathbf{x} + \int_{\{\mathbf{x}:\psi(\mathbf{x})=1\}} f_{\mathbf{X},Y}(\mathbf{x}, 0)\, d\mathbf{x}, \qquad (4.2)$$

which reduces to the right-hand side of equation (4.1) owing to the definition of $\eta(\mathbf{x})$. Since $0 \leq \eta(\mathbf{x}) \leq 1$, the right-hand side of equation (4.1) is minimized by

$$\psi_d(\mathbf{x}) = \begin{cases} 0 & \text{if } \eta(\mathbf{x}) \leq 1/2, \\ 1 & \text{if } \eta(\mathbf{x}) > 1/2, \end{cases} \qquad (4.3)$$

which defines the Bayes classifier. As a function of $\mathbf{X}$, $\eta(\mathbf{X})$ is a random variable dependent on $f_{\mathbf{X}}(\mathbf{x})$. Defining the *class conditional probability* of $Y$ given $\mathbf{x}$ by $P(Y = 1|\mathbf{x}) = \eta(\mathbf{x})$ gives it the intuitive form

$$\psi_d(\mathbf{x}) = \begin{cases} 0 & \text{if } P(Y = 1|\mathbf{x}) \leq 1/2, \\ 1 & \text{if } P(Y = 1|\mathbf{x}) > 1/2. \end{cases} \qquad (4.4)$$

$\psi_d(\mathbf{x})$ is defined to be 0 or 1 according to whether $Y$ is less or more likely to be 1 given $\mathbf{x}$.

It follows at once from equations (4.1) and (4.3) that the Bayes error is given by

$$\varepsilon_d = \int_{\{x:\eta(x)<1/2\}} \eta(\mathbf{x}) f_{\mathbf{x}}(\mathbf{x}) \, d\mathbf{x} + \int_{\{x:\eta(x)\geq1/2\}} (1 - \eta(\mathbf{x})) f_{\mathbf{x}}(\mathbf{x}) \, d\mathbf{x}. \qquad (4.5)$$

Considering the integration domains for the two integrals, the integral kernels can both be expressed as $\min(\eta(\mathbf{x}), 1 - \eta(\mathbf{x}))$. The integrals can then be combined to yield

$$\varepsilon_d = E[\min(\eta(\mathbf{x}), 1 - \eta(\mathbf{x}))] = \int_{\Re^d} \min\{\eta(\mathbf{x}), 1 - \eta(\mathbf{x})\} f_{\mathbf{X}}(\mathbf{x}) \, d\mathbf{x}, \qquad (4.6)$$

where the expectation in the first expression is with respect to the distribution of $\mathbf{X}$. The identity

$$\min(a, b) = \frac{a + b - |a - b|}{2} \qquad (4.7)$$

yields the error expression

$$\varepsilon_d = \frac{1}{2}(1 - E[|2\eta(\mathbf{X}) - 1|]). \qquad (4.8)$$

## 4.2  CLASSIFICATION RULES

The problem with the Bayes classifier is that we typically do not know the class conditional probabilities and therefore must design a classifier from sample data. An obvious approach would be to estimate the conditional probabilities from data, but often we do not have sufficient data to obtain good estimates. Moreover, as we will see, good classifiers can be obtained even when we lack sufficient data for satisfactory density estimation.

### 4.2.1  Consistent Classifier Design

The design of a classifier $\psi_n$ from a random sample

$$S_n = \{(\mathbf{X}_1, Y_1), (\mathbf{X}_2, Y_2), \ldots, (\mathbf{X}_n, Y_n)\} \qquad (4.9)$$

of vector-label pairs drawn from the feature-label distribution requires a *classification rule* that operates on random samples to yield a classifier. A *classification rule* is a mapping of the form $\Psi_n \colon [\Re^d \times \{0, 1\}]^n \to \mathcal{F}$, where $\mathcal{F}$ is the family of $\{0, 1\}$-valued functions on $\Re^d$. Given a sample $S_n$, we obtain a designed classifier $\psi_n = \Psi_n(S_n)$ according to the rule $\Psi_n$. To be completely formal, one might write $\psi(S_n; \mathbf{X})$ rather than $\psi_n(\mathbf{X})$; however, we will use the simpler notation, keeping in mind that $\psi_n$ derives from a classification rule applied to a feature-label sample.

Note that a classification rule is really a sequence of classification rules depending on $n$.

The Bayes error $\varepsilon_d$ is estimated by the error $\varepsilon_n$ of $\psi_n$. There is a *design cost*

$$\Delta_n = \varepsilon_n - \varepsilon_d, \tag{4.10}$$

$\varepsilon_n$ and $\Delta_n$ being sample-dependent random variables. The expected design cost is $E[\Delta_n]$, the expectation being relative to all possible samples. The expected error of $\psi_n$ is decomposed as

$$E[\varepsilon_n] = \varepsilon_d + E[\Delta_n]. \tag{4.11}$$

Asymptotic properties of a classification rule concern large samples (as $n \to \infty$). A rule is said to be *consistent* for a distribution of $(\mathbf{X}, Y)$ if $\Delta_n \to 0$ in the mean, meaning $E[\Delta_n] \to 0$ as $n \to \infty$. Since $\Delta_n \leq 1$ for all $n$, convergence to 0 in the mean is equivalent to convergence to 0 in probability: for any $\tau > 0$, $P(\Delta_n > \tau) \to 0$ as $n \to \infty$. Rewriting this, $P(\Delta_n \leq \tau) \to 1$ as $n \to \infty$. For a consistent rule, the expected design cost can be made arbitrarily small for a sufficiently large amount of data. Since rarely is there a good estimate of the distribution in pattern recognition, rules for which convergence is independent of the distribution are desirable. A classification rule is *universally consistent* if $\Delta_n \to 0$ in the mean for any distribution of $(\mathbf{X}, Y)$.

Consider the Bayes classifier defined by equation (4.3). If $\eta_n(\mathbf{x})$ is an estimate of $\eta(\mathbf{x})$ based on a sample $S_n$, then a reasonable classification rule is to define $\psi_n(\mathbf{x})$ according to equation (4.3) with $\eta_n(\mathbf{x})$ in place of $\eta(\mathbf{x})$. This approach is known as the *plug-in rule*. For this rule,

$$\Delta_n = \int_{\{\mathbf{x}: \psi_n(\mathbf{x}) \neq \psi_d(\mathbf{x})\}} |2\eta(\mathbf{x}) - 1| f_\mathbf{x}(\mathbf{x}) \, d\mathbf{x}. \tag{4.12}$$

By recognizing that $\psi_d(\mathbf{x}) \neq \psi_n(\mathbf{x})$ implies $|\eta(\mathbf{x}) - \eta_n(\mathbf{x})| \geq |\eta(\mathbf{x}) - 1/2|$ and integrating over $\Re^d$, it follows at once that the design cost possesses the $L_1$ bound

$$\Delta_n \leq 2 \int_{\Re^d} |\eta(\mathbf{x}) - \eta_n(\mathbf{x})| f_\mathbf{x}(\mathbf{x}) \, d\mathbf{x} = 2E[|\eta(\mathbf{X}) - \eta_n(\mathbf{X})|]. \tag{4.13}$$

Application of the Schwarz inequality yields a corresponding $L_2$ bound

$$\Delta_n \leq 2E[|\eta(\mathbf{X}) - \eta_n(\mathbf{X})|^2]^{1/2}. \tag{4.14}$$

Consequently, the plug-in rule is consistent if either $E[|\eta(\mathbf{X}) - \eta_n(\mathbf{X})|]$ or $E[|\eta(\mathbf{X}) - \eta_n(\mathbf{X})|^2]$ converges to 0 in probability as $n \to \infty$. For the $L_1$ bound, convergence in probability means that for any $\tau > 0$,

$$\lim_{n \to \infty} P(E[|\eta(\mathbf{X}) - \eta_n(\mathbf{X})|] > \tau) = 0, \tag{4.15}$$

where the probability is relative to the distribution of the sample. A similar statement applies to the $L_2$ bound.

Since the right-hand side of equation (4.14) is a constant, taking expectations on both sides yields

$$E[\Delta_n] \leq 2E[|\eta(\mathbf{X}) - \eta_n(\mathbf{X})|^2]^{1/2}. \qquad (4.16)$$

Hence, if $\eta_n$ converges to $\eta$ in the $L_2$ sense ($\eta_n$ is a consistent estimator of the density $\eta$), then the plug-in classifier is consistent. In this way, consistent density estimation yields consistent classifier estimation. However, as the next theorem shows, the bound is not tight. In fact, it shows that $E[\Delta_n]$ converges much faster than $E[|\eta(\mathbf{X}) - \eta_n(\mathbf{X})|^2]^{1/2}$, which in turn shows that classification is easier than density estimation in the sense that it requires fewer data.

**THEOREM 4.1 (Devroye et al., 1996)** *If $E[|\eta(\mathbf{X}) - \eta_n(\mathbf{X})|^2] \to 0$ as $n \to \infty$, then for the plug-in rule,*

$$\lim_{n \to \infty} \frac{E[\Delta_n]}{E[|\eta(\mathbf{X}) - \eta_n(\mathbf{X})|^2]^{1/2}} = 0. \qquad (4.17)$$

While universal consistency is useful for large samples, it has little consequence for small samples. One issue is whether the design cost $E[\Delta_n]$ can be made universally small for fixed $n$. It cannot be, because $n$ may be insufficient for some highly irregular distribution for $(\mathbf{X}, Y)$.

**THEOREM 4.2 (Devroye, 1982)** *For any $\tau > 0$, fixed n, and designed classifier $\psi_n$, there exists a distribution of $(\mathbf{X}, Y)$ such that $\varepsilon_d = 0$ and*

$$E[\varepsilon_n] > \frac{1}{2} - \tau. \qquad (4.18)$$

Even if a classifier is universally consistent, the rate at which $E[\Delta_n] \to 0$ is critical to application. If we desire a classifier whose expected error is within some tolerance of the Bayes error, consistency is not sufficient. Rather, we would like a statement of the following form: for any $\tau > 0$, there exists $n(\tau)$ such that $E[\Delta_{n(\tau)}] < \tau$ for any distribution of $(\mathbf{X}, Y)$. The next theorem shows that such a proposition is not possible. It states that, even if $\psi_n$ is universally consistent, the design error converges to zero arbitrarily slowly relative to all possible distributions.

**THEOREM 4.3 (Devroye, 1982)** *Let $\{a_n\}$ be a decreasing sequence of positive numbers with $a_1 \leq 1/16$. For any sequence of designed classifiers $\psi_n$, a distribution of $(\mathbf{X}, Y)$ exists such that $\varepsilon_d = 0$ and $E[\varepsilon_n] > a_n$.*

A large class of rules partition $\Re^d$ into a disjoint union of cells, and then $\psi_n(\mathbf{x})$ is defined to be 0 or 1 according to which is the majority among the labels in the cell containing $\mathbf{x}$. The cells may change with $n$ and may depend on the sample points, but they do not depend on the labels. Consistency can be shown for some of these by reducing the problem to one of plug-in consistency and applying the plug-in considerations for the $L_1$ bound. To obtain consistency for a distribution of $(\mathbf{X}, Y)$, we should expect two necessary properties for the partition: (1) it should be sufficiently fine to take into account local structure of the distribution, and (2) there should be enough labels in each cell so that the majority decision reflects well the theoretical decision based on the posterior distributions. In fact, these properties are sufficient. Given a point $\mathbf{x}$, let $U_n(\mathbf{x})$ denote the cell containing $\mathbf{x}$, where it may be that the cell depends on the sample points, and let $N_n(\mathbf{x})$ denote the number of sample points in $U_n(\mathbf{x})$. For any cell $U$, define the diameter of $U$, denoted by diam$[U]$, to be the supremum over all distances between points in $U$.

**THEOREM 4.4 (Devroye et al., 1996)** *A partition-based classification rule is consistent for a distribution of $(\mathbf{X}, Y)$ if (1) diam$[U_n(\mathbf{X})] \to 0$ in probability as $n \to \infty$, and (2) $N_n(\mathbf{X}) \to \infty$ in probability as $n \to \infty$.*

Implicit in this theorem is that the cell $U_n(\mathbf{X})$ is random, depending on the outcome of $\mathbf{X}$. The first condition means that $P(\text{diam}[U_n(\mathbf{X})] > \varepsilon) \to 0$ as $n \to \infty$ for any $\varepsilon > 0$, where the probability is relative to the distribution of $\mathbf{X}$. The second condition means that $P(N_n(\mathbf{X}) < k) \to 0$ as $n \to \infty$ no matter how large $k$ is. The proof involves showing that $E[|\eta(\mathbf{X}) - \eta_n(\mathbf{X})|] \to 0$ in probability, where

$$\eta_n(\mathbf{x}) = \frac{1}{N_n(\mathbf{x})} \sum_{\{i:\mathbf{x}_i \in U_n(\mathbf{x})\}} Y_i. \tag{4.19}$$

The theorem yields universal consistency if the conditions hold for any distribution of $(\mathbf{X}, Y)$.

To illustrate the theorem, we consider the *cubic histogram rule*. $\Re^d$ is partitioned into cubes of side length $r_n$. For each point $\mathbf{x} \in \Re^d$, $\psi_n(\mathbf{x})$ is defined to be 0 or 1 according to which is the majority among the labels for points in the cube containing $\mathbf{x}$. If the cubes are defined so that $r_n \to 0$ and $nr_n^d \to \infty$ as $n \to \infty$, then the conditions of the theorem are satisfied for any distribution of $(\mathbf{X}, Y)$ and the rule is universally consistent (Gordon and Olshen, 1978; Devroye et al., 1996). The cubes get smaller, but $n$ times the cube volume approaches infinity as $n \to \infty$.

A universal consistency theorem can be obtained via the plug-in $L_2$ bound for a general class of weighted-average estimators of the posterior distribution. For a sample $S_n$, as in equation (4.9), define the estimator

$$\eta_n(\mathbf{x}) = \sum_{i=1}^{n} W_{ni}(\mathbf{x}) Y_i, \tag{4.20}$$

where the nonnegative weight $W_{ni}(\mathbf{x})$ depends on the sample points and the weights sum to 1 for any $n$. By making the weights larger for sample points closer to $\mathbf{x}$, $\eta_n(\mathbf{x})$ provides a *local-weighted-average rule* and $\psi_n(\mathbf{x})$ is determined by a local weighted majority vote. If three conditions are satisfied by the weights, then a local-weighted-average rule is consistent. The first condition states that the total weight outside a ball of fixed radius from $\mathbf{x}$ must go to 0.  Specifically, for any $\delta > 0$,

$$\lim_{n \to \infty} E\left[ \sum_{\{i: \|\mathbf{x} - \mathbf{x}_i\| > \delta\}} W_{ni}(\mathbf{X}) \right] = 0. \tag{4.21}$$

Second, the expectation of the maximum of the weights must tend to 0:

$$\lim_{n \to \infty} E[\max\{W_{n1}(\mathbf{X}), W_{n2}(\mathbf{X}), \dots, W_{nn}(\mathbf{X})\}] = 0. \tag{4.22}$$

There is a third technical condition required for the proof: there exists a constant $c$ such that for every nonnegative measurable function $g$ for which $E[g(\mathbf{X})]$ is finite,

$$E\left[ \sum_{i=1}^{n} W_{ni}(\mathbf{X}) g(\mathbf{X}_i) \right] \leq cE[g(\mathbf{X})]. \tag{4.23}$$

**THEOREM 4.5 (Stone, 1977)** *If for any distribution of $\mathbf{X}$ the weights of a local-weighted-average rule satisfy the three preceding conditions, then the local-weighted-average rule is universally consistent.*

A classification rule is said to be *strongly consistent* for a distribution of $(\mathbf{X}, Y)$ if $\Delta_n \to 0$ almost surely as $n \to \infty$, meaning that $\Delta_n \to 0$ with probability 1. Specifically, for every training sequence $\{S_1, S_2, \dots\}$, except for a set $\mathcal{S}$ of training sequences, $\Delta_n \to 0$ as $n \to \infty$, and the probability of $\mathcal{S}$ is 0. Since almost sure convergence implies convergence in probability, strong consistency implies consistency. A rule is *universally strongly consistent* if $\Delta_n \to 0$ almost surely for any distribution of $(\mathbf{X}, Y)$. It is possible to derive consistency and strong consistency in a single argument. According to the Borel-Cantelli lemma, if

$$\sum_{n=1}^{\infty} P(\Delta_n > \tau) < \infty \tag{4.24}$$

for any $\tau > 0$, then $\Delta_n \to 0$ almost surely. Convergence of the sum implies $P(\Delta_n > \tau) \to 0$ as $n \to \infty$. The lemma is applied by bounding $P(\Delta_n > \tau)$ by a summable sequence. This is not easy.

### 4.2.2  Examples of Classification Rules

Recalling that the cubic histogram rule is universally consistent if the cubes are defined so that $r_n \to 0$ and $nr_n^d \to \infty$ as $n \to \infty$, under these same limit conditions, for any distribution of $(\mathbf{X}, Y)$ and any $\tau > 0$, there exists an integer $n(\tau)$ such that $n > n(\tau)$ implies that

$$P(\Delta_n > \tau) \leq 2e^{-n\tau^2/32}. \tag{4.25}$$

The Borel-Cantelli lemma shows that $\Delta_n \to 0$ almost surely. Since $n(\tau)$ depends on the distribution of $(\mathbf{X}, Y)$, the result gives strong consistency but not universal strong consistency. For discrete data the cubic histogram rule reduces to the plug-in rule if the cubes are sufficiently small, because then a cube contains at most one sample point and choosing the majority among the labels means choosing the majority label for the point and this is equivalent to using the relative frequency to estimate $\eta(\mathbf{x})$ and applying equation (4.3).

For the *nearest-neighbor* (NN) rule, $\psi_n$ is defined for each $\mathbf{x} \in \Re^d$ by letting $\psi_n(\mathbf{x})$ take the label of the sample point that is closest to $\mathbf{x}$. For the NN rule, no matter the distribution of $(\mathbf{X}, Y)$,

$$\varepsilon_d \leq \lim_{n \to \infty} E[\varepsilon_n] = E[2\eta(\mathbf{X})(1 - \eta(\mathbf{X}))] \leq 2\varepsilon_d(1 - \varepsilon_d) \leq 2\varepsilon_d \qquad (4.26)$$

(Cover and Hart, 1967; Stone, 1977). It follows that $\lim_{n \to \infty} E[\Delta_n] \leq \varepsilon_d$. Hence, the asymptotic expected cost of design is small if the Bayes error is small; however, this result does not give consistency.

For the *k-nearest-neighbor* (kNN) *rule*, $k$ odd, the $k$ points closest to $\mathbf{x}$ are selected and $\psi_n(\mathbf{x})$ is defined to be 0 or 1 according to which is the majority among the labels of these points. If $k = 1$, this gives the NN rule. We will not consider even $k$. The limit of $E[\varepsilon_n]$ as $n \to \infty$ can be expressed analytically, and various upper bounds exist. We state a simple one (Devroye et al., 1996):

$$\varepsilon_d \leq \lim_{n \to \infty} E[\varepsilon_n] \leq \varepsilon_d + \frac{1}{\sqrt{ke}}. \qquad (4.27)$$

It follows that

$$\lim_{n \to \infty} E[\Delta_n] \leq (ke)^{-1/2}. \qquad (4.28)$$

This does not give consistency, but it does show that the design cost becomes arbitrarily small for sufficiently large $k$ as $n \to \infty$. The kNN rule is seen to be a local-weighted-average rule by defining $W_{ni}(\mathbf{X}) = 1/k$ if $\mathbf{X}_i$ is one of the $k$ nearest neighbors of $\mathbf{X}$, and $W_{ni}(\mathbf{X}) = 0$ otherwise. Theorem 4.5 can be applied under the conditions that $k \to \infty$ and $k/n \to 0$ as $n \to \infty$ to yield universal consistency for the kNN rule (Stone, 1977).

Instead of taking the majority label among a predecided number of nearest neighbors as with the NN rule, the *moving-window rule* presets a distance and takes the majority label among all the sample points within that distance of $\mathbf{x}$.

The moving-window rule can be "smoothed" by giving more weight to sample points closer to $\mathbf{x}$. A histogram rule based on the majority between 0 and 1 labels is equivalent to the median of the labels, where in case of a tie a 0 is given. More generally, a weighted median of binary values is computed by adding up the weights associated with the 0-labeled and 1-labeled points separately and defining the output to be the larger sum. A *kernel rule* is constructed by defining a weighting kernel based on the distance of a sample point from $\mathbf{x}$ in conjunction with a smoothing factor. The *Gaussian* kernel is defined by

$$K_h(\mathbf{x}) = e^{-\|\mathbf{x}/h\|^2}, \qquad (4.29)$$

and the *Epanechnikov* kernel is defined by

$$K_h(\mathbf{x}) = 1 - ||\mathbf{x}/h||^2 \tag{4.30}$$

if $||\mathbf{x}|| \le h$ and $K_h(\mathbf{x}) = 0$ if $||\mathbf{x}|| > h$. If $\mathbf{x}$ is the point at which the classifier is being defined and $\mathbf{x}_k$ is a sample point, then the weighting is given by $K_h(\mathbf{x} - \mathbf{x}_k)$. Since the Gaussian kernel is never 0, all sample points are given some weight. The Epanechnikov kernel is 0 for sample points more than $h$ from $\mathbf{x}$, so that, as with the moving-window rule, only sample points within a certain radius contribute to the definition of $\psi_n(\mathbf{x})$. The moving-window rule is a special case of a kernel rule with the weights being 1 within a specified radius.

A kernel $K$ is said to be *regular* if it is nonnegative and there exists a ball $B_r$ of radius $r$ centered at the origin and a constant $\delta > 0$ such that $K \ge \delta$ on $B_r$ and

$$\int_{\Re^d} \sup_{\mathbf{y} \in \mathbf{x} + B_r} K(\mathbf{y}) \, d\mathbf{x} < \infty. \tag{4.31}$$

The three kernel rules mentioned previously are regular.  Applying the Borel-Cantelli lemma to the next theorem shows that regular kernel rules are strongly universally consistent.

**THEOREM 4.6 (Devroye and Krzyzak, 1989)** *Suppose $K$ is a regular kernel. If $h \to \infty$ in such a way that $nh^d \to \infty$ as $n \to \infty$, then there exists a constant $\rho$ depending on the dimension and the kernel such that for any distribution $(\mathbf{X}, Y)$ and every $\tau > 0$, there exists an integer $n_0$ such that $n \ge n_0$ implies*

$$P(\Delta_n > \tau) \le 2e^{-n\tau^2/32\rho^2}. \tag{4.32}$$

## 4.3  CONSTRAINED CLASSIFIERS

Previously we stated two properties of a partition-based rule that lead to consistency, in particular, to theorem 4.4: (1) the partition should be sufficiently fine to take into account local structure of the feature-label distribution, and (2) there should be enough labels in each cell so that the majority decision reflects well the theoretical decision based on the posterior distributions. A key issue with small samples is that the second criterion is not satisfied if the partition is fine. If the partition is too fine, one can end up with an extreme situation in which there is at most one point in each cell. Even if a rule is structured so that empty cells are joined with nearby nonempty cells, the result is a partition consisting of cells that each contain a single point, so that the majority decisions are quite meaningless, and a partition that perhaps yields a complex decision boundary whose complexity is unsupported by the data relative to the feature-label distribution. This situation exhibits the problem of *overfitting*. Relative to the sample data, a classifier may have small error; but relative to the feature-label distribution, the error may be severe. In the case of a partition-based rule, it might be better to use a coarse partition, have the

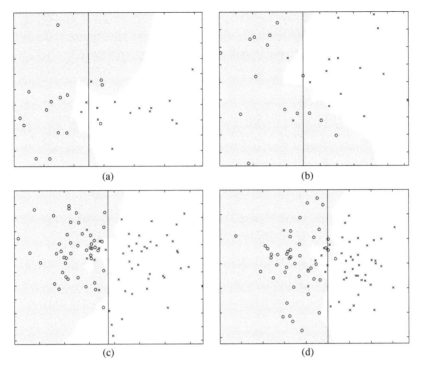

Figure 4.1. 3NN classification applied to two equal-variance circular Gaussian class conditional distributions. (a) For a 30-point sample; (b) for a second 30-point sample;
(c) for a 90-point sample; (d) for a second 90-point sample. (Dougherty et al.,
2005b.)

classifier make errors on the sample data, and have a smaller error relative to the
feature-label distribution than that resulting from a fine partition. The general point,
which will be a recurring theme, is that a classification rule should not cut up the
space in a manner too complex for the number of sample data available.

Overfitting is illustrated in figure 4.1, in which the 3NN rule is applied to two
equal-variance circular Gaussian class conditional distributions. Figure 4.1a and
4.1b show the 3NN classifier for two 30-point samples, and figure 4.1c and 4.1d
show the 3NN classifier for two 90-point samples. Note the greater overfitting
of the data for the 30-point samples, in particular, the greater difference in the two
30-point designed classifiers as compared to the difference between the two
90-point classifiers, and the closer the latter are to the Bayes classifier given by
the vertical line.

The problem of overfitting is not necessarily mitigated by applying an error
estimation rule to the designed classifier to see if it "actually" performs well since
when only a small number of data are available, error estimation rules tend to be
very imprecise, and the imprecision tends to be worse for complex classification
rules. Hence, a low error estimate may not be sufficient to overcome the large
expected design error resulting from using a complex classifier with a small sample.

Depending on the number of data available, we need to consider constrained classification rules. To formalize these considerations we need a measure of the capacity of a classifier to partition the feature space relative to sample size.

Constraining classifier design means restricting the functions from which a classifier can be chosen to a class $\mathcal{C}$. This leads to trying to find an optimal *constrained* classifier $\psi_{\mathcal{C}} \in \mathcal{C}$ having error $\varepsilon_{\mathcal{C}}$. Constraining the classifier can reduce the expected design error but at the cost of increasing the error of the best possible classifier. Since optimization in $\mathcal{C}$ is over a subclass of classifiers, the error $\varepsilon_{\mathcal{C}}$ of $\psi_{\mathcal{C}}$ typically exceeds the Bayes error unless the Bayes classifier happens to be in $\mathcal{C}$. This *cost of constraint (approximation)* is

$$\Delta_{\mathcal{C}} = \varepsilon_{\mathcal{C}} - \varepsilon_d. \tag{4.33}$$

A classification rule yields a classifier $\psi_{n,\mathcal{C}} \in \mathcal{C}$ with error $\varepsilon_{n,\mathcal{C}}$ and $\varepsilon_{n,\mathcal{C}} \geq \varepsilon_{\mathcal{C}} \geq \varepsilon_d$. The design error for constrained classification is

$$\Delta_{n,\mathcal{C}} = \varepsilon_{n,\mathcal{C}} - \varepsilon_{\mathcal{C}}. \tag{4.34}$$

For small samples, this can be substantially less than $\Delta_n$ depending on $\mathcal{C}$ and on the rule. The error of the designed constrained classifier is decomposed as

$$\varepsilon_{n,\mathcal{C}} = \varepsilon_d + \Delta_{\mathcal{C}} + \Delta_{n,\mathcal{C}}. \tag{4.35}$$

The expected error of the designed classifier from $\mathcal{C}$ can be decomposed as

$$E[\varepsilon_{n,\mathcal{C}}] = \varepsilon_d + \Delta_{\mathcal{C}} + E[\Delta_{n,\mathcal{C}}]. \tag{4.36}$$

The constraint is beneficial if and only if $E[\varepsilon_{n,\mathcal{C}}] < E[\varepsilon_n]$, which means that

$$\Delta_{\mathcal{C}} < E[\Delta_n] - E[\Delta_{n,\mathcal{C}}]. \tag{4.37}$$

That is, if the cost of constraint is less than the decrease in expected design cost, then the expected error of $\psi_{n,\mathcal{C}}$ is less than that of $\psi_n$. The dilemma: strong constraint reduces $E[\Delta_{n,\mathcal{C}}]$ at the cost of increasing $\varepsilon_{\mathcal{C}}$.

The matter can be graphically illustrated. For the discrete-data plug-in rule and the cubic histogram rule with fixed cube size, $E[\Delta_n]$ is nonincreasing, meaning that $E[\Delta_{n+1}] \leq E[\Delta_n]$. This means that the expected design error never increases as sample sizes increase, and it holds for any feature-label distribution. Such classification rules are called *smart*. They fit our intuition about increasing sample sizes. The nearest-neighbor rule is not smart because there exist distributions for which $E[\Delta_{n+1}] \leq E[\Delta_n]$ does not hold for all $n$. Now consider a consistent rule, constraint, and distribution for which $E[\Delta_{n+1}] \leq E[\Delta_n]$ and $E[\Delta_{n+1,\mathcal{C}}] \leq E[\Delta_{n,\mathcal{C}}]$. Figure 4.2 illustrates the design problem. The axes correspond to sample size and error. The horizontal dashed lines represent $\varepsilon_{\mathcal{C}}$ and $\varepsilon_d$; the decreasing solid lines represent $E[\varepsilon_{n,\mathcal{C}}]$ and $E[\varepsilon_n]$. If $n$ is sufficiently large, then $E[\varepsilon_n] < E[\varepsilon_{n,\mathcal{C}}]$; however, if $n$ is sufficiently small, then $E[\varepsilon_n] > E[\varepsilon_{n,\mathcal{C}}]$. The point $N_0$ at which the decreasing lines cross is the cutoff: for $n > N_0$, the constraint is detrimental; for $n < N_0$, it is beneficial. If $n < N_0$, then the advantage of the constraint is the difference between the decreasing solid lines.

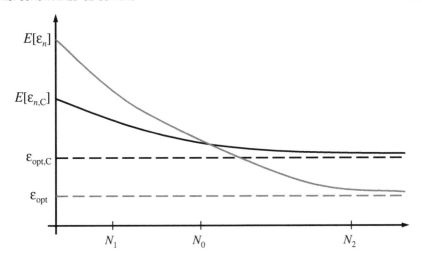

Figure 4.2. Errors for unconstrained and constrained classification as a function of sample size. (Dougherty, 2001.)

### 4.3.1 Shatter Coefficient

To examine the matter more closely, for class $C$, sample $S_n$, and classifier $\psi \in C$, define the *empirical error* $\hat{\epsilon}_n[\psi]$ for the sample to be the fraction of times that $\psi(\mathbf{X}_i) \neq Y_i$ for $(\mathbf{X}_i, Y_i) \in S_n$. It is the error rate on the sample. $S_n$ is a random sample for the distribution $(\mathbf{X}, Y)$ and the empirical error estimates $P(\psi(\mathbf{X}) \neq Y)$. An *empirical-error-rule classifier* $\hat{\psi}_{n,C}$ is a classifier in $C$ possessing minimal empirical error on the sample. Since $\psi_C$ is the classifier in $C$ minimizing the probability $P(\psi(\mathbf{X}) \neq Y)$ and $\hat{\epsilon}_n[\psi]$ estimates this probability, choosing a classifier with minimal empirical error appears to be a reasonable classification rule for optimization relative to $C$; in fact, the worth of an empirical-error-rule classifier depends upon its performance on the feature-label distribution. The empirical-error-rule classifier has empirical error $\hat{\epsilon}_n[\hat{\psi}_{n,C}]$ on the sample and error $\hat{\epsilon}_{n,C} = \epsilon[\hat{\psi}_{n,C}]$. Equations 4.34–4.36 hold with $\epsilon_{n,C}$ and $\Delta_{n,C}$ replaced by $\hat{\epsilon}_{n,C}$ and $\hat{\Delta}_{n,C}$, respectively.

If $C$ has finite cardinality $|C|$, $C$ contains the Bayes classifier, and the Bayes error is zero, meaning $\epsilon_C = \epsilon_d = 0$, then there exist simple bounds on $\hat{\epsilon}_{n,C} = \hat{\Delta}_{n,C}$: for any $n$ and any $\tau > 0$,

$$P(\hat{\epsilon}_{n,C} > \tau) \leq |C|e^{-n\tau} \tag{4.38}$$

and

$$E[\hat{\epsilon}_{n,C}] \leq \frac{1 + \log |C|}{n} \tag{4.39}$$

(Vapnik and Chervonenkis, 1974). It follows at once that $\hat{\Delta}_{n,C} \to 0$ in probability as $n \to \infty$.

A simple empirical-error inequality leads ultimately to much stronger results. By minimality, $\hat{\varepsilon}_n[\hat{\psi}_{n,\mathcal{C}}] \le \hat{\varepsilon}_n[\psi]$ for all $\psi \in \mathcal{C}$. Hence,

$$\hat{\varepsilon}_n[\hat{\psi}_{n,\mathcal{C}}] - \varepsilon_{\mathcal{C}} \le \sup_{\psi \in \mathcal{C}} |\hat{\varepsilon}_n[\psi] - \varepsilon[\psi]|. \tag{4.40}$$

Since $\hat{\psi}_{n,\mathcal{C}} \in \mathcal{C}$, $\hat{\varepsilon}_{n,\mathcal{C}} - \hat{\varepsilon}_n[\hat{\psi}_{n,\mathcal{C}}]$ has the same upper bound. Moreover, since

$$\hat{\varepsilon}_{n,\mathcal{C}} - \hat{\varepsilon}_n[\hat{\psi}_{n,\mathcal{C}}] = \varepsilon[\hat{\psi}_{n,\mathcal{C}}] - \hat{\varepsilon}_n[\hat{\psi}_{n,\mathcal{C}}], \tag{4.41}$$

the right-hand side also possesses the same upper bound. Taken together in conjunction with the triangle inequality, these bounds yield (Vapnik and Chervonenkis, 1974)

$$\hat{\Delta}_{n,\mathcal{C}} = \hat{\varepsilon}_{n,\mathcal{C}} - \hat{\varepsilon}_n[\hat{\psi}_{n,\mathcal{C}}] + \hat{\varepsilon}_n[\hat{\psi}_{n,\mathcal{C}}] - \varepsilon_{\mathcal{C}}$$

$$\le 2 \sup_{\psi \in \mathcal{C}} |\hat{\varepsilon}_n[\psi] - \varepsilon[\psi]|. \tag{4.42}$$

For any measurable set $A \subset \Re^d \times \{0, 1\}$, let $\nu[A] = P((\mathbf{X}, Y) \in A)$. Then

$$\varepsilon[\psi] = \nu[\{(\mathbf{x}, y) : \psi(\mathbf{x}) \ne y\}]$$

$$= \nu[(\mathcal{K}[\psi] \times \{0\}) \cup (\mathcal{K}[\psi]^c \times \{1\})], \tag{4.43}$$

where $\mathcal{K}[\psi] = \{\mathbf{x} : \psi(\mathbf{x}) = 1\}$ is called the *kernel* of $\psi$. Let $\nu_n[A]$ be the empirical measure of $A$ based on the sample $S_n$, meaning that $\nu_n[A]$ is the fraction of the sample points that fall in $A$. An expression similar to equation (4.43) holds for the empirical error $\hat{\varepsilon}_n[\psi]$ with $\nu_n$ in place of $\nu$. Hence,

$$\sup_{\psi \in \mathcal{C}} |\hat{\varepsilon}_n[\psi] - \varepsilon[\psi]| \le \sup_{A \in \mathcal{A}_{\mathcal{C}}} |\nu_n[A] - \nu[A]|, \tag{4.44}$$

where

$$\mathcal{A}_{\mathcal{C}} = \{\{(\mathbf{x}, y) : \psi(\mathbf{x}) \ne y\} : \psi \in \mathcal{C}\}$$

$$= \{(\mathcal{K}[\psi] \times \{0\}) \cup (\mathcal{K}[\psi]^c \times \{1\}) : \psi \in \mathcal{C}\}. \tag{4.45}$$

The supremum on the right-hand side of equation (4.44) can be taken for any class $\mathcal{A}$ of measurable sets, and we denote it by $\hat{\Delta}_n(\mathcal{A})$. When applying it to $\mathcal{A}_{\mathcal{C}}$, we denote it by $\hat{\Delta}_n(\mathcal{C})$. Taken together, the inequalities in equations (4.42) and (4.44) yield $\hat{\Delta}_{n,\mathcal{C}} \le 2\hat{\Delta}_n(\mathcal{C})$ and

$$P(\hat{\Delta}_{n,\mathcal{C}} > \tau) \le P(\hat{\Delta}_n(\mathcal{C}) > \tau/2). \tag{4.46}$$

Hence, to bound $P(\hat{\Delta}_{n,\mathcal{C}} > \tau)$, it is sufficient to bound $P(\hat{\Delta}_n(\mathcal{C}) > \tau/2)$.

Let $\mathcal{A}$ be any class of measurable sets in $\Re^d$. If $\{\mathbf{z}_1, \mathbf{z}_2, \ldots, \mathbf{z}_n\}$ is a point set in $\Re^d$, let

$$N_{\mathcal{A}}(\mathbf{z}_1, \mathbf{z}_2, \ldots, \mathbf{z}_n) = |\{\{\mathbf{z}_1, \mathbf{z}_2, \ldots, \mathbf{z}_n\} \cap A : A \in \mathcal{A}\}| \tag{4.47}$$

be the cardinality of the class of distinct subsets of $\{\mathbf{z}_1, \mathbf{z}_2, \ldots, \mathbf{z}_n\}$ created by intersection with sets in $\mathcal{A}$. The $n$th *shatter coefficient* of $\mathcal{A}$ is defined by

$$\xi(\mathcal{A}, n) = \max_{\mathbf{z}_1, \mathbf{z}_2, \ldots, \mathbf{z}_n} N_{\mathcal{A}}(\mathbf{z}_1, \mathbf{z}_2, \ldots, \mathbf{z}_n). \tag{4.48}$$

It is possible to have $\xi(\mathcal{A}, n) = 2^n$, in which case $\mathcal{A}$ is said to shatter $\{\mathbf{z}_1, \mathbf{z}_2, \ldots, \mathbf{z}_n\}$. If $\xi(\mathcal{A}, n) < 2^n$, then $\xi(\mathcal{A}, j) < 2^j$ for $j > n$.

As an illustration, consider the set $\mathcal{A}$ of half-lines in $\mathfrak{R}^1$. For a set consisting of $n$ points, the emanating point of a half-line can be smaller than any point, larger than any point, or between any two adjacent points. Each of these creates a distinct subset, and there are no other possibilities. Hence, $\xi(\mathcal{A}, n) = n + 1$. Now suppose $\mathcal{A}$ is the set of intervals in $\mathfrak{R}^1$. The endpoints of an interval can be such that the intersection in equation (4.47) is null, the interval contains a single point, or the interval contains more than one point. These three cases generate $1$, $n$, and $C(n, 2)$ sets, respectively, so that $\xi(\mathcal{A}, n) = 1 + n + C(n, 2)$.

**THEOREM 4.7 (Vapnik and Chervonenkis, 1971)** *For any probability measure $\nu$, set class $\mathcal{A}$, $n$, and $\tau > 0$,*

$$P(\hat{\Delta}_n(\mathcal{A}) > \tau) \leq 8\xi(\mathcal{A}, n)e^{-n\tau^2/8}. \tag{4.49}$$

Bounds on $P(\hat{\Delta}_n(\mathcal{C}) > \tau)$ follow at once by defining the shatter coefficient of a classifier class $\mathcal{C}$ to be $\Xi(\mathcal{C}, n) = \xi(\mathcal{A}_{\mathcal{C}}, n)$. From equation (4.46),

$$P(\hat{\Delta}_{n,\mathcal{C}} > \tau) \leq P(\hat{\Delta}_n(\mathcal{C}) > \tau/2) \leq 8\Xi(\mathcal{C}, n)e^{-n\tau^2/32}. \tag{4.50}$$

In general, if $Z$ is a nonnegative random variable such that $P(Z > \tau) \leq ce^{-2n\tau^2}$ for some constant $c$ and all $\tau > 0$, then $E[Z] \leq \sqrt{(\log ce)/2n}$. Hence, an immediate corollary of the Vapnik-Chervonenkis theorem is that

$$E[\hat{\Delta}_{n,\mathcal{C}}] \leq 8\sqrt{\frac{\log(8e\,\Xi(\mathcal{C}, n))}{2n}}. \tag{4.51}$$

It is not difficult to show that the shatter coefficient of $\mathcal{C}$ can be expressed via kernels: $\Xi(\mathcal{C}, n) = \xi(\mathcal{K}_{\mathcal{C}}, n)$, where $\mathcal{K}_{\mathcal{C}}$ is the class of all kernels for classifiers in $\mathcal{C}$ (Devroye et al., 1996). This emphasizes that the ability of a classifier class to separate points depends on the corresponding kernel class. This should be expected since there is a one-to-one relation between $\mathcal{K}_{\mathcal{C}}$ and $\mathcal{C}$.

### 4.3.2 VC Dimension

The shatter coefficient of a class of sets measures the extent to which the sets in the class can separate points for various point set sizes. The largest integer $n$ for which $\xi(\mathcal{A}, n) = 2^n$ is called the *Vapnik-Chervonenkis* (VC) *dimension* of a set class $\mathcal{A}$ and is denoted by $V_{\mathcal{A}}$. If $\xi(\mathcal{A}, n) = 2^n$ for all $n$, then $V_{\mathcal{A}} = \infty$. If its VC dimension is finite, then $\mathcal{A}$ is called a *VC class*. If $\mathcal{A}$ is finite, then $\xi(\mathcal{A}, k) \leq |\mathcal{A}|$ and $V_{\mathcal{A}} \leq \log_2 |\mathcal{A}|$.

Recalling the shatter coefficient of the set $\mathcal{A}$ of half-lines in $\mathfrak{R}^1$, $\xi(\mathcal{A}, n) = n + 1$. Thus, $\xi(\mathcal{A}, 1) = 2^1$, $\xi(\mathcal{A}, n) < 2^n$ for $n > 1$, and $V_\mathcal{A} = 1$. For the set $\mathcal{A}$ of intervals in $\mathfrak{R}^1$, $\xi(\mathcal{A}, n) = 1 + n + C(n, 2)$. Thus, $\xi(\mathcal{A}, 2) = 2^2$, $\xi(\mathcal{A}, n) < 2^n$ for $n > 2$, and $V_\mathcal{A} = 2$.

To show that $V_\mathcal{A} = n$, one can show that there exists a set with $n$ points shattered by $\mathcal{A}$ but that there does not exist a set containing $n + 1$ points shattered by $\mathcal{A}$. Using this approach, we can demonstrate that in $\mathfrak{R}^d$ the VC dimensions of the classes of quadrants and rectangles are $d$ and $2d$, respectively. For simplicity of notation, we restrict ourselves to $d = 2$, but the argument easily extends. Sets in the class $\mathcal{Q}$ of quadrants take the form $[-\infty, x_1] \times [-\infty, x_2]$. The two points $(-1, 1)$ and $(1, -1)$ are shattered by the quadrants with $(x_1, x_2) = (2, 2)$, $(0, 0)$, $(2, 0)$, and $(0, 2)$. To see that no three points can be shattered by $\mathcal{Q}$, let $\mathbf{z}_1$ be the point with the largest first coordinate, let $\mathbf{z}_2$ be the point with the largest second coordinate, and note that any quadrant that contains $\mathbf{z}_1$ and $\mathbf{z}_2$ must contain the third point. For the class $\mathcal{R}$ of rectangles, it is easy to see that the points $(-1, -1)$, $(1, 1)$, $(1, -1)$, and $(-1, 1)$ can be shattered by $\mathcal{R}$. For a set of five points, let $\mathbf{z}_1$ be the point with the largest first component, $\mathbf{z}_2$ be the point with the largest second component, $\mathbf{z}_3$ be the point with the smallest first component, and $\mathbf{z}_4$ be the point with the smallest second component. Any rectangle that contains $\mathbf{z}_1, \mathbf{z}_2, \mathbf{z}_3$, and $\mathbf{z}_4$ must contain the fifth point.

Various shatter coefficient bounds can be given in terms of the VC dimension. For instance, for any $n$,

$$\xi(\mathcal{A}, n) \leq \sum_{i=1}^{V_\mathcal{A}} \binom{n}{i} \tag{4.52}$$

(Vapnik and Chervonenkis, 1974). It follows immediately that $\xi(\mathcal{A}, n) \leq (n + 1)^{V_\mathcal{A}}$. It can also be shown that $\xi(\mathcal{A}, n) \leq n^{V_\mathcal{A}} + 1$ and that, for $V_\mathcal{A} > 2$, $\xi(\mathcal{A}, n) \leq n^{V_\mathcal{A}}$. In terms of classifier classes, the latter bound takes the form

$$\Xi(\mathcal{C}, n) \leq n^{V_\mathcal{C}}. \tag{4.53}$$

Thus, classifier shatter coefficients can be bounded in terms of the VC dimension of the class. Constraining the classifier class lowers the VC dimension.

From equations (4.51) and (4.42), for finite $V_\mathcal{C} > 2$,

$$E[\hat{\Delta}_{n,\mathcal{C}}] \leq 8\sqrt{\frac{V_\mathcal{C} \log n + 4}{2n}}. \tag{4.54}$$

Hence,

$$E[\hat{\Delta}_{n,\mathcal{C}}] = O\left(\sqrt{\frac{V_\mathcal{C} \log n}{n}}\right). \tag{4.55}$$

In fact, it can be shown that for $n\tau^2 \geq 64$,

$$E[\hat{\Delta}_{n,\mathcal{C}}] \leq \frac{8 + 32\sqrt{2V_\mathcal{C} \log(2^{12} V_\mathcal{C})}}{\sqrt{n}} \tag{4.56}$$

(Alexander, 1984). Taking $V_C$ out of $O(\sqrt{V_C \log n/n})$ shows that $E[\hat{\Delta}_{n,C}] = O(\sqrt{\log n/n})$, whereas the bound in equation (4.56) shows that $E[\hat{\Delta}_{n,C}] = O(\sqrt{1/n})$. There are two reasons for using $O(\sqrt{V_C \log n/n})$ convergence. First, it shows the role of the VC dimension. Second, the coefficient in the bound in equaton (4.56) is so huge that it is practically worse than the bound in equation (4.54).

While the $O(\sqrt{V_C \log n/n})$ convergence of $E[\hat{\Delta}_{n,C}]$ is encouraging relative to the reduction of design error for VC classes, there is another side to the story: the constraint cost of VC classes. Taken together with theorem 4.2 and equation (4.34), the bound in equation (4.54) shows that, for any $\tau > 0$, there exists a distribution of $(\mathbf{X}, Y)$ such that the Bayes error is 0 and

$$\frac{1}{2} - \frac{\tau}{2} - \varepsilon_C \le E[\hat{\Delta}_{n,C}] \le 8\sqrt{\frac{V_C \log n + 4}{2n}}. \tag{4.57}$$

For $n$ sufficiently large that the right-hand side is less than $\tau/2$, we obtain $\varepsilon_C \ge 1/2 - \tau$. Hence, any VC class can perform arbitrarily poorly.

Not only can the expected design cost be bounded above via the VC dimension, thereby guaranteeing performance, the VC dimension can also be used to bound the cost from below. In this case, minimax lower bounds show that the worst performance among a class of distributions exceeds some lower bound.

**THEOREM 4.8 (Devroye and Lugosi, 1995)** *Suppose $C$ has a finite VC dimension greater than $1$, $0 < \varepsilon_0 < 1/2$ and $\mathcal{D}$ is the class of all distributions of $(\mathbf{X}, Y)$ for which $\varepsilon_C = \varepsilon_0$. Then, for any classification rule,*

$$\sup_{(\mathbf{X},Y)\in\mathcal{D}} E[\hat{\Delta}_{n,C}] \ge \sqrt{\frac{\varepsilon_0(V_C - 1)}{24e^{16}n}} \tag{4.58}$$

*if $n \ge (V_C - 1)/2\varepsilon_0 \times \max(9, 1/(1 - 2\varepsilon_0)^2)$.*

This theorem gives a minimax lower bound for $E[\hat{\Delta}_{n,C}]$ as a constant times $n^{-1/2}$, while at the same time $E[\hat{\Delta}_{n,C}] = O(n^{-1/2})$. It also means that the sample size must significantly exceed the VC dimension to ensure good distribution-free performance.

Various bounds exist on the shatter coefficients and VC dimensions of classes formed from other classes. We summarize some of these. For a union of two classes,

$$\xi(\mathcal{A}_1 \cup \mathcal{A}_2, n) \le \xi(\mathcal{A}_1, n) + \xi(\mathcal{A}_2, n). \tag{4.59}$$

For any class $\mathcal{A}$, the complement class is defined by $\mathcal{A}^c = \{A^c : A \in \mathcal{A}\}$. Then, $\xi(\mathcal{A}^c, n) = \xi(\mathcal{A}, n)$. For classes $\mathcal{A}_1$ and $\mathcal{A}_2$, a binary set operation $*$ defines the class $\langle \mathcal{A}_1 * \mathcal{A}_2 \rangle = \{A_1 * A_2 : A_k \in \mathcal{A}_k\}$. For the operations of union, intersection, and Cartesian product,

$$\xi(\langle \mathcal{A}_1 * \mathcal{A}_2 \rangle, n) \le \xi(\mathcal{A}_1, n)\xi(\mathcal{A}_2, n). \tag{4.60}$$

Often there is no computationally feasible way to choose a classifier from $\mathcal{C}$ via the empirical error, in which case some algorithm can be used to approximate the best empirical error classifier. In general, for a classifier $\psi: \Re^d \to \{0, 1\}$, suppose the approximating algorithm yields the estimated classifier sequence $\{\hat{\phi}_{n,c}\}$ from the sample, $\{\tau_n\}$ and $\{\delta_n\}$ are positive sequences for which $\tau_n \to 0$ and $\delta_n \to 0$ as $n \to \infty$, and the empirical error for $\{\hat{\phi}_{n,c}\}$ gets close to the error of the optimal empirical error classifier in the sense that

$$P(\hat{\varepsilon}_n[\hat{\phi}_{n,c}] - \hat{\varepsilon}_{n,c} \le \tau_n) \ge 1 - \delta_n. \tag{4.61}$$

Then the Vapnik-Chervonenkis theorem can be modified to state

$$P(\varepsilon[\hat{\phi}_{n,c}] - \varepsilon_c > \tau) \le \delta_n + 8\Xi(\mathcal{C}, n)e^{-n(\tau - \tau_n)^2/128} \tag{4.62}$$

(Devroye et al., 1996).

## 4.4 LINEAR CLASSIFICATION

A classical way of constructing classifiers is to use parametric representation. The classifier is postulated to have a functional form $\psi(x_1, x_2, \ldots, x_d; a_0, a_1, \ldots, a_r)$, where the parameters $a_0, a_1, \ldots, a_r$ are to be determined by some estimation procedure based on the sample data. For parametric representation, we assume the labels to be $-1$ and $1$. The most basic functional form involves a linear combination of the coordinates of the observations. A binary function is obtained by thresholding. A *linear classifier* has the form

$$\psi(\mathbf{x}) = T\left[a_0 + \sum_{i=1}^{d} a_i x_i\right], \tag{4.63}$$

where $\mathbf{x} = (x_1, x_2, \ldots, x_d)$ and T thresholds at 0 and yields $-1$ or 1. A linear classifier divides the space into two half-spaces determined by the hyperplane defined by the parameters $a_0, a_1, \ldots, a_d$. The hyperplane is determined by the equation formed from setting the linear combination equal to 0. Using the dot product $\mathbf{a} \cdot \mathbf{x}$, which is equal to the sum in the preceding equation absent the constant term $a_0$, the hyperplane is defined by $\mathbf{a} \cdot \mathbf{x} = -a_0$. The classifier kernel is determined by the hyperplane. The VC dimension of the class of linear classifiers equals the VC dimension of the class of such half-spaces. The next theorem applies.

**THEOREM 4.9 (Cover, 1965)** *Let $\mathcal{G}$ be the linear space spanned by $m$ real-valued functions $g_1, g_2, \ldots, g_m$ on $\Re^d$, define $\mathbf{g} = (g_1, g_2, \ldots, g_m)$, and let $\mathcal{A}$ be the class of all zero-threshold sets of functions in $\mathcal{G}$, these being $\{\mathbf{z} : g(\mathbf{z}) \ge 0\}$, where $g \in \mathcal{G}$. If there exists a point set $\{\mathbf{z}_1, \mathbf{z}_2, \ldots, \mathbf{z}_n\} \subset \Re^d$ such that every $m$-element subset of $\{\mathbf{g}(\mathbf{z}_1), \mathbf{g}(\mathbf{z}_2), \ldots, \mathbf{g}(\mathbf{z}_n)\}$ is linearly independent, then*

$$\xi(\mathcal{A}, n) = 2\sum_{i=1}^{m-1} \binom{n-1}{i}. \tag{4.64}$$

Setting $n = m$ and $m + 1$ in the theorem yields $\xi(\mathcal{A}, m) = 2^m$ and $\xi(\mathcal{A}, m+1) = 2^{m+1} - 2$, respectively. Thus, $V_{\mathcal{A}} = m$. More generally,

$$\xi(\mathcal{A}, n) \leq 2(n-1)^{m-1} + 2. \tag{4.65}$$

Let $g_i(\mathbf{x}) = x_i$ for $i = 1, 2, \ldots, d$, and $g_{d+1}(\mathbf{x}) = 1$. Then $g \in \mathcal{G}$ if and only if $g$ is a linear combination of the form defining a linear classifier and the threshold sets in theorem 4.9 comprise the class of classifier kernels. Consequently, letting $\mathcal{L}$ denote the class of linear classifiers, $\Xi(\mathcal{L}, n)$ is given by equation (4.64) with $d$ replacing $m - 1$ as the upper limit of the sum,

$$\Xi(\mathcal{L}, n) \leq 2(n-1)^d + 2, \tag{4.66}$$

and $V_{\mathcal{L}} = d + 1$. Hence, $\Xi(\mathcal{L}, n) \leq n^{d+1}$ for $d > 1$. From the Vapnik-Chervonenkis theorem,

$$P(\hat{\Delta}_{n,\mathcal{L}} > \tau) \leq 8n^{d+1} e^{-n\tau^2/32}. \tag{4.67}$$

Numerous classification rules have been proposed for linear classifiers. Each finds parameters that hopefully define a linear classifier whose error is close to optimal. Sometimes, analysis of a design procedure depends on whether the sample data are *linearly separable*, meaning there exists a hyperplane such that the points with label $-1$ lie on one side of the hyperplane and the points with label 1 lie on the other side. Each design algorithm for linear classification is meant to achieve some advantage relative to other methods. Here we describe four methods, referring the reader to the literature for a more complete discussion of the many methods (Duda et al., 2001).

Before considering specific classification rules, we illustrate linear classification in the case of breast tumors from patients carrying mutations in one of the predisposing genes, BRCA1 or BRCA2, or from patients not expected to carry a hereditary predisposing mutation. Pathological and genetic differences appear to imply different but overlapping functions for BRCA1 and BRCA2, and in an early study involving expression-based classification, cDNA microarrays were used to show the feasibility of using differences in global gene expression profiles to separate BRCA1 and BRCA2 mutation-positive breast cancers (Hedenfalk et al., 2001). Based upon data from that study, figure 4.3 shows the BRCA1 tumors linearly separated from the BRCA2 tumors and sporadic tumors in the sample using genes KRT8 and DRPLA. What we can infer from this separation regarding classification between the tumor classes is a nontrivial issue concerning the method used to select the features KRT8 and DRPLA, the classification rule for constructing the separating line, how the error of the classifier is to be estimated, and the relationship between the estimated error based on the sample data and the true classifier error relative to the feature-label distribution.

### 4.4.1 Rosenblatt Perceptron

The classical *Rosenblatt perceptron algorithm* uses an iterative procedure to find a separating hyperplane for a linearly separable sample (Rosenblatt, 1962). Let

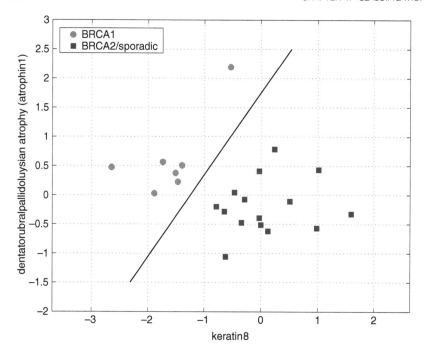

Figure 4.3. Linear classification of BRCA1 versus BRCA2 and sporadic patients. (Dougherty et al., 2005b.)

$\{(\mathbf{x}^1, y^1), (\mathbf{x}^2, y^2), \ldots, (\mathbf{x}^n, y^n)\}$ be a sample data set, $\mathbf{a}$ be a parameter vector defining the linear classifier, which in this context is called a *perceptron*, and $\{(\mathbf{x}_t, y_t)\}$ be a sequence of observation-label pairs formed by cycling repeatedly through the data. Proceed iteratively by forming a sequence $\mathbf{a}(0), \mathbf{a}(1), \mathbf{a}(2), \ldots$. Let $\mathbf{a}(0) = (0, 0, \ldots, 0)$. Given $\mathbf{a}(t)$, define $\mathbf{a}(t+1)$ by

$$\mathbf{a}(t+1) = \begin{cases} \mathbf{a}(t) & \text{if } \psi_t(\mathbf{x}_{t+1}) = y_{t+1}, \\ \mathbf{a}(t) + y_{t+1}\mathbf{x}_{t+1} & \text{if } \psi_t(\mathbf{x}_{t+1}) \neq y_{t+1}, \end{cases} \quad (4.68)$$

where $\psi_t$ is the perceptron formed using the parameter vector $\mathbf{a}(t)$. The parameter vector changes if and only if the label for time $t+1$ is not correctly given by applying the perceptron for time $t$ to the observation vector for time $t+1$. In finite time, the procedure produces a hyperplane that correctly separates the data. The time can be bounded in terms of the maximum norm of the observation vectors and the degree to which the 0-labeled and 1-labeled vectors are separated. For data sets that are not linearly separable, another method has to be employed to estimate the optimal perceptron. Gradient-based methods lead to consistent design under very general conditions.

### 4.4.2  Linear and Quadratic Discriminant Analysis

To this point the discussion has remained binary, with classification between $Y = 0$ and $Y = 1$. In this subsection and the next we will deviate because the matrix-based theory goes through without change and because the methods are commonly employed for multiclass discrimination.

The error of a binary classifier gives the probability of misclassification. This notion extends directly to a classifier $\psi$ taking values in $\{0, 1, \ldots, m\}$, its error being defined by

$$\varepsilon[\psi] = \sum_{k=0}^{m} P(Y = k, \xi(\mathbf{X}) \neq k)$$

$$= 1 - \sum_{k=0}^{m} P(Y = k, \xi(\mathbf{X}) = k)$$

$$= 1 - \sum_{k=0}^{m} \int_{R_k} f(\mathbf{x}, k) \, d\mathbf{x}, \qquad (4.69)$$

where $f(\mathbf{x}, k)$ is the feature-label density and $R_k = \{\mathbf{x} : \psi(\mathbf{x}) = k\}$. The error is minimized by choosing $\psi$ to maximize the latter sum, and this is accomplished if

$$R_k = \{\mathbf{x} : f(\mathbf{x}, k) \geq f(\mathbf{x}, j) \text{ for } j = 0, 1, \ldots, m\}. \qquad (4.70)$$

Hence, an optimal classifier is defined by $\psi_d(\mathbf{x}) = k$ if $f(\mathbf{x}, k) \geq f(\mathbf{x}, j)$ for $j = 0, 1, \ldots, m$. Relative to conditional probabilities, $\psi_d(\mathbf{x}) = k$ if $f(k|\mathbf{x}) \geq f(j|\mathbf{x})$ for $j = 0, 1, \ldots, m$ because $f(j|\mathbf{x}) = f(\mathbf{x}|j)/f(\mathbf{x})$ [assuming $f(\mathbf{x}) \neq 0$]. Vectors for which $\psi_d(\mathbf{x})$ is optimal define the decision boundaries in $\Re^d$. Here $\varepsilon_d = \varepsilon[\psi_d]$ is given by equation (4.69) using $R_k$ in equation (4.70).

It is common to view classification via *discriminant* functions: there are $m + 1$ functions, $d_0, d_1, \ldots, d_m$, and $\mathbf{x}$ is classified as $Y = k$ if $d_k(\mathbf{x}) \geq d_j(\mathbf{x})$ for $j = 0, 1, \ldots, m$. Misclassification error is minimized if we let $f(\mathbf{x}, k) = f(\mathbf{x}|k) f(k)$ be the discriminant. Since the order relation among discriminant functions is unaffected by any strictly increasing function, taking the logarithm yields the equivalent discriminant

$$d_k(\mathbf{x}) = \log f(\mathbf{x}|k) + \log f(k). \qquad (4.71)$$

If the conditional densities $f(\mathbf{x}|0), f(\mathbf{x}|1), \ldots, f(\mathbf{x}|n)$ are normally distributed, then, for $k = 0, 1, \ldots, m$, $f(\mathbf{x}|k)$ is of the form

$$f(\mathbf{x}|k) = \frac{1}{\sqrt{(2\pi)^n \det[\mathbf{K}_k]}} \exp\left[-\frac{1}{2}(\mathbf{x} - \mathbf{u}_k)^t \mathbf{K}_k^{-1}(\mathbf{x} - \mathbf{u}_k)\right], \qquad (4.72)$$

where $\mathbf{K}_k$ and $\mathbf{u}_k$ are the covariance matrix and mean vector, respectively. Inserting $f(\mathbf{x}|k)$ into equation (4.71), dropping the constant terms, and multiplying by the factor 2 (which has no effect on discrimination) yields

$$d_k(\mathbf{x}) = -(\mathbf{x} - \mathbf{u}_k)^t \mathbf{K}_k^{-1}(\mathbf{x} - \mathbf{u}_k) - \log(\det[\mathbf{K}_k]) + 2\log f(k). \qquad (4.73)$$

The form of this equation shows that the decision boundaries $d_k(\mathbf{x}) = d_j(\mathbf{x})$ are quadratic. The equation characterizes *quadratic discriminant analysis* (QDA).

If all conditional densities possess the same covariance matrix $\mathbf{K}$, then $\log(\det [\mathbf{K}_k])$ can be dropped from $d_k(\mathbf{x})$ and

$$d_k(\mathbf{x}) = -(\mathbf{x} - \mathbf{u}_k)^t \mathbf{K}^{-1}(\mathbf{x} - \mathbf{u}_k) + 2\log f(k), \tag{4.74}$$

which is a linear function of $\mathbf{x}$ and produces hyperplane decision boundaries. The discriminant characterizes *linear discriminant analysis* (LDA).

If, besides a common covariance matrix, we assume that the conditional random vectors are uncorrelated with covariance matrix $\mathbf{K} = \sigma^2 \mathbf{I}$, then the preceding equation reduces to

$$d_k(\mathbf{x}) = -\sigma^{-2}||\mathbf{x} - \mathbf{u}_k|| + 2\log f(k). \tag{4.75}$$

This form does not require estimation of the covariance matrix; however, uncorrelatedness is a strong assumption. If we go one step further and assume that the classes possess equal prior probabilities $f(k)$, then the logarithm term can be dropped to obtain $d_k(\mathbf{x}) = -\sigma^{-2}||\mathbf{x} - \mathbf{u}_k||$ and $\mathbf{x}$ is assigned to class $Y = k$ if its distance to the mean vector for $f(\mathbf{x}|k)$ is minimal.

Equations (4.73) and (4.74) for quadratic and linear discriminant analysis are derived under the Gaussian assumption but in practice can perform well as long as the class conditional densities are not too far from Gaussian and there are sufficient data to obtain good estimates of the relevant covariance matrices, the point being that the QDA and LDA classification rules involve finding the sample covariance matrices and sample means. Owing to the greater number of parameters to be estimated for QDA as opposed to LDA, one can proceed with smaller samples with LDA than with QDA; moreover, LDA appears to be more robust relative to the underlying assumptions than QDA (Wald and Kronmal, 1977).

### 4.4.3  Linear Discriminants Based on Least-Squares Error

An alternative to linear discriminant analysis for obtaining a linear classifier when the conditional densities are not Gaussian or the feature vectors do not possess a common covariance matrix is to take a least-squares approach. There will be $m + 1$ discriminant functions of the form

$$d_k(\mathbf{x}) = \sum_{i=0}^{d} a_{ki} x_i, \tag{4.76}$$

where $\mathbf{x} = (x_0, x_1, \ldots, x_n)^t$, the nonhomogeneous term arising by setting $x_0 = 1$. Letting

$$\mathbf{A} = \begin{pmatrix} a_{00} & a_{01} & \cdots & a_{0d} \\ a_{10} & a_{11} & \cdots & a_{1d} \\ \vdots & \vdots & \ddots & \vdots \\ a_{m0} & a_{m1} & \cdots & a_{md} \end{pmatrix} \tag{4.77}$$

and $\mathbf{d}(\mathbf{x}) = (d_0(\mathbf{x}), d_1(\mathbf{x}), \ldots, d_m(\mathbf{x}))^t$, the discriminant functions can be expressed as

$$\mathbf{d}(\mathbf{x}) = \mathbf{A}\mathbf{x}. \qquad (4.78)$$

An observed vector $\mathbf{x}$ is assigned to $Y = k$ if $d_k(\mathbf{x})$ is maximum among discriminant values.

Instead of finding a matrix $\mathbf{A}$ to minimize misclassification error among all possible weight matrices, we can take a sum-of-squares error approach based on target values for the discriminant functions and then select $\mathbf{A}$ to minimize the sum-of-squares error. Since $\mathbf{x}$ is classified as $Y = k$ if $d_k(\mathbf{x})$ is maximum among discriminant values, for $k = 1, 2, \ldots, m$, we choose target values $b_{k1}, b_{k2}, \ldots, b_{kn}$ for $d_k$ corresponding to the sample pairs $(\mathbf{x}_1, y_1), (\mathbf{x}_2, y_2), \ldots, (\mathbf{x}_n, y_n)$ according to $b_{kj} = 1$ if $j = k$ and $b_{kj} = 0$ if $j \neq k$. This reflects the goal that $d_k(\mathbf{x}_j) = 1$ if $j = k$ and $d_k(\mathbf{x}_j) = 0$ if $j \neq k$. Define the *sum-of-squares error* by

$$
\begin{aligned}
e &= \sum_{k=0}^{m} \sum_{j=1}^{n} |d_k(\mathbf{x}_j) - b_{kj}|^2 \\
&= \sum_{k=0}^{m} \sum_{j=1}^{n} \left( \sum_{i=0}^{d} a_{ki} x_{ji} - b_{kj} \right)^2, \qquad (4.79)
\end{aligned}
$$

where $\mathbf{x}_j = (x_{j0}, x_{j1}, \ldots, x_{jn})^t$, with $x_{j0} = 1$. To minimize $e$ with respect to the coefficients, take partial derivatives with respect to $a_{ki}$ and set $\partial e / \partial a_{ki} = 0$ to obtain the equations

$$\sum_{j=1}^{n} \left( \sum_{l=0}^{d} a_{kl} x_{jl} - b_{kj} \right) x_{ji} = 0 \qquad (4.80)$$

for $k = 0, 1, \ldots, m$ and $i = 0, 1, \ldots, d$. Letting

$$\mathbf{H} = \begin{pmatrix} x_{10} & x_{11} & \cdots & x_{1d} \\ x_{20} & x_{21} & \cdots & x_{2d} \\ \vdots & \vdots & \ddots & \vdots \\ x_{n0} & x_{n1} & \cdots & x_{nd} \end{pmatrix} \qquad (4.81)$$

and

$$\mathbf{B} = \begin{pmatrix} b_{01} & b_{02} & \cdots & b_{0n} \\ b_{11} & b_{12} & \cdots & b_{1n} \\ \vdots & \vdots & \ddots & \vdots \\ b_{m1} & b_{m2} & \cdots & b_{mn} \end{pmatrix}, \qquad (4.82)$$

the system in equation (4.80) can be expressed as

$$\mathbf{H}^t \mathbf{H} \mathbf{A}^t = \mathbf{H}^t \mathbf{B}^t, \qquad (4.83)$$

so that the least squares weights are determined via the pseudoinverse of $\mathbf{H}$ by

$$\mathbf{A}^t = \mathbf{H}^+ \mathbf{B}^t = (\mathbf{H}'\mathbf{H})^{-1}\mathbf{H}'\mathbf{B}^t. \tag{4.84}$$

Existence of the pseudoinverse $\mathbf{H}^+$ is ensured if $\mathbf{H}'\mathbf{H}$ is invertible, which is itself ensured if the columns of $\mathbf{H}$ are linearly independent. If the sample is sufficiently large that $n$ significantly exceeds $d$, then the columns of $\mathbf{H}$ can be expected to be linearly independent. If $\mathbf{H}$ is square, then $\mathbf{H}^+ = \mathbf{H}^{-1}$ and equation (4.84) reduces to $\mathbf{A}^t = \mathbf{H}^{-1}\mathbf{B}^t$ or $\mathbf{B}^t = \mathbf{H}\mathbf{A}^t$. Componentwise, this means that

$$b_{kj} = \sum_{i=0}^{d} a_{ki} x_{ji} \tag{4.85}$$

and $e = 0$. If $\mathbf{H}$ is square and singular, or if $r < n$, then the least-squares system is underdetermined and there exists a class of coefficient matrices all giving zero sum-of-squares error. Along with the invertible case, these solutions are not consequential owing to the paucity of training observations.

For binary classification, the least-squares solution can be viewed in terms of the Wiener filter by expressing the error in equation (4.1) as a mean square error,

$$\varepsilon[\psi] = E[|Y - \psi(\mathbf{X})|^2]. \tag{4.86}$$

If we take an optimization perspective relative to the feature-label distribution and apply the Wiener filter theory, then the optimal nonhomogeneous mean-square-error linear estimator of $Y$ based on $\mathbf{X}$ is given by a weight vector $\mathbf{a} = (a_0, a_1, \dots, a_d)^t$ determined by the autocorrelation matrix for the augmented vector $\mathbf{X}^{(1)} = (1, \mathbf{X}^t)^t$ and the cross-correlation vector for $\mathbf{X}^{(1)}$ and $Y$. These are given by $\mathbf{R}_{\mathbf{X}^{(1)}} = E[\mathbf{X}^{(1)}(\mathbf{X}^{(1)})^t]$ and $E[\mathbf{X}^{(1)}Y]$, respectively. According to the Wiener theory, the optimal weight vector is given by $\mathbf{a} = \mathbf{R}_{\mathbf{X}^{(1)}}^{-1} E[\mathbf{X}^{(1)}Y]$ if $\mathbf{R}_{\mathbf{X}^{(1)}}$ is nonsingular (which we assume). A linear classifier is defined by $T[\mathbf{a}^t \mathbf{X}^{(1)}]$, where T is the zero-threshold function.

Application requires estimation of the autocorrelation matrix and the cross-correlation vector. The most commonly employed estimations are the sample autocorrelation matrix $\hat{\mathbf{R}}_{\mathbf{X}^{(1)}}$ and the sample cross-correlation vector $\hat{E}[\mathbf{X}^{(1)}Y]$. These are expressed by

$$\hat{\mathbf{R}}_{\mathbf{X}^{(1)}} = \frac{1}{n}\mathbf{H}'\mathbf{H}, \tag{4.87}$$

$$\hat{E}[\mathbf{X}^{(1)}Y] = \frac{1}{n}\mathbf{H}'\mathbf{y}, \tag{4.88}$$

where $\mathbf{y} = (y_1, y_2, \dots, y_n)^t$. Hence, the coefficient vector for the Wiener filter is given by

$$\mathbf{a} = \mathbf{H}^+ \mathbf{y}. \tag{4.89}$$

In the binary setting equation (4.84) takes the form

$$\mathbf{A}^t = \mathbf{H}^+(\mathbf{b}_0, \mathbf{b}_1), \tag{4.90}$$

where $(\mathbf{b}_0, \mathbf{b}_1)$ is a two-column matrix whose columns are the first two columns of $\mathbf{B}^t$. Write $\mathbf{H}^+$ in terms of its rows using the vectors $\mathbf{h}_0, \mathbf{h}_1, \ldots, \mathbf{h}_d$. In this way,

$$\mathbf{H}^+ = \begin{pmatrix} \mathbf{h}_0^t \\ \mathbf{h}_1^t \\ \vdots \\ \mathbf{h}_d^t \end{pmatrix}. \tag{4.91}$$

If the classifier $\psi_w$ is designed according to the thresholded Wiener filter, then $\psi_w(\mathbf{x}) = 1$ for $\mathbf{x} = (x_1, x_2, \ldots, x_n)^t$ if and only if

$$\sum_{k=0}^{d} \mathbf{h}_k^t \mathbf{y} x_k > 0. \tag{4.92}$$

If $\psi_d$ is designed according to least squares, then $\psi_d(\mathbf{x}) = 1$ if and only if

$$\sum_{k=0}^{d} \mathbf{h}_k^t (\mathbf{b}_1 - \mathbf{b}_0) x_k > 0. \tag{4.93}$$

Since $\mathbf{y} = \mathbf{b}_1 - \mathbf{b}_0$, the two classifiers are identical.

### 4.4.4 Support Vector Machines

The *support vector machine* (SVM) provides another method for designing linear classifiers (Vapnik, 1998). Figure 4.4 shows a linearly separable data set and three hyperplanes (lines). The outer lines pass through points in the sample data, and the third, called the *maximal-margin hyperplane* (MMH) is equidistant between the outer lines. It has the property that the distance from it to the nearest $-1$-labeld vector is equal to the distance from it to the nearest 1-labeled vector. The vectors closest to it are called *support vectors*. The distance from the MMH to any support vector is called the *margin*. The matter is formalized by recognizing that differently labeled sets are separable by the hyperplane $\mathbf{u} \cdot \mathbf{x} = c$, where $\mathbf{u}$ is a unit vector and $c$ is a constant, if $\mathbf{u} \cdot \mathbf{x}_k > c$ for $y_k = 1$ and $\mathbf{u} \cdot \mathbf{x}_k < c$ for $y_k = -1$. For any unit vector $\mathbf{u}$, the margin is defined by

$$\rho(\mathbf{u}) = \frac{1}{2} \left( \min_{\{\mathbf{x}_k : y_k = 1\}} \mathbf{u} \cdot \mathbf{x}_k - \max_{\{\mathbf{x}_k : y_k = -1\}} \mathbf{u} \cdot \mathbf{x}_k \right). \tag{4.94}$$

The MMH, which is unique, can be found by solving the following quadratic optimization problem: among the set of all vectors $\mathbf{v}$ for which there exists a constant $b$ such that

$$\begin{aligned} \mathbf{v} \cdot \mathbf{x}_k + b &\geq 1 & \text{if } y_k = 1, \\ \mathbf{v} \cdot \mathbf{x}_k + b &\leq -1, & \text{if } y_k = -1, \end{aligned} \tag{4.95}$$

find the vector of minimum norm, $||\mathbf{v}||$. If $\mathbf{v}_0$ satisfies this optimization problem, then the vector defining the MMH and the margin are given by $\mathbf{u}_0 = \mathbf{v}_0/||\mathbf{v}_0||$ and $\rho(\mathbf{u}_0) = ||\mathbf{v}_0||^{-1}$, respectively.

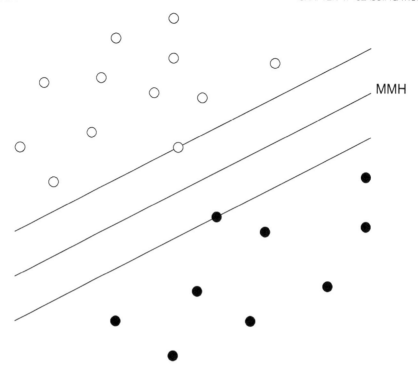

Figure 4.4.  Support vector hyperplane. (Dougherty and Attoor, 2002.)

We can approach this optimization via the method of Lagrange multipliers. If we rewrite the constraint in equation (4.95) as

$$y_k(\mathbf{v} \cdot \mathbf{x}_k + b) \geq 1 \tag{4.96}$$

for $k = 1, 2, \ldots, d$, then finding the MMH involves minimizing the quadratic form $||\mathbf{v}_0||^2$ subject to the constraints in equation (4.96). Using Lagrange multipliers, this minimization involves finding the saddle point of the Lagrange function,

$$L(\mathbf{v}, b, \mathbf{a}) = \frac{||\mathbf{v}||^2}{2} - \sum_{k=1}^{n} a_k(y_k(\mathbf{v} \cdot \mathbf{x}_k + b) - 1), \tag{4.97}$$

where $\mathbf{a} = (a_1, a_2, \ldots, a_n)^t$ is the vector of (nonnegative) Lagrange multipliers. Taking partial derivatives with respect to $\mathbf{v}$ and $b$, setting these equal to 0, and solving yields

$$\mathbf{v} = \sum_{k=1}^{n} y^k a_k \mathbf{x}_k, \tag{4.98}$$

$$\sum_{k=1}^{n} y^k a_k = 0. \tag{4.99}$$

Substituting into $L(\mathbf{v}, b, \mathbf{a})$ yields the function

$$L(\mathbf{a}) = \sum_{k=1}^{n} a_k + \frac{1}{2} \sum_{i,j=1}^{n} y_i y_j a_i a_j (\mathbf{x}_i \cdot \mathbf{x}_j), \tag{4.100}$$

where $L$ is now reduced to a function of $\mathbf{a}$. Maximizing $L(\mathbf{a})$ over $a_1, a_2, \ldots, a_n$ yields the solution

$$\mathbf{v}_0 = \sum_{k=1}^{n} y_k a_k^0 \mathbf{x}_k \tag{4.101}$$

for the optimizing values $a_1^0, a_2^0, \ldots, a_n^0$, and $b_0$ is found by maximizing the margin. The optimal hyperplane is defined by

$$0 = \mathbf{x} \cdot \mathbf{v}_0 + b_0 = \sum_{k=1}^{n} y^k a_k^0 \mathbf{x} \cdot \mathbf{x}_k + b_0. \tag{4.102}$$

Note that $L(\mathbf{a})$ and the hyperplane depend only on the vector inner products, not on the vectors themselves. Moreover, the form of the hyperplane is greatly simplified because it can be shown that the nonzero values of $a_1^0, a_2^0, \ldots, a_n^0$ correspond to the support vectors among $\mathbf{x}_1, \mathbf{x}_2, \ldots, \mathbf{x}_n$.

If the sample is not linearly separable, then one has two choices: try to find a good linear classifier or find a nonlinear classifier. In the first case, the preceding method can be modified (Vapnik, 1998). A second approach is to map the sample points into a higher-dimensional space, find a hyperplane in that space, and then map back into the original space. We will restrict our attention here to the latter approach. The basic idea is to map $\mathbf{x}_1, \mathbf{x}_2, \ldots, \mathbf{x}_n$ from $\Re^d$ into the Hilbert space $l_2$, which consists of all infinite numerical sequences $\mathbf{w} = \{w_1, w_2, \ldots\}$ such that

$$||\mathbf{w}||^2 = \sum_{k=1}^{\infty} w_k^2 \tag{4.103}$$

is finite (here considering only real-valued sequences) and the inner product is given by

$$\mathbf{w}_1 \cdot \mathbf{w}_2 = \sum_{k=1}^{\infty} w_{1k} w_{2k}. \tag{4.104}$$

The mapping procedure depends on Mercer's theorem.

**THEOREM 4.10 (Mercer)** *If $K(\mathbf{u}, \mathbf{v})$ is a real-valued, continuous, symmetric function on a compact set $C \times C \subset \Re^{2d}$ such that for all functions $g \in L_2(C)$,*

$$\int_C \int_C K(\mathbf{u}, \mathbf{v}) g(\mathbf{u}) g(\mathbf{v}) \, d\mathbf{u} \, d\mathbf{v} > 0, \tag{4.105}$$

*then there exist functions $z_1, z_2, \ldots$ defined on $C$ and positive constants $c_1, c_2, \ldots$*
*such that*

$$K(\mathbf{u}, \mathbf{v}) = \sum_{k=1}^{\infty} c_k z_k(\mathbf{u}) z_k(\mathbf{v}). \tag{4.106}$$

Assuming that $\mathbf{x}_1, \mathbf{x}_2, \ldots, \mathbf{x}_n \in C$, the desired mapping $C \to l_2$ is defined by

$$z(\mathbf{u}) = (\sqrt{c_1}\mathbf{z}_1(\mathbf{u}), \sqrt{c_2}\mathbf{z}_2(\mathbf{u}), \sqrt{c_3}\mathbf{z}_3(\mathbf{u}), \ldots). \tag{4.107}$$

According to the definition of the inner product in $l_2$, for any $\mathbf{u}, \mathbf{v} \in C$,

$$z(\mathbf{u}) \cdot z(\mathbf{v}) = K(\mathbf{u}, \mathbf{v}). \tag{4.108}$$

Map the data to $z(\mathbf{x}_1), z(\mathbf{x}_2), \ldots, z(\mathbf{x}_n) \in l_2$ and consider the new feature-label set $\{(z(\mathbf{x}_1), y_1), (z(\mathbf{x}_2), y_2), \ldots, (z(\mathbf{x}_n), y_n)\}$, where the labels are those associated with the original points. If the new points are linearly separable in $l_2$, then we can apply the original procedure to find the separating hyperplane, except that now the optimization problem in equation (4.100) takes the form

$$L(\mathbf{a}) = \sum_{k=1}^{n} a_k + \frac{1}{2} \sum_{i,j=1}^{n} y_i y_j a_i a_j K(\mathbf{x}_i, \mathbf{x}_j) \tag{4.109}$$

because the inner product $\mathbf{x}_i \cdot \mathbf{x}_j$ in equation (4.100) is replaced by $z(\mathbf{x}_i) \cdot z(\mathbf{x}_j)$. The separating hyperplane in $l_2$ defines a nonlinear decision boundary in $\mathfrak{R}^d$. If the points are not separable in $l_2$, then one can apply the non-linear-separable procedure.

The choice of the kernel $K(\mathbf{u}, \mathbf{v})$ is crucial, and different choices have been applied. A popular choice is the *polynomial support vector machine*, in which case

$$K(\mathbf{u}, \mathbf{v}) = [(\mathbf{u} \cdot \mathbf{v}) + 1]^d. \tag{4.110}$$

For the polynomial SVM, the coordinates for the mapping $z(\mathbf{u})$ consist of all products $u_{i_1} u_{i_2} \cdots u_{i_d}$ among the coordinates of $\mathbf{u}$.

### 4.4.5 Representation of Design Error for Linear Discriminant Analysis

Historically, there has been a tendency to focus on sample-size and dimensionality issues for quadratic or linear classification (El-Sheikh and Wacker, 1980; Raudys and Pikelis, 1980). Here we consider approximate analytic expressions for the design error $\Delta_{n,C}$ when $C$ is based on a parametric representation, with particular attention paid to linear discriminant analysis (Fukunaga and Hayes, 1989).

We consider a classifier of the form $\psi(\mathbf{x}; \mathbf{a})$, where $\mathbf{a} = (a_1, a_1, \ldots, a_r)^t$ is a parameter to be determined by some estimation procedure based on a sample. Hence, the designed classifier is of the form $\psi(\mathbf{x}; \hat{\mathbf{a}})$, where $\hat{\mathbf{a}} = (\hat{a}_1, \hat{a}_2, \ldots, \hat{a}_r)^t$. Since the optimal classifier is defined via a coefficient vector $\mathbf{a}_d$, $\Delta_{n,C}$ depends on

$\Delta \mathbf{a} = \hat{\mathbf{a}} - \mathbf{a}_d$, which implicitly depends on the sample size $n$. The basic approximation, obtained by Fourier transform methods, is given by

$$E[\Delta_{n,c}] \approx \frac{1}{4\pi} \int_{-\infty}^{\infty} \int_{\Re^d} \sum_{i=1}^{r} \sum_{k=1}^{r} \left( \frac{\partial^2 \psi(\mathbf{x}; \mathbf{a})}{\partial a_i \partial a_k} + j\omega \frac{\partial \psi(\mathbf{x}; \mathbf{a})}{\partial a_i} \frac{\partial \psi(\mathbf{x}; \mathbf{a})}{\partial a_k} \right)$$

$$\times E[\Delta a_i \Delta a_k] e^{j\omega \psi(\mathbf{x};\mathbf{a})} [f(\mathbf{x}, 0) - f(\mathbf{x}, 1)] d\mathbf{x} d\omega, \quad (4.111)$$

where $f(\mathbf{x}, y)$ is the feature-label density. The effect of the sample size appears in the term $E[\Delta a_i \Delta a_k]$. The expression inside the integral depends on the form of the classifier relative to the parameter vector, which contributes to the partial derivatives and the classification rule since the latter impacts $E[\Delta a_i \Delta a_k]$. The approximation is good only when the deviations $|a_i - \hat{a}_i|$ are small and therefore cannot be assumed good for small-sample estimation.

For linear discriminant analysis with Gaussian conditional distributions $N(\mathbf{0}, \mathbf{I})$ and $N(\mathbf{m}, \mathbf{I})$, equal label probabilities, and estimation using the sample moments, a fairly straightforward (but tedious) evaluation of the partial derivatives reduces the general approximation to

$$E[\Delta_{n,c}] \approx \frac{1}{n} \frac{e^{-\mathbf{m}'\mathbf{m}/8}}{4\sqrt{2\pi \mathbf{m}'\mathbf{m}}} \left[ d^2 + \left( 1 + \frac{\mathbf{m}'\mathbf{m}}{2} \right) d + \left( \frac{(\mathbf{m}'\mathbf{m})^2}{16} - \frac{\mathbf{m}'\mathbf{m}}{2} - 1 \right) \right].$$

$$(4.112)$$

The accuracy of this expression depends on the sample size and the number of features. Because $E[\Delta_{n,c}]$ is approximately of the order $d^2/n$, the sample size must significantly exceed the square of the number of features as the latter grows. Quantitatively, the approximation yields $E[\Delta_{n,c}] \approx O(d^2/n)$. This approximation is quite consistent with the VC dimension, which in this case is $d$. Indeed, according to the bound in equation (4.54), $E[\hat{\Delta}_{n,c}] = O(\sqrt{d \log n/n})$, and for sufficiently large $n$,

$$\frac{d^2}{n} \leq \frac{d \log n}{n} = \sqrt{\frac{d \log n}{n}} \sqrt{\frac{d \log n}{n}} \leq \sqrt{\frac{d \log n}{n}}. \quad (4.113)$$

Not only is $O(d^2/n)$ bounded by $O(\sqrt{d \log n/n})$, but in fact convergence is much faster in the present setting, with

$$\lim_{n \to \infty} \frac{d^2/n}{\sqrt{d \log n/n}} = 0. \quad (4.114)$$

This demonstrates the looseness of the distribution-free VC bound in comparison to convergence in the case of a simple mixture of Gaussian distributions.

### 4.4.6 Distribution of a QDA Sample-Based Discriminant

For quadratic discriminant analysis, the class decision depends on the discriminants defined in equation (4.73). In the two-class problem with equal prior class probabilities $f(0) = f(1) = 0.5$, the two discriminants, $d_0$ and $d_1$, can be combined to provide the unified discriminant,

$$Q(\mathbf{x}) = (\mathbf{x} - \mathbf{u}_0)^t \mathbf{K}_0^{-1}(\mathbf{x} - \mathbf{u}_0) - (\mathbf{x} - \mathbf{u}_1)^t \mathbf{K}_1^{-1}(\mathbf{x} - \mathbf{u}_1) + \log \frac{\det \mathbf{K}_0}{\det \mathbf{K}_1}, \qquad (4.115)$$

with $\mathbf{x}$ being put into class 1 if $Q(\mathbf{x}) > 0$. Using parameter estimation, the estimate $\hat{Q}$ of $Q$ is obtained from a random sample by replacing the means and covariance matrices by their respective sample means and sample covariance matrices. The unbiasedness of these estimators ensures good estimation for large samples but not for small samples. For the latter, it is beneficial to know the distribution of $\hat{Q}$ in terms of the numbers of observations, $n_0$ and $n_1$, of each class in the sample and the model parameters $\mathbf{u}_0$, $\mathbf{u}_1$, $\mathbf{K}_0$, and $\mathbf{K}_1$.

The theorem concerning the distribution of $\hat{Q}$ involves a number of mutually independent random variables: $X_0$ and $X_1$ possess chi-square distributions with $(n_0 - d)$ and $(n_1 - d)$ degrees of freedom, respectively; $Z_{01}, Z_{02}, \ldots, Z_{0d}, Z_{11}, Z_{12}, \ldots, Z_{1d}$ possess standard normal distributions; and $F_1, F_2, \ldots, F_{d-1}$ possess F distributions with $F_k$ having $(n_0 - k, n_1 - k)$ degrees of freedom.

A number of different parameters are involved. Decompose the matrix $\mathbf{K}_0^{-1/2} \mathbf{K}_1 \mathbf{K}_0^{-1/2}$ by

$$\mathbf{K}_0^{-1/2} \mathbf{K}_1 \mathbf{K}_0^{-1/2} = \mathbf{H} \mathbf{D} \mathbf{H}^t, \qquad (4.116)$$

where $\mathbf{H}$ is an orthogonal matrix and $\mathbf{D}$ is a diagonal matrix with the eigenvalues $\lambda_1, \lambda_2, \ldots, \lambda_d$ of $\mathbf{K}_0^{-1} \mathbf{K}_1$ running down its main diagonal. Define the column vector

$$\mathbf{u} = (u_1, u_2, \ldots, u_d)^t = \mathbf{H}^t \mathbf{K}_0^{-1/2}(\mathbf{u}_1 - \mathbf{u}_0). \qquad (4.117)$$

For $k = 1, 2, \ldots, d$, define the parameters

$$w_{0k} = \left( \frac{n_0 n_1}{(n_0 + 1)(n_1 + \lambda_k)} \right)^{1/2},$$

$$w_{1k} = \left( \frac{n_0 n_1 \lambda_k}{(\lambda_k n_0 + 1)(n_1 + 1)} \right)^{1/2},$$

$$c_{0k} = (1 + \lambda_k n_1^{-1})^{-1/2} u_k, \qquad (4.118)$$

$$c_{1k} = (\lambda_k + n_0^{-1})^{-1/2} u_k,$$

and the random variables

$$V_1 = \frac{(n_1 - 1)(n_1 + 1)}{n_1 X_1},$$

$$V_{1k} = \frac{(n_0 - 1)(\lambda_k + n_0^{-1})}{X_0},$$

$$V_0 = \frac{(n_0 - 1)(n_0 + 1)}{n_0 X_0}, \qquad (4.119)$$

$$V_{0k} = \frac{(n_1 - 1)(\lambda_k^{-1} + n_1^{-1})}{X_1}.$$

The stochastic representations for the conditional distributions of $\hat{Q}$ given $Y = 0$ and $\hat{Q}$ given $Y = 1$, describing the random variables $\hat{Q}_0$ and $\hat{Q}_1$, respectively, are given in the next theorem.

**THEOREM 4.11 (McFarland and Richards, 2002)** *In terms of the preceding description of the QDA discriminant, the estimated discriminant possesses the representations*

$$\hat{Q}_1 = \sum_{k=1}^{d} [V_{1k}(w_{1k}Z_{1k} + (1 - w_{1k}^2)^{1/2}Z_{0k} + c_{1k})^2 - V_1 Z_{1k}^2]$$

$$+ \log \frac{(n_1 - 1)^d X_0}{(n_0 - 1)^d X_1} - \log(\det \mathbf{K}_0^{-1} \mathbf{K}_1) + \sum_{k=1}^{d-1} \log\left(\frac{n_0 - k}{n_1 - k} F_k\right); \quad (4.120)$$

$$\hat{Q}_0 = \sum_{k=1}^{d} [V_0 Z_{1k}^2 - V_{0k}(w_{0k}Z_{1k} + (1 - w_{0k}^2)^{1/2}Z_{0k} + c_{0k})^2]$$

$$+ \log \frac{(n_1 - 1)^d X_0}{(n_0 - 1)^d X_1} - \log(\det \mathbf{K}_0^{-1} \mathbf{K}_1) + \sum_{k=1}^{d-1} \log\left(\frac{n_0 - k}{n_1 - k} F_k\right). \quad (4.121)$$

The error of classification by $\hat{Q}$ is given by

$$\varepsilon[\hat{Q}] = \frac{1}{2}[P(\hat{Q}(\mathbf{X}) \leq 0 | \mathbf{X} \in C_1) + P(\hat{Q}(\mathbf{X}) > 0 | \mathbf{X} \in C_0)]. \quad (4.122)$$

This can be estimated via Monte Carlo simulation using the representation in the theorem.

## 4.5 NEURAL NETWORK CLASSIFIERS

As a class, neural networks are universally strongly consistent, but they require a large number of neurons to be close to the Bayes classifier, and this in turn requires very large samples. As a classifier a (feed-forward) two-layer *neural network* has the form

$$\psi(\mathbf{x}) = T\left[c_0 + \sum_{i=1}^{k} c_i \sigma[\psi_i(\mathbf{x})]\right], \quad (4.123)$$

where T thresholds at $1/2$, $\sigma$ is a sigmoid function, and

$$\psi_i(\mathbf{x}) = a_{i0} + \sum_{j=1}^{d} a_{ij}x_j. \quad (4.124)$$

The sigmoid function is nondecreasing, with limits $-1$ and $+1$ at $-\infty$ and $\infty$, respectively. Each operator in the sum in equation (4.123) is a *neuron*. These form the *hidden layer*. We consider neural networks with the threshold sigmoid [$\sigma(x) = -1$ if $x \leq 0$ and $\sigma(x) = 1$ if $x > 0$]. $\mathcal{N}_k$ denotes this class. $\mathcal{L}_k$ denotes the class of functions of the form given in equation (4.123) prior to thresholding by T. Our focus is on the error decomposition

$$\hat{\varepsilon}_{n,\mathcal{N}_k} = \varepsilon_d + \Delta_{\mathcal{N}_k} + \hat{\Delta}_{n,\mathcal{N}_k}. \tag{4.125}$$

Our first concern is convergence of the constraint cost $\Delta_{\mathcal{N}_k}$. We consider this from a general perspective. Given a sequence of real-valued function classes $\mathcal{F}_k$ on $\mathfrak{R}^d$, a sequence of classifier classes $\mathcal{C}_k$ is formed by thresholding. Conditions are needed on $\mathcal{F}_k$ to ensure convergence of $\Delta_{\mathcal{C}_k}$ as $k \to \infty$. The denseness of the $\mathcal{F}_k$ in larger function spaces with respect to the $L_\infty$ and $L_1$ norms is key. By definition, $\mathcal{F}_k$ becomes dense in the normed space $\mathcal{H}$ if for every $g \in \mathcal{H}$,

$$\lim_{k \to \infty} \inf_{\phi \in \mathcal{F}_k} \|g - \phi\| = 0. \tag{4.126}$$

First consider the $L_1$ norm. For $\psi = T[\phi]$, equation (4.13) yields

$$\varepsilon[\psi] - \varepsilon_d \leq 2E[|\psi(\mathbf{x}) - \eta(\mathbf{x})|] = 2 \int_{\mathfrak{R}^d} |\psi(\mathbf{x}) - \eta(\mathbf{x})| f_{\mathbf{x}}(\mathbf{x}) \, d\mathbf{x}. \tag{4.127}$$

Hence, $\Delta_{\mathcal{C}_k} \to 0$ as $k \to \infty$ if there exists a sequence $\{\phi_k\}$ in $\mathcal{F}_k$ such that $2E[|\phi_k(\mathbf{x}) - \eta(\mathbf{x})|] \to 0$ as $k \to \infty$. This is assured if the classes $\mathcal{F}_k$ become dense in $L_1[f_{\mathbf{x}}]$. In fact, the function class $\mathcal{L}_k$ generating the class $\mathcal{N}_k$ of neural networks does become dense in $L_1[f_{\mathbf{x}}]$ for any distribution of $\mathbf{X}$, so that we can conclude that $\Delta_{\mathcal{N}_k} \to 0$ as $k \to \infty$ (Hornik, 1991).

For the $L_\infty$ norm, if for any $[\mathbf{a}, \mathbf{b}]^d$ the classes $\mathcal{F}_k$ become dense in the space $C([\mathbf{a}, \mathbf{b}]^d)$ of continuous functions on $[\mathbf{a}, \mathbf{b}]^d$, then it is not difficult to show that $\Delta_{\mathcal{C}_k} \to 0$ as $k \to \infty$ for any distribution of $(\mathbf{X}, Y)$. For $\delta > 0$, there exists a cube $[\mathbf{a}, \mathbf{b}]^d$ such that

$$P(\mathbf{X} \notin [\mathbf{a}, \mathbf{b}]^d) < \delta \tag{4.128}$$

and $\gamma \in C([\mathbf{a}, \mathbf{b}]^d)$ such that $|\gamma| \leq 1$ and

$$E[|\gamma(\mathbf{x}) - \eta(\mathbf{x})|] < \delta. \tag{4.129}$$

Because the $\mathcal{F}_k$ become dense in $C([\mathbf{a}, \mathbf{b}]^d)$, there exist $k$ and $\phi \in \mathcal{F}_k$ such that $\|\phi - \gamma\|_\infty < \delta$. For $\psi = T[\phi]$, equation (4.13) and the triangle inequality yield

$$\varepsilon[\psi] - \varepsilon_d \leq 2E[|\gamma(\mathbf{X}) - \eta(\mathbf{X})| I_{\{\mathbf{X} \in [\mathbf{a}, \mathbf{b}]^d\}}] + 2\delta$$

$$\leq 2E[|\gamma(\mathbf{X}) - \phi(\mathbf{X})| I_{\{\mathbf{X} \in [\mathbf{a}, \mathbf{b}]^d\}}] + 2E[|\phi(\mathbf{X}) - \eta(\mathbf{X})| I_{\{\mathbf{X} \in [\mathbf{a}, \mathbf{b}]^d\}}] + 2\delta \tag{4.130}$$

$$\leq 2\|\phi - \gamma\|_\infty + 2E[|\phi(\mathbf{X}) - \eta(\mathbf{X})| I_{\{\mathbf{X} \in [\mathbf{a}, \mathbf{b}]^d\}}] + 2\delta < 6\delta.$$

It can be shown via the Stone-Weierstrass approximation theorem that the function class $\mathcal{L}_k$ becomes dense relative to the $L_\infty$ norm in the space of continuous functions on any cube $[\mathbf{a}, \mathbf{b}]^d$ in $\mathfrak{R}^d$ for any distribution of $(\mathbf{X}, Y)$. Thus, $\Delta_{\mathcal{N}_k} \to 0$ as

$k \to \infty$ for any distribution of $(\mathbf{X}, Y)$ (Cybenko, 1989; Funahashi, 1989; Hornik et al., 1989).

Regarding estimation error, the VC dimension of $\mathcal{N}_k$ is bounded below by $dk$ if $k$ is even and by $d(k+1)$ if $k$ is odd (Baum, 1988). Thus, good distribution-free performance requires the sample size to greatly exceed $kd$. It is often counterproductive to try to obtain a good classifier by using large numbers of variables and neurons since this creates a requirement for a very large sample. For instance, if $d = k = 10$, then the VC dimension is bounded below by 100, and setting $V_C = 100$ and $n = 5000$ in equation (4.54) yields a bound exceeding 1. In fact, the VC dimension can be bounded from above and below:

$$d \min \left( k, \frac{2^d}{d^2/2 + d + 1} \right) + 1 \le V_{\mathcal{N}_k} \le 2(kd + 2k + 1) \log_2[e(kd + 2k + 1)]$$

(4.131)

(Bartlett, 1993). Strong universal consistency follows from convergence of the constraint error and this bound.

**THEOREM 4.12 (Farago and Lugosi, 1993)** *Let $\hat{\psi}_{n,\mathcal{N}_k}$ be the k-node neural network with threshold sigmoid that minimizes the empirical error. If $k \to \infty$ such that $(k \log n)/n \to 0$ as $n \to \infty$, then $\hat{\psi}_{n,\mathcal{N}_k}$ is universally strongly consistent, meaning $\hat{\varepsilon}_{n,\mathcal{N}_k} \to \varepsilon_d$ with probability 1 for all distributions of $(\mathbf{X}, Y)$.*

*Proof.* The theorem follows from the error decomposition in equation (4.125). We have seen that $\Delta_{\mathcal{N}_k} \to 0$ as $k \to \infty$ for any distribution of $(\mathbf{X}, Y)$. By the Vapnik-Chervonenkis theorem and equation (4.131),

$$P(\hat{\Delta}_{n,\mathcal{N}_k} > \tau) \le 8 \Xi(\mathcal{N}_k, n) e^{-n\tau^2/32} \le 8n^{2(kd+2k+1) \log_2[e(kd+2k+1)]} e^{-n\tau^2/32}. \quad (4.132)$$

With the assumption on the convergence of $(k \log n)/n$, the expression on the right is summable and the theorem follows by the Borel-Cantelli lemma. □

The preceding results are for the threshold sigmoid. In fact, there exists a sigmoid that is monotonically increasing, continuous, concave on $(0, \infty)$, and convex on $(-\infty, 0)$ and for which the VC dimension of $\mathcal{N}_k$ is infinite for $k \ge 8$ (Devroye et al., 1996).

The denseness of the class of neural networks relative to both $L_1$ and $L_\infty$ is theoretically important, but there are limitations to its practical usefulness. Not only do we not know the function whose approximation is desired, but even if we knew the function and how to find the necessary coefficients, a close approximation can require an extremely large number of model parameters. As the number of model parameters grows, use of the model for classifier design becomes increasingly intractable from the perspectives of both computation and statistical estimation of the model parameters. Given the neural network structure, the task is to estimate the optimal weights. Since the number of hidden units must be kept relatively small, there is no assurance that the optimal neural network of the prescribed form closely approximates the Bayes classifier, nor do we necessarily know the degree to which it is suboptimal. In any event, given the form, an algorithm is required to estimate the parameters from the data.

## 4.6 CLASSIFICATION TREES

The cubic histogram rule partitions the space without reference to the actual data. One can instead partition the space based on the data, either with or without reference to the labels. Tree classifiers are a common way of performing data-dependent partitioning. Since any tree can be transformed into a binary tree, we need consider only binary classification trees. A tree is constructed recursively based on some criteria. If $S$ represents the set of all data, then it is partitioned according to some rule into $S = S_1 \cup S_2$. There are four possibilities: (1) $S_1$ is partitioned into $S_1 = S_{11} \cup S_{12}$ and $S_2$ is partitioned into $S_2 = S_{21} \cup S_{22}$; (2) $S_1$ is partitioned into $S_1 = S_{11} \cup S_{12}$ and partitioning of $S_2$ is terminated; (3) $S_2$ is partitioned into $S_2 = S_{21} \cup S_{22}$ and partitioning of $S_1$ is terminated; and (4) partitioning of both $S_1$ and $S_2$ is terminated. In the last case, the partition is complete; in any of the others, it proceeds recursively until all the branches end in termination, at which point the leaves on the tree represent the partition of the space. On each cell (subset) in the final partition, the designed classifier is defined to be 0 or 1 according to which is the majority among the labels of the points in the cell.

As an example of a classification tree, consider the median tree. Using one of the coordinates, say, the first, split the observation vectors into two groups based on the median value of the selected coordinates in the sample. If the first coordinate is used, then the $x_1$-axis is split at the median $\tilde{x}_1$, thereby partitioning the data into $S = S_1 \cup S_2$: for $\mathbf{x} = (x_1, x_2, \ldots, x_d) \in S$, $\mathbf{x} \in S_1$ if $x_1 \le \tilde{x}_1$, and $\mathbf{x} \in S_2$ if $x_1 > \tilde{x}_1$. The second level of the tree is constructed by partitioning $S_1$ and $S_2$ according to the median values of the second coordinate among the points in $S_1$ and $S_2$, respectively, where the point selected for the first level is not considered. The partitioning is continued to some chosen level $k$. If the level goes beyond the dimension $d$ of the space, then one recycles through coordinates again. Note that the median tree does not depend on the labels of the sample data. Median tree classifiers are consistent if $k \to \infty$ in such a way that $n/k2^k \to \infty$, as long as $\mathbf{X}$ possesses a density (Devroye et al., 1996).

A wide variety of classification trees whose leaves are rectangles in $\Re^d$ can be obtained by perpendicular splits. At each stage of growing the tree, a decision to split a rectangle $R$ is made according to a coordinate decision of the form $X_i^j \le \alpha$, where $\mathbf{X}^j = (X_1^j, X_2^j, \ldots, X_d^j)$ is a sample point in $R$. At each stage there are also two collections of rectangles $\mathcal{R}_0$ and $\mathcal{R}_1$ determined by majority vote of the labels, so that $R \in \mathcal{R}_1$ if and only if the majority of labels for points in $R$ have value 1. The 0 and 1 decision regions are determined by the unions of rectangles in $\mathcal{R}_0$ and $\mathcal{R}_1$, respectively. A final classification tree, and therefore the designed classifier, depends on the splitting criterion, the choice of $\alpha$, and a stopping criterion. Two desirable attributes of a stopping criterion are that the leaf nodes (final rectangles) be small in number so that the complexity of the classifier will not be too great for the number of data, and that the labels in each final rectangle not be evenly split, thereby increasing the likelihood that the majority label accurately reflects the distribution in the rectangle. A rectangle is said to be *pure* relative to a particular sample if all the labels corresponding to points in the rectangle possess the same label.

### 4.6.1 Classification and Regression Trees

These considerations lead to a method of splitting that goes under the name *classification and regression trees* (CART) and is based on the notion of an "impurity" function. For any rectangle $R$, let $N_0(R)$ and $N_1(R)$ be the numbers of 0-labeled and 1-labeled points, respectively, in $R$, and let

$$N(R) = N_0(R) + N_1(R) \tag{4.133}$$

be the total number of points in $R$. The *impurity* of $R$ is defined by

$$\kappa(R) = \psi(p_R, 1 - p_R), \tag{4.134}$$

where $p_R = N_0(R)/N(R)$ is the proportion of 0 labels in $R$ and $\psi(p, 1 - p)$ is a nonnegative function such that

1. $\psi(0.5, 0.5) \geq \psi(p, 1 - p)$ for any $p \in [0, 1]$,
2. $\psi(0, 1) = \psi(1, 0) = 0$,
3. as a function of $p$, $\psi(p, 1 - p)$ increases in $p \in [0, 0.5]$ and decreases in $p \in [0, 0.5]$.

Several observations follow from the definition of $\psi$: (1) $\kappa(R)$ is maximum when the proportions of 0-labeled and 1-labeled points in $R$ are equal (corresponding to maximum impurity); (2) $\kappa(R) = 0$ if $R$ is pure; and (3) $\kappa(R)$ increases for greater impurity.

We mention three possible choices for $\psi$:

1. $\psi_e(p, 1 - p) = -p \log p - (1 - p) \log(1 - p)$ *(entropy impurity)*,
2. $\psi_g(p, 1 - p) = p(1 - p)$ *(Gini impurity)*,
3. $\psi_m(p, 1 - p) = \min(p, 1 - p)$ *(misclassification impurity)*.

The origins of these three impurities lie in the definition of $\kappa(R)$: $\psi_e(p, 1 - p)$ provides an entropy estimate, $\psi_g(p, 1 - p)$ provides a variance estimate for a binomial distribution, and $\psi_m(p, 1 - p)$ provides a misclassification rate estimate.

A splitting regimen is determined by the manner in which a split will cause an overall decrease in impurity. Let $i$ be a coordinate, $\alpha$ be a real number, $R$ be a rectangle to be split along the $i$th coordinate, $R^i_{\alpha, -}$ be the subrectangle resulting from the $i$th coordinate being less than or equal to $\alpha$, and $R^i_{\alpha, +}$ be the subrectangle resulting from the $i$th coordinate being greater than $\alpha$. Define the *impurity decrement* by

$$\Delta_i(R, \alpha) = \kappa(R) - \frac{N(R^i_{\alpha, -})}{N(R)} \kappa(R^i_{\alpha, -}) - \frac{N(R^i_{\alpha, +})}{N(R)} \kappa(R^i_{\alpha, +}). \tag{4.135}$$

A good split will result in impurity reductions in the subrectangles. In computing $\Delta_i(R, \alpha)$, the new impurities are weighted by the proportions of points going into the subrectangles. CART proceeds iteratively by splitting a rectangle at $\hat{\alpha}$ on the $i$th coordinate if $\Delta_i(R, \alpha)$ is maximized for $\alpha = \hat{\alpha}$. There are two possible splitting strategies: (1) the coordinate $i$ is given, and $\Delta_i(R, \alpha)$ is maximized over all $\alpha$ and $R$; (2) the coordinate is not given, and $\Delta_i(R, \alpha)$ is maximized over all $i$, $\alpha$, and $R$.

Various stopping strategies are possible—for instance, stopping when maximization of $\Delta_i(R, \alpha)$ yields a value below a preset threshold. One can also continue to grow the tree until all leaves are pure and then prune.

In addition to there being distributions where no splitting impurity-based procedure yields a consistent classification rule, the behavior of CART can be pathological. Consider the feature-label distribution for two variables with the masses of the 0-label and 1-label distributions being uniformly distributed over the regions

$$T_1 = [-1, 0] \times [0, 1] \cup [0, 1] \times [0, -1], \qquad (4.136)$$

$$T_2 = [-1, 0] \times [-1, 0] \cup [0, 1] \times [0, 1], \qquad (4.137)$$

respectively, and the mass of the full distribution uniform over $T_1 \cup T_2$ (Devroye et al., 1996). For a very large sample, the empirical probability $p$ is essentially equal to the distributional probability. Then, no matter how we split the square (from either coordinate), $p = 1 - p = 0.5$.

### 4.6.2 Strongly Consistent Rules for Data-Dependent Partitioning

Having described CART, we note that a host of ways to build binary classification trees are described in the literature. Rather than describe more trees in detail, we consider a general consistency theorem that applies to classification rules determined by data-dependent partitioning and then show how it applies to binary classification trees in particular.

Before stating the theorem we need to introduce some concepts related to partitions. $\mathcal{P} = \{A_1, A_2, \ldots\}$ is a countable partition of $\mathfrak{R}^d$ if the intersection of any two sets in $\mathcal{P}$ is null and the union of all sets in $\mathcal{P}$ equals $\mathfrak{R}^d$. For any $\tau > 0$, let $\mathcal{P}^{(\tau)}$ be the restriction of $\mathcal{P}$ to the closed ball $B_\tau$ of radius $\tau$ centered at the origin:

$$\mathcal{P}^{(\tau)} = \{A_1 \cap B_\tau, A_2 \cap B_\tau, \ldots\}, \qquad (4.138)$$

which is a partition of $B_\tau$. We assume that the number of sets in $\mathcal{P}^{(\tau)}$ is finite for any $\tau < \infty$. Thus, the collection $\mathcal{A}(\mathcal{P}^{(\tau)})$ of all sets in $\mathfrak{R}^d$ formed by unions of sets in $\mathcal{P}^{(\tau)}$ is finite. Now suppose $\mathcal{F}$ is a (perhaps infinite) family of partitions and let $\mathcal{F}^{(\tau)}$ be the family of partitions $\mathcal{P}^{(\tau)}$ of $B_\tau$ resulting from the partitions $\mathcal{P} \in \mathcal{F}$. Finally, let $\mathcal{D}(\mathcal{F}^{(\tau)})$ be the collection of all sets in $\mathcal{A}(\mathcal{P}^{(\tau)})$ for some $\mathcal{P}_\tau \in \mathcal{F}_\tau$ (equivalently, $\mathcal{P} \in \mathcal{F}$). $A \in \mathcal{D}(\mathcal{F}^{(\tau)})$ if and only if there exists a partition $\mathcal{P}_A \in \mathcal{F}$ for which $A$ is a union of sets in $\mathcal{P}_A^{(\tau)}$. Our main interest is with the shatter coefficient $\xi(\mathcal{D}(\mathcal{F}^{(\tau)}), n)$.

We now describe the general formulation for data-dependent partitioning using a random sample $S_n$. There is a rule $\pi_n$ yielding a random partition $\mathcal{P}_n = \pi_n(S_n)$, and a classifier is defined according to majority decision among the labels for sample points in each set of the partition. The rule $\pi_n$ induces a family $\mathcal{F}_n$ of nonrandom partitions composed of all realizations of the random partition (which themselves result from all realizations of the random sample). A sequence $\{\pi_n\}$ of partitioning rules yields the classifier $\psi_n$ according to majority decision. The next theorem gives strong consistency under two conditions.

**THEOREM 4.13 (Lugosi and Nobel, 1996)** *Let* $\mathbf{X}_1, \mathbf{X}_2, \ldots$ *be independent and identically distributed random vectors on* $\mathfrak{R}^d$ *with associated distribution (probability measure)* $\mu$, $\{\pi_n\}$ *be a sequence of partitioning rules, and* $\mathcal{F}_n$ *be the family of partitions induced by* $\pi_n$. *Suppose that for any* $\tau < \infty$,

$$\lim_{n \to \infty} \frac{\log \xi(\mathcal{D}(\mathcal{F}_n^{(\tau)}))}{n} = 0 \qquad (4.139)$$

*and, for any ball* $B_\tau$ *and* $\gamma > 0$,

$$\lim_{n \to \infty} \mu(\{\mathbf{x} : \mathrm{diam}(A_n(\mathbf{x}) \cap B_\tau) > \gamma\}) = 0 \text{ (with probability 1)}, \qquad (4.140)$$

*where* $A_n(\mathbf{x})$ *is the set in* $\mathcal{D}(\mathcal{F}_n^{(\tau)})$ *containing* $\mathbf{x}$. *Then the classification rule associated with* $\{\pi_n\}$ *is strongly consistent.*

The first condition limits the shattering ability of the partitioning rules relative to the sample size, and the second ensures that the majority decisions determining the classification rule are localized relative to the probability distribution. This theorem can be at once applied to binary classification trees formed so that each node splits its associated set by a hyperplane.

**THEOREM 4.14 (Lugosi and Nobel, 1996)** *If every cell of a binary tree classifier formed by hyperplane splitting contains at least* $k_n$ *points, where* $k_n / \log n \to \infty$ *as* $n \to \infty$, *and the second condition in theorem 4.13 is satisfied, then the classification rule is strongly consistent.*

*Proof.* The proof gives the flavor of how the first condition regarding the shatter coefficient pertains to the manner of splitting. First, the number of ways that $n$ points can be split into two sets by hyperplanes is bounded by $n^{d+1}$. Second, since each cell contains at least $k_n$ points, the number of hyperplane splits is bounded by $n/k_n$. Hence, the number of ways $n$ points can be partitioned by the classification rule is bounded by $(n^{d+1})^{n/k_n+1} = 2^{(d+1)(n/k_n+1)\log n}$. Consequently,

$$\xi(\mathcal{D}(\mathcal{F}_n^{(\tau)})) \leq 2^{n/k_n+1} 2^{(d+1)(n/k_n+1)\log n} \qquad (4.141)$$

and, under the condition $k_n / \log n \to \infty$ as $n \to \infty$,

$$
\begin{aligned}
\frac{\log \xi(\mathcal{D}(\mathcal{F}_n^{(\tau)}))}{n} &\leq \frac{n/k_n + 1}{n} + \frac{(d+1)(n/k_n + 1)\log n}{n} \\
&= \left(\frac{1}{k_n} + \frac{1}{n}\right) + \frac{(d+1)\log n}{n} + \frac{\log n}{k_n} \to 0.
\end{aligned} \qquad (4.142)
$$

$\square$

## 4.7  ERROR ESTIMATION

Error estimation is critical for classification because the error of a classifier determines its worth. Thus, the precision of error estimation is extremely important—
indeed, on epistemological grounds, one could argue that it is the most important
issue in classification (Dougherty and Braga-Neto, 2006). It is complicated because
the precision of an error estimator depends on the classification rule, feature-label
distribution, number of features, and sample size. Error estimation is especially
problematic in gene expression-based classification owing to performance degradation with small samples.

   If a classifier $\psi_n$ is designed from a random sample $S_n$, then the error of the
classifier relative to a particular sample is given by

$$\varepsilon_n = E_F[|Y - \psi_n(\mathbf{X})|], \tag{4.143}$$

where the expectation is taken relative to the feature-label distribution $F$ (as indicated by the notation $E_F$). The expected error of the classifier over all samples of
size $n$ is given by

$$E[\varepsilon_n] = E_{F_n} E_F[|Y - \psi_n(\mathbf{X})|], \tag{4.144}$$

where the outer expectation is with respect to the joint distribution of the sample
$S_n$. In practice, the feature-label distribution is unknown and the expected error
must be estimated.

   If there is an abundance of sample data, then it can be split into *training* data and
*test* data. A classifier is designed on the training data, and its estimated error is the
proportion of errors it makes on the test data. We denote this test data error estimate
by $\tilde{\varepsilon}_{n,m}$, where $m$ is the number of sample pairs in the test data. It is unbiased in the
sense that $E[\tilde{\varepsilon}_{n,m}|S_n] = \varepsilon_n$. Moreover, the conditional distribution of $m\tilde{\varepsilon}_{n,m}$ given
$S_n$ is binomial with parameters $m$ and $\varepsilon_n$. Hence, $m\tilde{\varepsilon}_{n,m}$ has conditional variance

$$E[(m\tilde{\varepsilon}_{n,m} - m\varepsilon_n)^2|S_n] = m\varepsilon_n(1 - \varepsilon_n). \tag{4.145}$$

Factoring out $m$ yields

$$E[(\tilde{\varepsilon}_{n,m} - \varepsilon_n)^2|S_n] = \frac{\varepsilon_n(1 - \varepsilon_n)}{m} \leq \frac{1}{4m}, \tag{4.146}$$

which tends to zero as $m \to \infty$. The problem with using both training data and
test data is that one would like to use all the data for design, especially when there
is a small number of data. Thus, we would like to use the same data for training
(design) and testing.

### 4.7.1  Resubstitution

One approach is to use all sample data to design a classifier $\psi_n$ and estimate $\varepsilon_n$
by applying $\psi_n$ to the same data. The *resubstitution estimate* $\bar{\varepsilon}_n$ is the fraction of

errors made by $\psi_n$ on the sample (training) data:

$$\bar{\varepsilon}_n = \frac{1}{n} \sum_{i=1}^{n} |Y_i - \psi_n(\mathbf{X}_i)|. \tag{4.147}$$

To appreciate the problem with resubstitution, consider the plug-in rule for discrete data. For any vector $\mathbf{x}$, let $n(\mathbf{x})$ be the number of occurrences of $\mathbf{x}$ in the sample data, $n(Y = 1|\mathbf{x})$ be the number of times $\mathbf{x}$ has label 1, and $P_n(Y = 1|\mathbf{x}) = n(Y = 1|\mathbf{x})/n(\mathbf{x})$. There are three possibilities: (1) $\mathbf{x}$ is observed in training, $n(Y = 1|\mathbf{x}) > n(\mathbf{x})/2$, $P_n(Y = 1|\mathbf{x}) > 1/2$, and $\psi_n(\mathbf{x}) = 1$; (2) $\mathbf{x}$ is observed in training, $n(Y = 1|\mathbf{x}) \leq n(\mathbf{x})/2$, $P_n(Y = 1|\mathbf{x}) \leq 1/2$, and $\psi_n(\mathbf{x}) = 0$; or (3) $\mathbf{x}$ is not observed in training and $\psi_n(\mathbf{x})$ is defined by a convention. Each $\mathbf{x}$ in the first category contributes $n(Y = 0|\mathbf{x})$ errors. Each $\mathbf{x}$ in the second category contributes $n(Y = 1|\mathbf{x})$ errors. For a small sample, the points in the third category contribute nothing to $\bar{\varepsilon}_n$ but may contribute substantially to $\varepsilon_n$. Moreover, there may be many vectors in the first and second categories observed only once. It is not surprising then that for *multinomial discrimination*, in which the components of each observation $\mathbf{x}$ are in the range $\{0, 1, \dots, r\}$, $E[\bar{\varepsilon}_n] \leq \varepsilon_d$ and therefore

$$E[\bar{\varepsilon}_n] \leq \varepsilon_d \leq E[\varepsilon_n]. \tag{4.148}$$

For fixed-partition histogram rules, the resubstitution error is biased low, meaning $E[\bar{\varepsilon}_n] \leq E[\varepsilon_n]$, and its variance is bounded by the sample size $\mathrm{Var}[\bar{\varepsilon}_n] \leq 1/n$. For small samples, the bias can be severe. It typically improves for large samples. Indeed, for fixed partitions, meaning those that are independent of sample size and of the data, $E[\bar{\varepsilon}_n]$ is monotonically increasing. The next theorem bounds the mean square error for resubstitution error estimation.

**THEOREM 4.15 (Deveroye et al., 1996)** *For a fixed-partition histogram rule having at most $q$ cells,*

$$E[|\bar{\varepsilon}_n - \varepsilon_n|^2] \leq \frac{6q}{n}. \tag{4.149}$$

This preceding inequality indicates decent performance if the number of cells is not too large; however, as the next theorem indicates, even for fixed partitions the situation can be bad if the number of cells is large.

**THEOREM 4.16 (Devroye and Wagner, 1976)** *For every $n$ there exists a fixed-partition rule and a feature-label distribution such that*

$$E[|\bar{\varepsilon}_n - \varepsilon_n|^2] \geq \frac{1}{4}. \tag{4.150}$$

While we have with justification concentrated on the low bias of resubstitution estimation, we note that there exist consistent classification rules and distributions for which the resubstitution estimator is high-biased, $E[\bar{\varepsilon}_n] \geq E[\varepsilon_n] \geq \varepsilon_d$.

### 4.7.2 Cross-validation

With cross-validation, classifiers are designed from parts of the sample, each is tested on the remaining data, and $\varepsilon_n$ is estimated by averaging the errors. In *k-fold cross-validation*, $S_n$ is partitioned into $k$ folds $S_{(i)}$, for $i = 1, 2, \ldots, k$, where for simplicity we assume that $k$ divides $n$. Each fold is left out of the design process and used as a test set, and the estimate is the overall proportion of errors committed on all folds:

$$\hat{\varepsilon}_{n,k} = \frac{1}{n} \sum_{i=1}^{k} \sum_{j=1}^{n/k} |Y_j^{(i)} - \psi_{n,i}(\mathbf{X}_j^{(i)})|, \tag{4.151}$$

where $\psi_{n,i}$ is designed on $S_n - S_{(i)}$ and $(\mathbf{X}_j^{(i)}, Y_j^{(i)})$ is a sample point in $S_{(i)}$. The process can be repeated, where several cross-validated estimates are computed, using different partitions of the data into folds, and the results averaged. In *stratified cross-validation*, the classes are represented in each fold in the same proportion as in the original data. A $k$-fold cross-validation estimator is unbiased as an estimator of $E[\varepsilon_{n-n/k}]$: $E[\hat{\varepsilon}_{n,k}] = E[\varepsilon_{n-n/k}]$.

In *leave-one-out estimation* ($n$-fold cross-validation), a single observation is left out, $n$ classifiers are designed from sample subsets formed by leaving out one sample pair, each is applied to the left-out pair, and the leave-one-out estimator $\hat{\varepsilon}_n^{\text{loo}}$ is $1/n$ times the number of errors made by the $n$ classifiers. $\hat{\varepsilon}_n^{\text{loo}}$ is an unbiased estimator of $E[\varepsilon_{n-1}]$, namely, $E[\hat{\varepsilon}_n^{\text{loo}}] = E[\varepsilon_{n-1}]$. Approximate unbiasedness is important, but a key concern is the variance of the estimator for small $n$. Performance depends on the classification rule.

**THEOREM 4.17 (Rodgers and Wagner, 1978)** *For the kNN rule with randomized tie breaking,*

$$E[|\hat{\varepsilon}_n^{\text{loo}} - \varepsilon_n|^2] \le \frac{6k+1}{n}. \tag{4.152}$$

Given that $\hat{\varepsilon}_n^{\text{loo}}$ is approximately an unbiased estimator of $\varepsilon_n$, this inequality bounds the variance of $\hat{\varepsilon}_n^{\text{loo}} - \varepsilon_n$. Although an upper bound does not say how bad the situation is, but only how bad it can at most be, it can be instructive to look at its order of magnitude. For $k = 1$ and $n = 175$, upon taking the square root, this bound only ensures that the standard deviation of $\hat{\varepsilon}_n^{\text{loo}} - \varepsilon_n$ is less than 0.2. Applying Chebyshev's inequality to equation (4.152) yields a convergence-in-probability-type bound:

$$P(|\hat{\varepsilon}_n^{\text{loo}} - \varepsilon_n| > \delta) \le \frac{6k+1}{n\delta^2}. \tag{4.153}$$

**THEOREM 4.18 (Devroye et al., 1996)** *For a fixed-partition histogram rule,*

$$E[|\hat{\varepsilon}_n^{\text{loo}} - \varepsilon_n|^2] \le \frac{1 + 6e^{-1}}{n} + \frac{6}{\sqrt{\pi(n-1)}}. \tag{4.154}$$

Comparing theorems 4.15 and 4.18 for fixed-partition histogram rules, $\sqrt{n-1}$ for leave-one-out estimation as opposed to $n$ in the denominator for resubstitution shows greater variance for $\hat{\varepsilon}_n^{loo}$. There is a certain tightness to this bound. For any partition there is a distribution for which

$$E[|\hat{\varepsilon}_n^{loo} - \varepsilon_n|^2] \geq \frac{1}{e^{1/12}4\sqrt{2\pi n}}. \tag{4.155}$$

Performance can be very bad for small $n$. Unbiasedness comes with increased variance.

To appreciate the difficulties inherent in leave-one-out bounds, we will simplify them in a way that makes them more favorable to precise estimation. The performance of $\hat{\varepsilon}_n^{loo}$ guaranteed by equation (4.154) becomes better if we lower the bound. A bound lower than the one in equation (4.154) is $(1.8)/\sqrt{n-1}$. The corresponding standard-deviation bounds for $n = 50$ and $100$ exceed $0.5$ and $0.435$, respectively. These are essentially useless. The minimum worst-case-performance bound in equation (4.155) would be better if it were lower. A bound lower than the one given is $(0.35)/\sqrt{n}$. The corresponding standard-deviation bounds for $n = 50$ and $100$ exceed $0.11$ and $0.09$, respectively.

Cross-validation is likely the most popular error estimator in the microarray literature. Unfortunately, it is often used with neither justification nor mention of its serious shortcomings in small-sample settings. While not suffering from severe low bias like resubstitution, its large variance in small-sample settings makes its use problematic. According to the unbiasedness of leave-one-out estimation, $E[\hat{\varepsilon}_n^{loo}] = E[\varepsilon_{n-1}]$, we have $E[\hat{\varepsilon}_n^{loo} - \varepsilon_n] \approx 0$. Thus, the expected difference between the error estimator and the error is approximately 0. But we are not interested in the expected difference between the error estimator and the error; rather, we are interested in the precision of the error estimator in estimating the error. Our concern is the expected deviation $E[|\hat{\varepsilon}_n^{loo} - \varepsilon_n|]$ and unless the cross-validation variance is small, which it is not for small samples, this expected deviation will not be small. As will be demonstrated shortly by considering the deviation distributions for several error estimators, cross-validation generally does not perform well for small samples (Braga-Neto and Dougherty, 2004a).

The difficulty with cross-validation is illustrated in figure 4.5, which shows decision regions obtained from CART. The top part of the figure shows the classifier designed from the sample data, and the remaining parts show a few *surrogate* classifiers designed from the sample after a point (circle) has been removed. Note how different the surrogates are from the designed classifier. The leave-one-out estimator is obtained from counting the errors of all the surrogates, many of which have little relation to the actual classifier whose error we desire. Although cross-validation does not perform well for small samples, it can be beneficial for modest-sized samples in which one wishes to use all the data for design so as to obtain the best possible classifier. For larger samples, the variances of the cross-validation estimators get smaller and better performance is achieved.

### 4.7.3 Bootstrap

Bootstrap methodology is a general resampling strategy that can be applied to error estimation (Efron, 1979). It is based on the notion of an *empirical distribution $F^*$*,

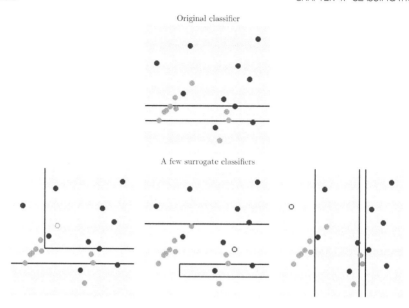

Figure 4.5. The CART classifier and a few leave-one-out surrogate classifiers. (Dougherty et al., 2005b.)

that puts mass on each of the $n$ sample points. A *bootstrap sample* $S_n^*$ from $F^*$ consists of $n$ equally likely draws with replacement from the original sample $S_n$. Some points may appear multiple times, whereas others may not appear at all. The probability that a given point will not appear in $S_n^*$ is $(1 - 1/n)^n \approx e^{-1}$. Hence, a bootstrap sample of size $n$ contains on average $(1 - e^{-1})n \approx (0.632)n$ of the original points. The basic bootstrap estimator, the *bootstrap zero estimator* (Efron, 1983), mimics equation (4.143) with respect to the empirical distribution and is defined by

$$\hat{\varepsilon}_n^b = E_{F^*}[|Y - \psi_n^*(\mathbf{X})| : (\mathbf{X}, Y) \in S_n - S_n^*], \qquad (4.156)$$

where $S_n$ is fixed. The classifier is designed on the bootstrap sample and tested on the points that are left out. In practice, the expectation is approximated by a sample mean based on $B$ independent replicates, $S_{n,b}^*$, $b = 1, 2, \ldots, B$:

$$\hat{\varepsilon}_n^b = \frac{\displaystyle\sum_{b=1}^{B} \sum_{i=1}^{n} |Y_i - \psi_{n,b}^*(\mathbf{X}_i)| I_{(\mathbf{X}_i, Y_i) \in S_n - S_{n,b}^*}}{\displaystyle\sum_{b=1}^{B} \sum_{i=1}^{n} I_{(\mathbf{X}_i, Y_i) \in S_n - S_{n,b}^*}}. \qquad (4.157)$$

A value of $B$ between 25 and 200 has been recommended. A variance-reducing technique, called balanced bootstrap resampling, is often employed (Chernick, 1999). With it, each sample is made to appear exactly $B$ times in the computation.

The bootstrap zero estimator tends to be a high-biased estimator of $E[\varepsilon_n]$, since the number of points available for design is on average only $0.632n$. The $.632$ *bootstrap estimator* tries to correct this bias by calculating a weighted average of the zero and resubstitution estimators (Efron, 1983). It is defined by

$$\hat{\varepsilon}_n^{b632} = (1 - 0.632)\bar{\varepsilon}_n + 0.632\hat{\varepsilon}_n^b. \tag{4.158}$$

The *bias-corrected bootstrap estimator* tries to correct for resubstitution bias. It is defined by

$$\hat{\varepsilon}_n^{b632,\text{cor}} = \bar{\varepsilon}_n + \frac{1}{B} \sum_{b=1}^{B} \sum_{i=1}^{n} \left( \frac{1}{n} - P_{i,b}^* \right) |Y_i - \psi_{n,b}^*(\mathbf{X}_i)|, \tag{4.159}$$

where $P_{i,b}^*$ is the proportion of times that $(\mathbf{X}_i, Y_i)$ appears in the bootstrap sample $S_{n,b}^*$. The estimator adds to the resubstitution estimator the bootstrap estimate of its bias.

### 4.7.4 Bolstering

The resubstitution estimator is given by

$$\bar{\varepsilon}_n = E_{F^*}[|Y - \psi_n(\mathbf{X})|]. \tag{4.160}$$

Relative to $F^*$, no distinction is made between points near or far from the decision boundary. If one spreads the probability mass at each point of the empirical distribution, then variation is reduced because points near the decision boundary will have more mass on the other side than points far from the decision boundary. Relative to the decision, more confidence is attributed to points far from the decision boundary than to points near it.

To take advantage of this observation, consider a density function $f_i^\diamond, i = 1, 2, \ldots, n$, called a *bolstering kernel*, and define the bolstered error distribution $F^\diamond$ by

$$F^\diamond(\mathbf{x}) = \frac{1}{n} \sum_{i=1}^{n} f_i^\diamond(\mathbf{x} - \mathbf{x}_i). \tag{4.161}$$

The *bolstered resubstitution* estimator is obtained by replacing $F^*$ by $F^\diamond$ in equation (4.160):

$$\bar{\varepsilon}_n^\diamond = E_{F^\diamond}[|Y - \psi_n(\mathbf{X})|] \tag{4.162}$$

(Braga-Neto and Dougherty, 2004b).

Bolstering shares some similarities with the smoothing technique for linear discriminant analysis, where the basic idea is to reduce the variance of error-counting estimators by means of a smoothing function that plays a role similar to that of the bolstering kernels (Glick, 1978; Snapinn and Knoke, 1985). Bolstered error estimation can be seen as smoothed estimation in a few special cases; however, the

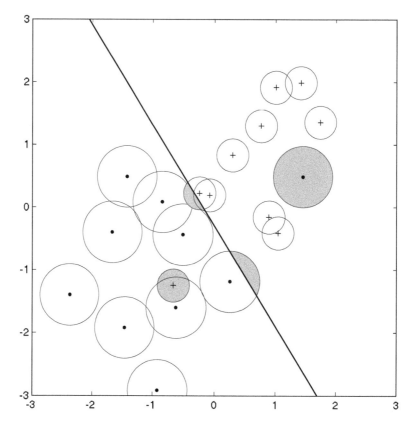

Figure 4.6. Bolstering error for linear classification.

smoothing technique does not have a general direct geometric interpretation and can be cumbersome to formulate for multiple classes, and it is not easily extended to classifiers other than linear discriminant analysis (Tutz, 1985; Devroye et al., 1996).

A computational expression for the bolstered resubstitution estimator is given by

$$\bar{\varepsilon}_n^{\diamond} = \frac{1}{n} \sum_{i=1}^{n} I_{y_i=0} \int\limits_{\{\mathbf{x}:\psi_n(\mathbf{x})=1\}} f_i^{\diamond}(\mathbf{x}-\mathbf{x}_i)\,d\mathbf{x} + I_{y_i=1} \int\limits_{\{\mathbf{x}:\psi_n(\mathbf{x})=0\}} f_i^{\diamond}(\mathbf{x}-\mathbf{x}_i)\,d\mathbf{x}.$$

(4.163)

The integrals are the error contributions made by the data points according to whether $y_i = 0$ or $y_i = 1$. The bolstered resubstitution estimate is equal to the sum of all the error contributions divided by the number of points. If the classifier is linear, then the decision boundary is a hyperplane and it is usually possible to find analytic expressions for the integrals, for instance, for Gaussian bolstering; otherwise, Monte Carlo integration can be employed. Figure 4.6 illustrates the error for linear classification when the bolstering kernels are uniform circular distributions.

A key issue is choosing the amount of bolstering. Since the purpose of bolstering is to improve error estimation in the small-sample setting, we do not want to use bolstering kernels that require complicated inferences. Hence, we consider zero-mean, spherical bolstering kernels with covariance matrices of the form $\sigma_i \mathbf{I}$. The choice of the parameters $\sigma_1, \sigma_2, \cdots, \sigma_n$ determines the variance and bias properties of the corresponding bolstered estimator. If $\sigma_1 = \sigma_2 = \cdots = \sigma_n = 0$, then there is no bolstering and the bolstered estimator reduces to the original estimator. As a general rule, larger $\sigma_i$ lead to lower-variance estimators, but after a certain point this advantage is offset by increasing bias. There are different methods to choose $\sigma_1, \sigma_2, \ldots, \sigma_n$. Here we focus on a simple nonparametric sample-based method applicable in small-sample settings.

We wish to select $\sigma_1, \sigma_2, \ldots, \sigma_n$ to make the bolstered resubstitution estimator nearly unbiased. One can think of $(\mathbf{X}, Y)$ in equation (4.143) as a random test point. Given $Y = y$, the expectation can be viewed as considering the error for a test point at a "true mean distance" $\delta(y)$ from the data points belonging to class $Y = y$. One reason why resubstitution is optimistically biased is that the test points are all at distance zero from the data set. We can view the selection of $\sigma_1, \sigma_2, \ldots, \sigma_n$ as finding the amount of spread that makes the test points as close as possible to the true mean distance to the data points. The true mean distance can be estimated by the sample-based estimate

$$\hat{\delta}(y) = \frac{\sum_{i=1}^{n} \min_{j \neq i}\{||\mathbf{x}_i - \mathbf{x}_j||\} I_{y_i=y}}{\sum_{i=1}^{n} I_{y_i=y}}. \tag{4.164}$$

This estimate is the mean minimum distance between points belonging to class $Y = y$.

To arrive at estimates of $\sigma_1, \sigma_2, \ldots, \sigma_n$, let $D_i$ be the random variable giving the distance to the origin of a randomly selected point from a unit-variance bolstering kernel and let $F_i$ be the probability distribution function for $D_i$. For the class $Y = y$, a value of $\sigma_y$ is to be chosen so that the median distance of a test point to the origin is equal to the estimate of the mean distance, the result being that half of the test points will be farther from the center than $\hat{\delta}(y)$ and the other half will be nearer. Hence, $\sigma_y$ is a solution of the equation $\sigma_y F_i^{-1}(0.5) = \hat{\delta}(y)$, and the estimated standard deviations for the bolstering kernels are given by

$$\sigma_i = \frac{\hat{\delta}(y_i)}{F_i^{-1}(0.5)}. \tag{4.165}$$

This value depends on the number of variables and, as the number of points in the sample increases, $\sigma_i$ decreases and less bias correction is introduced by the bolstered resubstitution.

When resubstitution is heavily low-biased, it may not be good to spread incorrectly classified data points because that increases the optimism of the error estimate (low bias). Bias is reduced by letting $\sigma_i = 0$ (no bolstering) for incorrectly classified points. The *semibolstered resubstitution* estimator results from not bolstering (no spread) incorrectly classified points. Bolstering can be applied to any error-counting estimation procedure. *Bolstered leave-one-out* estimation involves bolstering the estimates on the surrogate classifiers.

### 4.7.5 Error Estimator Performance

Large simulation studies involving the empirical distribution of $\hat{\varepsilon}_n - \varepsilon_n$ have been conducted to examine small-sample error estimator performance (Braga-Neto and Dougherty, 2004a; Braga-Neto and Dougherty, 2004b). We provide some results for 3NN and CART in which the error estimator $\hat{\varepsilon}_n$ is resubstitution (resub), 10-fold cross-validation with 10 repetitions (cv10r), .632 bootstrap (b632), or bolstered resubstitution with Gaussian bolstering kernels. The number of bootstrap samples is $B = 100$, which makes the number of designed classifiers the same as for cv10r. The simulations considered here use data from a microarray-based classification study that analyzes a large number of microarrays prepared with RNA from breast tumor samples from each of 295 patients (van de Vijver et al., 2002). Using a 70-gene prognosis profile, a prognosis signature based on gene expression is proposed that correlates with patient survival data and other existing clinical measures. Of the 295 microarrays, 115 belong to the "good-prognosis" class and 180 belong to the "poor-prognosis" class.

The simulation cases discussed here use log-ratio gene expression values associated with the top 5 genes, as ranked by a correlation-based measure (van't Veer et al., 2002). For each case, 1000 observations of size $n = 20$ and $n = 40$ are drawn independently from the pool of 295 microarrays. Sampling is stratified in the sense that half of the sample points are drawn from each of the two prognosis classes. The true error for each observation of size $n$ is approximated by a holdout estimator whereby the $(295 - n)$ sample points not drawn are used as the test set (a good approximation to the true error given the large test sample). This allows computation of the empirical deviation distribution for each error estimator using the considered classification rules. Note that because the observations are not independent, there is a degree of inaccuracy in the computation of the deviation distribution; however, for sample sizes $n = 20$ and $n = 40$ out of a pool of 295 sample points, the amount of overlap between samples is small. Figures 4.7 and 4.8 display plots of the empirical deviation distributions for LDA and CART, respectively, obtained by fitting beta densities to the raw data. Better performance is indicated by a narrow distribution centered close to 0.

In the overall set of experiments, not simply the ones shown here, resubstitution, leave-one-out, and even 10-fold cross-validation are generally outperformed by bootstrap and bolstered estimators. Bolstered resubstitution is very competitive with bootstrap, in some cases beating it. For LDA, the best estimator overall is bolstered resubstitution. For 3NN and CART, which are classifiers known to overfit in small-sample settings, the situation is not as clear. For 3NN, bolstered resubstitution fails in correcting the bias of resubstitution for $d = 5$ despite having small variance. For CART, the bootstrap estimator is affected by the extreme low-biasedness of resubstitution. In this case, bolstered resubstitution performs quite well, but the best overall estimator is semibolstered resubstitution. Increasing the sample size improves the performance of all the estimators considerably,

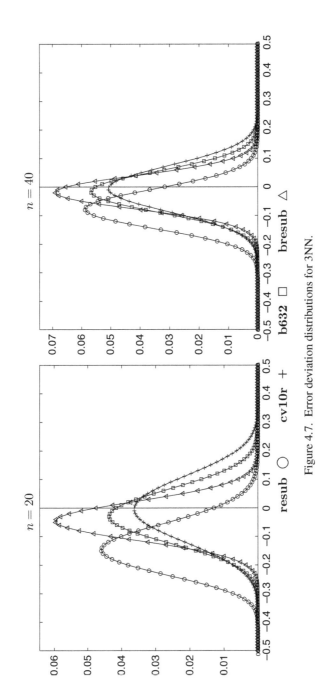

Figure 4.7. Error deviation distributions for 3NN.

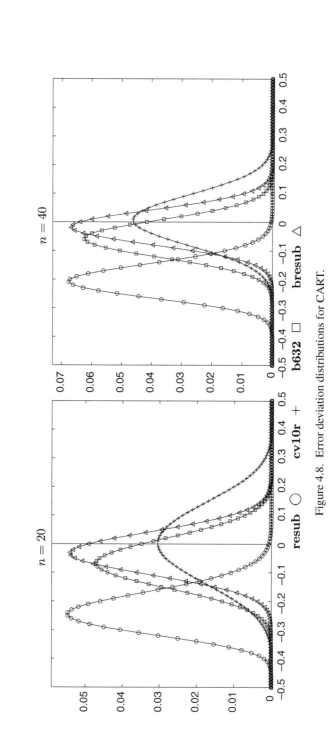

Figure 4.8. Error deviation distributions for CART.

but the general relative trends still hold with few exceptions. When considering these kinds of results, it must be kept in mind that specific performance advantages depend on the classification rule.

Computation time can be critical for error estimation, in particular, when examining tens of thousands of potential feature sets. Resubstitution is the fastest estimator. Leave-one-out is fast for a small number of samples, but its performance quickly degrades with an increasing number of samples. The 10-fold cross-validation and the bootstrap estimator are the slowest estimators. Bolstered resubstitution can be tens of times faster than the bootstrap estimator.

To highlight the need for caution when comparing error estimators, consider the often held opinion that cross-validation is "always" superior to resubstitution. In fact, this need not be true. In the case of the histogram rule for multinomial discrimination, there exist exact analytic formulations for the mean square error $E[|\hat{\varepsilon}_n - \varepsilon_n|^2]$ for both resubstitution and leave-one-out cross-validation, and these exact formulations are used to demonstrate the better performance of resubstitution when the number of cells is small (Braga-Neto and Dougherty, 2005). Rather than just give anecdotal examples for different distributions, the original study utilizes a parametric Zipf model, which is a power law discrete distribution where the parameter controls the difficulty of classification, and evaluates the root mean square as a function of the expected true error computed for a number of distinct models of the parametric Zipf model. For instance, for $n = 40$ and $b = 8$ cells, their performances are virtually the same (resubstitution is slightly better) for $E[\varepsilon_n] \leq 0.3$, which practically means equivalent performance in any situation where there is acceptable discrimination. For $b = 4$, resubstitution outperforms leave-one-out across the entire error range. For $b = 16$, the complexity of the model in comparison to the sample size is such that resubstitution is very low-biased, and leave-one-out has a better performance. If we fix the expected error at a value giving modest discrimination, $E[\varepsilon_n] = 0.2$, and consider different sample sizes, then resubstitution outperforms leave-one-out for $b \leq 7$, $b \leq 8$, and $b \leq 9$ at $n = 20$, $n = 40$, and $n = 60$, respectively. If we think of the Boolean model for gene regulation, then $b = 2, 4, 8,$ and 16 correspond to connectivity 1, 2, 3, and 4, respectively. Since in practice connectivity is often bounded by 3, this means that, relative to leave-one-out, resubstitution can provide equivalent estimation of prediction error at practical error levels for 3 predictors and better prediction error at all error levels for 1 and 2 predictors. Given the need to sometimes estimate the errors of tens of thousands of predictor functions, there is an enormous computational benefit in using resubstitution besides the better prediction for 1 and 2 predictors.

## 4.7.6 Feature Set Ranking

When choosing among a collection of potential feature sets for classification, estimating the errors of designed classifiers is a key issue, and it is natural to order the potential feature sets according to the error rates of their corresponding classifiers. Hence, it is important to apply error estimators that provide rankings that better correspond to rankings produced by true errors. This is especially important in small-sample settings, where all feature selection algorithms are subject to significant errors and feature selection can be viewed as finding a list of potential feature sets and not as trying to find a best feature set. For instance, in phenotype classification based on gene expression, feature selection can be viewed as gene selection, finding sets of genes whose expressions can be used for phenotypic discrimination; indeed, gene selection in the context of small samples can be viewed as an exploratory methodology.

We consider two performance measures concerning how well feature ranking using error estimators agrees with feature ranking based on true errors. Since our main interest is in finding good feature sets, say, the best $K$ feature sets, we wish to compare the rankings of the best estimate-based feature sets with those of the $K$ best based on true errors. We make this comparison for feature sets whose true performances attain certain levels. For $t > 0$, let $\mathcal{G}_t^K$ be the collection of all feature sets of a given size whose true errors are less than $t$, where $\mathcal{G}_t^K$ is defined only if there exists at least $K$ feature sets with true error less than $t$. Rank the best $K$ feature sets according to their true errors and rank all the feature sets in $\mathcal{G}_t^K$ according to their estimated errors, with rank 1 corresponding to the lowest error. We then have two ranks for each of the $K$ best feature sets: $k$ (true) and $k^*$ (estimated) for all feature sets in $\mathcal{G}_t^K$. In case of ties, the rank is equal to the mean of the ranks. In using the parameter $t$ we are reflecting two interests: (1) how well truly good feature sets are ranked (Braga-Neto et al., 2004); and (2) the pragmatic issue of having to rank feature sets based on error estimates without necessarily having any a priori restriction on the goodness of the feature sets being considered (Sima et al., 2005).

For feature discovery, a key interest is whether truly important features appear on the list of important feature sets based on error estimation. This is the list we obtain from data analysis, and good classification depends on discovering truly good classifying feature sets. The first performance statistic counts the number of feature sets among the top $K$ feature sets that also appear in the top $K$ using the error estimator

$$R_1^K(t) = \sum_{k=1}^{K} I_{k^* \leq K},\qquad(4.166)$$

where $I_A$ denotes the indicator function. For this measure, higher scores are better. Since $k^*$ is the estimate-based rank of the $k$th true-ranked feature set among the feature sets in $\mathcal{G}_t^K$ and since we consider only feature sets in $\mathcal{G}_t^K$, the larger $t$, the larger the collection of ranks $k^*$ and the greater the possibility that erroneous feature sets will appear among the top $K$, thereby resulting in a smaller value of $R_1^K(t)$. The curve of $R_1^K(t)$ flattens out for increasing $t$, which is reflective of the fact that, as we consider ever poorer feature sets, their effect on the top ranks becomes negligible owing to the fact that inaccuracy in the measurement of their errors is not sufficient to make them confuse the ranking of the best feature sets.

A second performance metric measures the mean absolute deviation in the ranks for the $K$ best features sets,

$$R_2^K(t) = \sum_{k=1}^{K} |k - k^*|.\qquad(4.167)$$

For this measure, lower scores are better. By analogy with $R_1^K(t)$, the larger $t$, the larger the collection of ranks $k^*$ and the greater the possible deviation between $k$ and $k^*$. As with $R_1^K(t)$, the curve for $R_2^K(t)$ flattens out as $t$ increases.

To apply the two measures, consider two 20-dimensional unit-variance spherical Gaussian class conditional distributions with means at $\delta\mathbf{a}$ and $-\delta\mathbf{a}$, where $\mathbf{a} = (a_1, a_2, \ldots, a_n)$, $|\mathbf{a}| = 1$, and $\delta > 0$ is a separation parameter. The Bayes classifier is a hyperplane perpendicular to the axis joining the means. The best feature set of size $k$ corresponds to the $k$ largest parameters among $\{a_1, a_2, \ldots, a_n\}$. We consider all feature sets of size 3, for each sample of size 30 we obtain the LDA, 3NN, and CART classifiers, and for each of these we obtain the true error from the distribution and estimated errors based on resubstitution, cross-validation, bootstrap, and bolstering (Sima et al., 2005). Figure 4.9 shows graphs obtained by averaging these measures over many samples. In the cases studied, cross-validation is generally substantially poorer than .632 bootstrap, whereas bolstered estimators are generally better.

## 4.8 ERROR CORRECTION

If we are interested in the ability of a feature set to discriminate between classes, then the actual measure of interest is the Bayes error, not the error of the designed classifier, because if we had sufficient data, we would estimate the Bayes classifier and its error. This section provides a correction term to subtract from the error of the designed classifier to obtain a better approximation of the Bayes error. The correction applies to binomial classifiers. Since these are the same as Boolean predictors, the correction applies to the binary coefficient of determination and its application to PBNs. A key property of the correction is that it is conservative, meaning that the corrected expected error is lower than the uncorrected expected error but still greater than the Bayes error. The correction applies to the actual error, not to error estimation.

Let us rewrite equation (4.36) as

$$E[\varepsilon_n] = \varepsilon_d + \xi_n, \qquad (4.168)$$

where $\varepsilon_n$ is the error of the designed classifier, whether constrained or not, and $\xi_n$ is the sum of the constraint cost, which may be null, and the expected design cost. If our interest is in the Bayes error, then we are motivated to find a correction $\tau_n$ and to approximate $\varepsilon_d$ by

$$\omega_n = E[\varepsilon_n] - \tau_n. \qquad (4.169)$$

If $\tau_n > 0$, then $\omega_n < E[\varepsilon_n]$ and the positive bias of the approximation is reduced. It is important that the correction be conservative, meaning $\omega_n \geq \varepsilon_d$; otherwise, it would lead us to believe that the Bayes error is less than it actually is. We desire a correction factor $\tau_n$ such that $0 \leq \tau_n \leq \xi_n$.

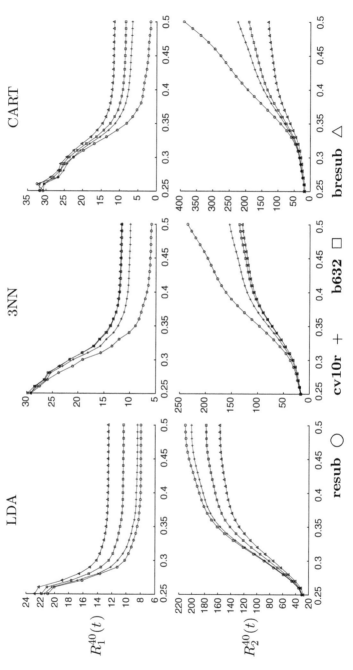

Figure 4.9. Performance curves for feature set ranking.

Consider a Boolean function on binary $d$-vectors, $\psi: \{0, 1\}^d \to \{0, 1\}$. Let $\{\mathbf{x}_1, \mathbf{x}_2, \ldots, \mathbf{x}_m\}$ be the set of all binary $d$-vectors, $m = 2^d$. Given a binary feature-label distribution $(\mathbf{X}, Y)$, for $i = 1, 2, \ldots, m$, let $r_i$ be the probability of observing $\mathbf{x}_i$ and let $p_i = P(Y = 1|\mathbf{x}_i)$. The distribution is determined by the probability vector

$$\mathbf{p} = (r_1, r_2, \ldots, r_m, p_1, p_2, \ldots, p_m)^t. \tag{4.170}$$

The Bayes error is given by

$$\varepsilon_d(\mathbf{p}) = \sum_{i=1}^{m} r_i \rho_i, \tag{4.171}$$

where $\rho_i = \min\{p_i, 1 - p_i\}$. Equation (4.168) applies directly, with $\xi_n(\mathbf{p})$ being a function of $\mathbf{p}$.

Given a random sample of size $n$, we define the classifier $\psi_n$ by the plug-in rule, so that $\psi_n(\mathbf{x}_i) = 1$ if $(\mathbf{x}_i, 1)$ appears in the sample more than $(\mathbf{x}_i, 0)$ does. If $\mathbf{x}_i$ is not observed in the sample, then for the purpose of our analysis we will randomly define $\psi_n(\mathbf{x}_i)$ as 0 or 1 with equal probability. It is possible to obtain an explicit representation for the expected error increase $\xi_n(\mathbf{p})$.

**THEOREM 4.19 (Brun et al., 2003)** *The expected error increase for a model $\mathbf{p}$ and sample size $n$ is given by*

$$\xi_n(\mathbf{p}) = \sum_{\{i: p_i < 0.5\}} (0.5 - \rho_i)c_1(\rho_i, r_i, n) + \sum_{\{i: p_i \geq 0.5\}} (0.5 - \rho_i)c_2(\rho_i, r_i, n), \tag{4.172}$$

*with*

$$c_1(\rho_i, r_i, n) = 2r_i(1 - r_i)^n \left[ \frac{1}{2} + \sum_{k=1}^{n} C_k^n \left( \frac{r}{1-r} \right)^k (1 - B_{k,\rho_i}(\lfloor k/2 \rfloor)) \right], \tag{4.173}$$

$$c_2(\rho_i, r_i, n) = 2r_i(1 - r_i)^n \left[ \frac{1}{2} + \sum_{k=1}^{n} C_k^n \left( \frac{r}{1-r} \right)^k B_{k,1-\rho_i}(\lfloor k/2 \rfloor) \right], \tag{4.174}$$

*where $C_k^n$, $B_{k,\rho}$, and $\lfloor a \rfloor$ denote the binomial coefficient for $n$ objects taken $k$ at a time, the probability distribution function for the binomial distribution with $k$ trials and probability $\rho$ of success, and the largest integer in $a$, respectively.*

The coefficients $c_1$ and $c_2$ in equation (4.173) depend on only the probabilities $r_i$ and the error contributions $\rho_i$. Whether $p_i$ is less than or greater than 0.5 determines the use of $c_1$ or $c_2$ in the sum. This lets us find a lower bound $\tau_n(\mathbf{p})$ of $\xi_n(\mathbf{p})$ that depends on only $r_i$ and $\rho_i$ and therefore does not depend on whether $\psi_d(\mathbf{x}_i) = 0$ or $\psi_d(\mathbf{x}_i) = 1$ for the Bayes classifier $\psi_d$. Since we want a lower bound for $\xi_n(\mathbf{p})$, we can use the minimum value between $c_1$ and $c_2$ for each $\mathbf{x}_i$. Let

$$c(\rho_i, r_i, n) = \min\{c_1(\rho_i, r_i, n), c_2(\rho_i, r_i, n)\}. \tag{4.175}$$

Define the correction factor for the parameter (model) $\mathbf{p}$ by

$$\tau_n(\mathbf{p}) = \sum_{i=1}^{m} (0.5 - \rho_i) c(\rho_i, r_i, n). \tag{4.176}$$

Then, $\tau_n(\mathbf{p}) \leq \xi_n(\mathbf{p})$ and $\omega_n(\mathbf{p}) \geq \varepsilon_d$. Thus, the corrected approximation is conservative.

The tightness of the lower bound depends on the difference between $c_1$ and $c_2$. If the difference is large, then the bound is loose, and vice versa. The difference between $c_1$ and $c_2$ is small for small $n$ and small $\rho_i$. In this case, the bound is near the real value for $\xi_n(\mathbf{p})$.

To compute $\tau_n(\mathbf{p})$ in equation (4.176) requires $r_i$ and $\rho_i$ for $i = 1, 2, \ldots, m$. Thus, there are $2m$ parameters to be estimated. For small samples, we employ a model with fewer parameters. The model assumes the probabilities $r_i$ are normalized random variables arising from a gamma distribution with shape parameter $\kappa$ (varying) and scale parameter $\beta = 1$, and $\rho_i = \rho$ for $i = 1, 2, \ldots, m$. Normalization enforces the probability requirement

$$r_1 + r_2 + \cdots + r_m = 1, \tag{4.177}$$

and the resulting distribution is a multivariate Dirichlet distribution (Wilks, 1962). The model parameters are the error contribution $\rho$ and the shape parameter $\kappa$. The model is summarized by $\rho_i = \rho$ for $i = 1, 2, \ldots, m$ and

$$r_i = s_i \left( \sum_{i=1}^{m} s_i \right)^{-1}, \tag{4.178}$$

where $s_i$ is gamma-distributed with parameters $\kappa$ and 1. The model corresponds to a class of distributions determined by $\mathbf{p}$ in equation (4.170). The probabilities $r_1, r_2, \ldots, r_m$ are specified only up to selection via equation (4.178), and the conditional probabilities $p_1, p_2, \ldots, p_m$ are specified only up to the requirement that $p_i = \rho$ or $p_i = 1 - \rho$.

For any $\kappa$, the probabilities $r_i$ can take different values. Thus, $\tau_n(\mathbf{p})$ and $c(\rho_i, r_i, n)$ are random variables. We define $\tau_n(\rho, \kappa)$ to be the expected value of $\tau_n(\mathbf{p})$ relative to the distribution of $r_1, r_2, \ldots, r_n$. From equation (4.176),

$$\tau_n(\rho, \kappa) = \sum_{i=1}^{m} (0.5 - \rho_i) E[c(\rho_i, r_i, n)]. \tag{4.179}$$

Since the correction is a function of the model, not the particular distribution, the corrected estimate for the error of the optimal classifier in equation (4.169) becomes

$$\omega_n(\mathbf{p}) = E[\varepsilon_n](\mathbf{p}) - \tau_n(\rho, \kappa). \tag{4.180}$$

Monte Carlo simulation can be used to obtain $\tau_n(\rho, \kappa)$.

Several issues need to be addressed with this kind of correction factor. Is it too conservative? That is, does it give significant correction ? Simulations show that it does provide significant correction for small samples even if we diminish the correction factor in the face of uncertainty regarding $\rho$ and $\kappa$, whose estimation is a major issue. Moreover, although it is conservative relative to the expectation of $\varepsilon_n$, is it sufficiently conservative relative to reasonable amounts of variation for $\varepsilon_n$ so as to only infrequently yield a low estimation of the Bayes error? Last, but certainly not least, how does the correction perform when used in conjunction with error estimation? We leave these matters to the original study (Brun et al., 2003).

## 4.9 ROBUST CLASSIFIERS

A fundamental aspect of any classifier is the degree to which its performance degrades when it is applied to distributions different from the one for which it has been designed. Qualitatively, a classifier is said to be *robust* when its performance degradation is acceptable for distributions close to the one for which it has been designed. Robustness is crucial for application because, once in practice, a classifier will surely be applied in nondesign settings.

In the context of signal processing, robust filter design has been posed in the following way: Given that the joint distribution determining the optimal filter is uncertain in the sense that it can be in different states, what is the best state at which to optimize the filter so as to optimize filter performance relative to the set of all possible states? It is implicit that a criterion be posited for optimization across the state (parameter) space. The design of optimal robust filters was first treated from a minimax perspective, which does not employ distributional information regarding the states: find a filter having the best worst performance over all states (Kuznetsov, 1976; Kassam and Lim, 1977; Poor, 1980; Verdu and Poor, 1984). Optimal robust filtering is treated in a Bayesian framework by assuming a prior distribution for the states. In this context, robustness analysis is relative to the prior distribution (Grygorian and Dougherty, 2001a,b).

In addition to the design of optimal robust filters in signal processing, there is a long history of robust statistics (Huber, 1981). In particular, we mention the influence function, which measures the effect that an infinitesimal contamination at a point has on an estimator (Hempel et al., 1986). The influence function has been adapted within the context of signal processing to measure the effect of an infinitesimal contamination on the output distribution of a filter (Peltonen et al., 2001). Our interest here, however, is strictly in defining optimization criteria for classifiers, not in simply measuring the effects of differences in the design and application distributions, although we naturally have to consider such differences. For instance, parameterized contamination ipso facto results in parameterized distributional differences and therefore fits into the general framework presented here.

In analyzing robustness, we will limit our discussion to the theoretical case in which we possess full distributional knowledge. Practically, it is important to treat robustness in the context of sample-based classifier design, especially in the case of small samples, where suboptimality of designed classifiers is a central issue.

Contenting ourselves with covering the basic theory here, we leave sample-based analysis to the literature (Dougherty et al., 2005a).

### 4.9.1  Optimal Robust Classifiers

To define robustness, consider a parameterized feature-label distribution $F_\mathbf{a}(\mathbf{x}, Y)$, where the *state* $\mathbf{a}$ is a parameter vector corresponding to our uncertainty regarding the distribution. Classification involves the two class conditional distributions $F_\mathbf{a}(\mathbf{x}|0)$ and $F_\mathbf{a}(\mathbf{x}|1)$. The error of a classifier $\psi$ is $\varepsilon_\mathbf{a}[\psi] = P_\mathbf{a}(\psi(\mathbf{X}) \neq Y)$, the subscript indicating that the error is state-dependent. If classifier $\psi_\mathbf{b}$ is optimal relative to the distribution $F_\mathbf{b}$, then $\varepsilon_\mathbf{a}[\psi_\mathbf{b}] \geq \varepsilon_\mathbf{a}[\psi_\mathbf{a}]$ because $\psi_\mathbf{b}$ can perform no better for the distribution $F_\mathbf{a}$ than can the optimal classifier $\psi_\mathbf{a}$ for $F_\mathbf{a}$. If the classifier is robust, then the inequality should not be too great when $\mathbf{b}$ is close to $\mathbf{a}$. The *robustness* of the optimal classifier $\psi_\mathbf{b}$ relative to the distribution $F_\mathbf{a}$ is defined by

$$\kappa(\mathbf{a}; \mathbf{b}) = \varepsilon_\mathbf{a}[\psi_\mathbf{b}] - \varepsilon_\mathbf{a}[\psi_\mathbf{a}]. \tag{4.181}$$

It is immediate that $\kappa(\mathbf{a}; \mathbf{b}) \geq 0$ and $\kappa(\mathbf{a}; \mathbf{a}) = 0$. Robustness is a function of two parameters. By fixing one or the other, it can be looked at in two different ways. Fixing $\mathbf{a}$, it measures the error increase resulting from using classifiers designed for distributions $F_\mathbf{b}$ on distribution $F_\mathbf{a}$. Fixing $\mathbf{b}$, it measures the error increase resulting from applying the classifier designed on system $F_\mathbf{b}$ to distributions $F_\mathbf{a}$. Here $\kappa(\mathbf{a}; \mathbf{b})$ is not symmetric with respect to the parameters since except in special circumstances $\kappa(\mathbf{b}; \mathbf{a}) \neq \kappa(\mathbf{a}; \mathbf{b})$. In the special case where $\mathbf{a} = a$ is a single variable, $\kappa(a; b)$ defines a robustness surface that is zero along the diagonal $b = a$. By analogy with the criteria in signal processing for robust filters, we define three families of robust classifiers.

For minimax robust analysis we do not assume a probability structure on the set $\mathcal{A}$ of states. The minimax approach is to define the best classifier to be one whose worst performance over $\mathcal{A}$ is best among all the classifiers in a class $\mathcal{C}$. Hence, a *minimax robust classifier* is defined by

$$\psi_0 \in \arg \inf_{\psi \in \mathcal{C}} \sup_{\mathbf{a} \in \mathcal{A}} \varepsilon_\mathbf{a}[\psi]. \tag{4.182}$$

In general, a solution is not guaranteed. Suppose there exists a state $\underline{\mathbf{a}} \in \mathcal{A}$ such that

$$\varepsilon_{\underline{\mathbf{a}}}[\psi_{\underline{\mathbf{a}}}] = \max_{\mathbf{a} \in \mathcal{A}} \varepsilon_\mathbf{a}[\psi_\mathbf{a}]. \tag{4.183}$$

This means that among all optimal classifiers for states in $\mathcal{A}$, the one having the greatest error is for the distribution $F_{\underline{\mathbf{a}}}$, which is called a *least favorable distribution*. A least favorable distribution is said to be a *saddle-point solution* to the optimization of equation (4.182) if

$$\varepsilon_\mathbf{a}[\psi_{\underline{\mathbf{a}}}] \leq \varepsilon_{\underline{\mathbf{a}}}[\psi_{\underline{\mathbf{a}}}] \leq \varepsilon_{\underline{\mathbf{a}}}[\psi] \tag{4.184}$$

for all $\mathbf{a} \in \mathcal{A}$ and all $\psi \in \mathcal{C}$. The right inequality holds because $\psi_{\underline{\mathbf{a}}}$ is optimal for $F_{\underline{\mathbf{a}}}$, and the left inequality says that $\psi_{\underline{\mathbf{a}}}$ performs at least as well for any distribution as it does for the least favorable distribution, for which it is optimal. If a least favorable distribution is a saddle-point solution, then $\psi_{\underline{a}}$ is a minimax robust classifier because, according to the left inequality, for it the supremum in equation (4.182) is $\varepsilon_{\underline{a}}[\psi_{\underline{a}}]$ and, according to the right inequality, the infimum in equation (4.182) is attained for $\psi_{\underline{a}}$. Finding a minimax robust classifier is reduced to finding a least favorable distribution that is a saddle-point solution. Besides the difficulty in finding a least favorable distribution that is a saddle-point solution, a basic problem with minimax robustness is that it is conservative and uses no knowledge of the distribution of the parameter vector. It gives great weight to states at which a classifier performs badly, and these states may occur rarely.

From a Bayesian approach, the expected error increase from using the classifier $\psi_b$ that is optimal for distribution $F_b$ instead of the classifier $\psi_a$ that is optimal relative to the state $\mathbf{a}$ is the expectation of $\kappa(\mathbf{a}; \mathbf{b})$ relative to the density $f(\mathbf{a})$ for the random vector $\mathbf{a}$:

$$\kappa(\mathbf{b}) = E_{\mathbf{a}}[\kappa(\mathbf{a}; \mathbf{b})] = \int_{-\infty}^{\infty} \int_{-\infty}^{\infty} \cdots \int_{-\infty}^{\infty} \kappa(\mathbf{a}; \mathbf{b}) f(\mathbf{a}) \, d\mathbf{a}, \qquad (4.185)$$

where $E_{\mathbf{a}}$ denotes expectation relative to the states. Here $\kappa(\mathbf{b})$ defines the *mean robustness*, and any minimizing value of $\mathbf{b}$ is a *maximally robust state*. The expected error from using $\psi_b$ is given by

$$\varepsilon(\mathbf{b}) = E_{\mathbf{a}}[\varepsilon_{\mathbf{a}}[\psi_b]] = E_{\mathbf{a}}[\kappa(\mathbf{a}; \mathbf{b}) + \varepsilon_{\mathbf{a}}[\psi_{\mathbf{a}}]] = \kappa(\mathbf{b}) + E_{\mathbf{a}}[\varepsilon_{\mathbf{a}}[\psi_{\mathbf{a}}]], \qquad (4.186)$$

where $E_{\mathbf{a}}[\varepsilon_{\mathbf{a}}[\psi_{\mathbf{a}}]]$ is the expected error of the optimal classifier relative to the distribution for which it is optimal. If $\mathbf{b}^*$ is a maximally robust state, then the optimal classifier for state $\mathbf{b}^*$, called a *Bayesian robust classifier*, performs best relative to the performance across all states. Practically, one would like to design a classifier at a maximally robust state, or at least at a state $\mathbf{b}$ for which $\kappa(\mathbf{b})$ is close to minimal.

Rather than design the optimal classifier for a maximally robust state, one can take a distributionwide approach and define a single *global classifier* that takes into account the distribution of the parametric mass. A simple way to do this is to use the optimal classifier for $E[\mathbf{a}]$. This is a naive approach in which one designs a single optimal classifier at an "average state of nature." $E[\mathbf{a}]$ is not typically maximally robust, although it may be. The mean takes into account the parameter distribution but not the characteristics determining the optimal classifier. Here we take into account the expectation of these characteristics. If the feature-label distributions are characterized by parameters, such as the mean vectors and covariance matrices, or if a constrained classifier depends on only certain characteristics of the feature-label distribution, then equation (4.181) can be written in the form

$$\kappa(\mathbf{a}; \mathbf{b}) = \varepsilon[Q_{X,Y,\mathbf{a}}; \psi_b] - \varepsilon[Q_{X,Y,\mathbf{a}}; \psi_{\mathbf{a}}], \qquad (4.187)$$

where the relevant characteristics are provided in functional form by $Q_{X,Y\mathbf{a}}$ and $\varepsilon[Q_{X,Y,\mathbf{a}}; \psi_{\mathbf{a}}]$ denotes classifier error. $Q_{X,Y\mathbf{a}}$ can be averaged to obtain a mean characteristic function,

$$Q_{X,Y;\bullet} = \int\limits_{-\infty}^{\infty} Q_{X,Y;\mathbf{a}} f(\mathbf{a}) \, d\mathbf{a}, \tag{4.188}$$

and the global classifier is defined as the optimal classifier relative to $Q_{X,Y;\bullet}$. Once a global classifier $\psi_\bullet$ is defined, it cannot outperform the state-specific optimal classifier $\psi_\mathbf{a}$ for the distribution $F_\mathbf{a}$. We denote the increase in error from applying $\psi_\bullet$ instead of $\psi_\mathbf{a}$ for $F_\mathbf{a}$ by $\kappa(\mathbf{a}; \bullet)$. The expected error increase from application of the global classifier is the constant

$$\kappa(\bullet) = E[\kappa(\mathbf{a}; \bullet)] = \int\limits_{-\infty}^{\infty} \kappa(\mathbf{a}; \bullet) f(\mathbf{a}) \, d\mathbf{a}. \tag{4.189}$$

### 4.9.2 Performance Comparison for Robust Classifiers

We examine robustness using spherical 10-dimensional (10-feature) Gaussian class conditional distributions $N(\mathbf{0}, \Sigma_0)$ and $N(\mathbf{1}, \Sigma_1)$ with means $\mathbf{0} = (0, 0, \ldots, 0)$ and $\mathbf{1} = (1, 1, \ldots, 1)$ and standard deviations $\sigma_0$ and $\sigma_1$, respectively. We consider three settings: uncertainty in the mean of the first variable of $N(\mathbf{1}, \Sigma_1)$ with $\sigma_0 = \sigma_1$; uncertainty in the mean of the first variable of $N(\mathbf{1}, \Sigma_1)$ with $\sigma_0 \neq \sigma_1$; and uncertainty in the variance with fixed means $\mathbf{0}$ and $\mathbf{1}$. In all cases, the offset (state variable) $b$ is uniformly distributed over $[-1, 1]$. For each case we plot six error curves as functions of the state: (1) $\varepsilon_b[\psi_b]$, the error of the state-specific optimal classifier for state $b$ at state $b$; (2) $E_a[\varepsilon_a[\psi_b]]$, the expected error of the optimal classifier for state $b$ across all states; (3) $\varepsilon_b[\psi_{b^*}]$, the error of the Bayesian robust classifier at state $b$; (4) $\varepsilon_b[\psi_\bullet]$, the error of the global classifier at state $b$; (5) $E_a[\varepsilon_a[\psi_{b^*}]]$, the expected error of the Bayesian robust classifier across all states; and (6) $E_a[\varepsilon_a[\psi_\bullet]]$, the expected error of the global classifier across all states. In the offset distribution considered, $\mu_b = 0$, so the global classifier is simply the optimal classifier in the absence of uncertainty, which would not be the case if the mean of the offset distribution were not 0.

For the offset of the first-variable mean with $\sigma_0 = \sigma_1 = 1$, the optimal classifier is determined by LDA. The error $\varepsilon_a[\psi_b]$ and the robustness surface $\kappa(a; b)$ are shown in figures 4.10 and 4.11, respectively. Error curves are shown in figure 4.12, where we see that $E[b] = 0$ and the maximally robust state is $-0.10$. The error $\varepsilon_b[\psi_b]$ of the state-specific optimal classifier is at a maximum of 0.5 for state $b = -1$ and falls monotonically to almost 0 for state $b = 1$, thus reflecting complete identification of the class conditional distributions at $b = -1$ and their increasing separation for increasing $b$. The expected error $E_a[\varepsilon_a[\psi_b]]$ of the state-specific optimal classifier drops very fast for states near $-1$ as the conditional distributions begin to separate, reaches a minimum at the maximally robust state, and increases thereafter.

In comparing the error $\varepsilon_b[\psi_{b^*}]$ of the Bayesian robust classifier at state $b$ with the error, $\varepsilon_b[\psi_b]$ of the state-specific optimal classifier for state $b$ at state $b$, $\varepsilon_b[\psi_b] \leq \varepsilon_b[\psi_{b^*}]$, as must be the case. Thus, if one is certain of the state at which the classifier is to be applied, it is better to use the optimal classifier for this state. Given this

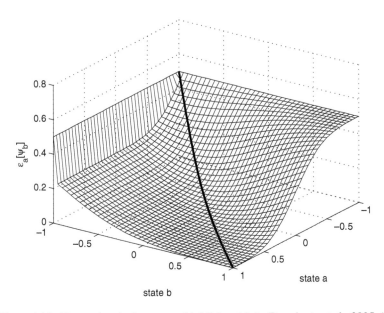

Figure 4.10. Uncertainty in the mean with LDA: $\varepsilon_a[\psi_b]$. (Dougherty et al., 2005a.)

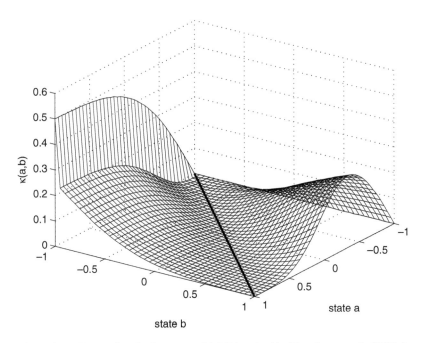

Figure 4.11. Uncertainty in the mean with LDA: $\kappa(a; b)$. (Dougherty et al., 2005a.)

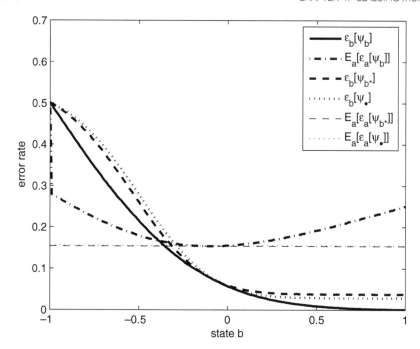

Figure 4.12. Error curves for uncertainty in the mean with LDA. (Dougherty et al., 2005a.)

state, say $b$, the cost of applying the Bayesian robust classifier is $\varepsilon_b[\psi_{b^*}] - \varepsilon_b[\psi_b]$; however, in the face of uncertainty, the gain in applying the Bayesian robust classifier is $E_a[\varepsilon_a[\psi_b]] - E_a[\varepsilon_a[\psi_{b^*}]]$. This gain is always positive and is substantial for states far from the maximally robust state. The global filter is always outperformed by the Bayesian robust filter and substantially so for certain states.

The potential pathological behavior of the minimax filter is clearly demonstrated. Specifically, the optimal classifier is linear for all states, and no matter at which state the classifier is designed, the maximum error for this classifier occurs at state $b = -1$ and is equal to 0.5. Thus, all states are minimax states and the maximum error of "the" minimax classifier is 0.5, thereby completely negating the reason for using the minimax classifier in the first place.

Now consider uncertainty in the mean of the first variable with $\sigma_0 = 1$ and $\sigma_1 = 2$. Because the covariance matrices are unequal, the optimal classifier is determined by QDA. The six curves are shown in figure 4.13. Many of the comments regarding the LDA case in figure 4.12 apply here; however, there are some noteworthy differences. When $b = -1$, so that the means are identical, except for the expectations of the errors for the Bayesian robust and global filters, the error for equal variances is 0.5, but these errors are much less for $\sigma_0 = 1$ and $\sigma_1 = 2$, and they are unequal. This is because the mass of $N(\mathbf{1}, \mathbf{\Sigma}_1)$ is much less concentrated about the mean than the mass of $N(\mathbf{1}, \mathbf{\Sigma}_0)$. Key to robustness is that $E_a[\varepsilon_a[\psi_b]]$, the expected error of the optimal classifier for state $b$ across all states, is almost flat for $b \leq 0.2$. The maximally robust state is $b = -0.2$; however, designing the classifier at any state

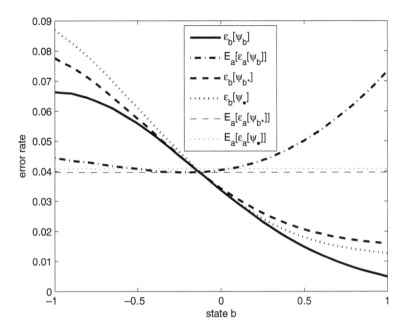

Figure 4.13. Error curves for uncertainty in the mean with QDA. (Dougherty et al., 2005a.)

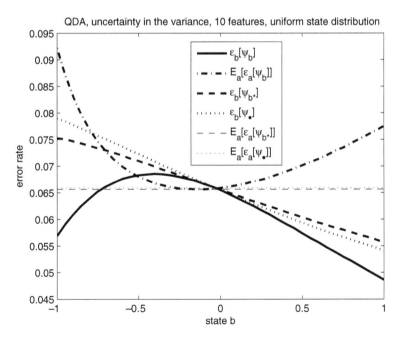

Figure 4.14. Error curves for uncertainty in the variance with QDA. (Dougherty et al., 2005a.)

$b \leq 0.2$ will do about as well. In particular, design at the mean state $b = 0$ yields a very robust classifier.

Finally, consider spherical Gaussian distributions $N(\mathbf{0}, \boldsymbol{\Sigma}_0)$ and $N(\mathbf{1}, \boldsymbol{\Sigma}_1)$ with diagonal $\boldsymbol{\Sigma}_0$ and $\boldsymbol{\Sigma}_1$ having variances $\sigma_0^2 = 1$ and $\sigma_1^2 = 2$. The variance for $N(\mathbf{1}, \boldsymbol{\Sigma}_1)$ is uncertain with uniform variance offset $b$ on $[-1, 1]$. The curves are shown in figure 4.14. Two points should be noted. First, $\varepsilon_b[\psi_b]$, the error of the state-specific optimal classifier for state $b$ at state $b$, has a maximum at $b = -0.4$. Second, as $b$ moves away from the maximally robust state, the gain $E_a[\varepsilon_a[\psi_b]] - E_a[\varepsilon_a[\psi_{b*}]]$ of applying the Bayesian robust classifier increases dramatically.

# Bibliography

Alexander K. (1984) Probability inequalities for empirical processes and a law of the iterated logarithm. *Ann Probab* 4: 1041–67.

Bartlett P. (1993) Lower bounds on the Vapnik-Chervonenkis dimension of multi-layer threshold networks, *Proceeding of the Sixth Annual ACM Conference on Computational Learning Theory*, Association for Computing Machinery, New York, pp. 144–50.

Baum E. (1988) On the capabilities of multilayer perceptrons. *Complexity*, 4: 193–215.

Braga-Neto UM, Hashimoto R, Dougherty ER, Nguyen DV, Carroll RJ. (2004) Is cross-validation better than resubstitution for ranking Genes? *Bioinformatics* 20(2): 253–58.

Braga-Neto UM, Dougherty ER. (2004a) Is cross-validation valid for small-sample microarray classification? *Bioinformatics* 20(3): 374–80.

Braga-Neto UM, Dougherty ER. (2004b) Bolstered error estimation. *Pattern Recognition* 37(6): 1267–81.

Braga-Neto U, Dougherty ER. ( 2005) Exact performance measures and distributions of error estimators for discrete classifiers. *Pattern Recognition* 38(11): 1799–814.

Brun M, Sabbagh D, Kim S, Dougherty ER. (2003) Corrected small-sample estimation of the Bayes error. *Bioinformatics* 19(8): 944–51.

Chernick M. (1999) *Bootstrap Methods: A Practitioner's Guide*, John Wiley & Sons, New York.

Cover T. (1965) Geometrical and statistical properties of systems of linear inequalities with applications in pattern recognition. *IEEE Trans Electron Comput* 14: 326–34.

Cover T, Hart P. (1967) Nearest-neighbor pattern classification. *IEEE Trans Inform. Theory* 13: 21–7.

Cybenko G. (1989) Approximation by superposition of sigmoidal functions. *Math Control Signals Systems* 2: 303–14.

Devroye L. (1982) Necessary and sufficient conditions for the almost every-where convergence of nearest neighbor regression function estimates. *Zeitschrift fur Wahrscheinlichkeitstheorie und verwandte Gebiete* 61: 467–81.

Devroye L, Wagner T. (1976) A Distribution-free performance bound on error estimation. *IEEE Trans Inform Theory* 22: 586–7.

Devroye L, Lugosi G. (1995) Lower bounds in pattern recognition and learning. *Pattern Recognition* 28: 1011–8.

Devroye L, Gyorfi L, Lugosi G. (1996) *A Probabilistic Theory of Pattern Recognition*, Springer-Verlag, New York.

Devroye L, Krzyzak A. (1989) An equivalence theorem for $L_1$ convergence of the kernel regression estimate. *Stat Planning Inference* 23: 71–82.

Dougherty ER. (2001) Small sample issues for microarray-based classification. *Comp Funct Genomics* 2: 28–34.

Dougherty ER, Attoor SN. (2002) Design issues and comparison of methods for microarray-based classification. *Computational and Statistical Approaches to Genomics*. Zhang W, Shmulevich I, eds., Kluwer Academic Publishers, New York.

Dougherty ER, Braga-Neto U. (2006) Epistemology of computational biology: mathematical models and experimental prediction as the basis of their validity. *Biol Syst* 14(1): 65–90.

Dougherty ER, Hua J, Xiong Z, Chen Y. (2005a) Optimal robust classifiers. *Pattern Recognition* 38: 1520-32.

Dougherty ER, Datta A, Sima C. (2005b) Research Issues in Genomic Signal Processing. *IEEE Signal Process Mag* 22(6): 46-68.

Duda RO, Hart PE, Stork DG. (2001) *Pattern Classification*, John Wiley & Sons, New York.

Efron B. (1979) Bootstrap methods: another look at the jacknife. *Ann Stat* 7: 1–26.

Efron B. (1983) *The Jacknife, the Bootstrap, and Other Resampling Plans*, SIAM NSF-CBMS, Monograph 38, SIAM, Philadelphia, PA.

El-Sheikh TS, Wacker AG. (1980) Effect of dimensionality and estimation on the performance of Gaussian classifiers. *Pattern Recognition* 12: 115–26.

Farago A, Lugosi G. (1993) Strong universal consistency of neural network classi-fiers. *IEEE Trans Inform Theory* 39: 1146–51.

Fukunaga K, Hayes RR. (1989) Effects of sample size in classifier design. *IEEE Trans Pattern Anal Mach Intell* 11(8): 873–85.

Funahashi K. (1989) On the approximate realization of continuous mappings by neural networks. *Neural Netw* 2: 183–92.

Glick N. (1978) Additive estimators for probabilities of correct classification. *Pattern Recognition* 10: 211–22.

Gordon L, Olshen R. (1978) Asymptotically efficient solutions to the classification problem. *Ann Stat* 6: 525-33.

Grygorian AM, Dougherty ER. (2001a) Design and analysis of robust optimal binary filters in the context of a prior distribution for the states of nature. *Math Imaging Vision* 11: 239–54.

Grygorian AM, Dougherty ER. (2001b) Bayesian robust optimal linear filters. *Signal Process* 81: 2503–21.

Hedenfalk I, Duggan D, Chen Y, Radmacher M, Bittner M, Simon R, Meltzer P, Gusterson B, Esteller M, Raffeld M, Yakhini Z, Ben-Dor A, Dougherty E, Kononen J, Bubendorf L, Fehrle W, Pittaluga S, Gruvverger S, Loman N, Johannsson O, Olsson H, Wifond B, Sauter G, Kallioniemi OP, Borg A, Trent J. (2001) Gene expression profiles in hereditary breast cancer. *N Engl J Med* 344(8): 539–48.

Hempel FR, Rousseeuw PJ, Ronchetti EM, Stahel WA. (1986) *Robust Statistics: The Approach Based on Influence Functions*, John Wiley & Sons, New York.

Hornik K. (1991) Approximation capabilities of multi-layer feed-forward networks. *Neural Netw* 4: 251–7.

Hornik K, Stinchombe M, White H. (1989) Multilayer feedforward networks are universal approximators. *Neural Netw* 2: 359–66.

Huber PJ. (1981) *Robust Statistics*, John Wiley & Sons, New York.

Kassam SA, Lim TI. (1977) Robust Wiener filters. *Franklin Institute* 304: 171-85.

Lugosi G, Nobel A. (1996) Consistency of data-driven histogram methods for density estimation and classification. *Ann Stat* 24: 687–706.

McFarland HR, Richards D. (2002) Exact misclassification probabilities for plug-in normal quadratic discriminant functions. II. The heterogeneous case. *Multivariate Anal* 82: 299–330.

Peltonen S, Kuosmanen P, Astola J. (2001) Output distributional influence function. *IEEE Trans Signal Proces* 49: 1953–60.

Poor HV. (1980) On robust Wiener filtering. *IEEE Trans Automat Control* 26: 531–6.

Raudys S, Pikelis V. (1980) On dimensionality, sample size, classification error, and complexity of classification algorithms in pattern recognition. *IEEE Trans Pattern Anal Mach Intell* 2(3): 242–52.

Rodgers W, Wagner T. (1978) A finite sample distribution-free performance bound for local discrimination rules. *Ann Stat* 6: 506–14.

Rosenblatt F. (1962) *Prinicples of Neurodynamics: Perceptrons and the Theory of Brain Mechanisms*, Spartan, Washington, DC.

Sima C, Braga-Neto U, Dougherty ER. (2005a) Superior feature-set ranking for small samples using bolstered error estimation. *Bioinformatics* 21(7): 1046–54.

Snapinn S, Knoke J. (1985) An evaluation of smoothed classification error-rate estimators. *Technometrics* 27(2) 199–206.

Stone C. (1977) Consistent nonparametric regression. *Ann Stat* 5: 595–645.

Tutz G. (1985) Smoothed additive estimators for non-error rates in multiple discrimination analysis. *Pattern Recognition* 18(2): 151–9.

van de Vijver MJ, He YD, van't Veer LJ, Dai H, Hart AAM, Voskuil DW, Schreiber GJ, Peterse JL, Roberts C, Marton MJ, et al. (2002) Gene expression signature as a predictor of survival in breast cancer. *N Engl J Med* 347: 1999–2009.

van't Veer LJ, Dai H, van de Vijver MJ, He YD, Hart AAM, Mao M, Peterse HL, van der Kooy K, Marton MJ, Witteveen AT, et al. (2002). Gene expression profiling predicts clinical outcome of breast cancer. *Nature* 415: 530–6.

Vapnik VN. (1998) *Statistical Learning Theory*, John Wiley & Sons, New York.

Vapnik VN, Chervonenkis A. (1974) *Theory of Pattern Recognition*, Nauka, Moscow.

Vapnik VN, Chervonenkis A. (1971) On the uniform convergence of relative frequencies of events to their probabilities. *Theory Probab Appl* 16: 264–80.

Verdu S, Poor HV. (1984) On minimax robustness: a general approach and applications. *IEEE Trans Inform Theory* 30: 328–40.

Wald PW, Kronmal RA. (1977) Discriminant functions when covariances are unequal and sample sizes are moderate. *Biometrics* 33: 479–84.

Wilks SS. (1962) *Mathematical Statistics*. John Wiley & Sons, New York.

# Chapter Five

## Regularization

Thus far we have taken the perspective that a collection of features is given, sample data are obtained, a classifier based on the features is designed from the data via a classification rule, and the error of the classifier is estimated from the data. The key design issue has been the relation between the classification rule and the sample size. The feature set and the sample data are taken as given, and the designer selects a classification rule. In this chapter, we consider alterations to this paradigm. First, we do not assume that the feature set is given; instead, it may be obtained from a larger set of potential features or it may result from a transformation of the features. Second, we do not assume that a classification rule is to be applied directly to the sample data; rather, the data may be supplemented or completely replaced by synthetic data. Third, we do not assume that a classification rule is chosen independently of the data; instead, based on the data, a classification rule may be automatically selected from a family of classification rules. We are not really changing directions; indeed, these adaptations are being made to address the same key design issue, the relation between the classification rule and the sample size. Reducing the number of features constrains the classifier; supplementing the data increases the sample size; and not fixing the classification rule allows a data-driven adjustment between classifier complexity and sample size.

## 5.1 DATA REGULARIZATION

Rather than design a classifier precisely according to some classification rule when the sample is small, it can be beneficial to *regularize* the sample data or the parameters estimated from the data, where by regularization we mean some alteration of the data or modification of the estimation rule for the parameters.

### 5.1.1 Regularized Discriminant Analysis

Even with linear constraint, the expected estimation error $E[\Delta_{n,C}]$ can be severe for very small samples. For quadratic (and linear) discriminant analysis, the key is the estimated covariance matrix $\hat{\mathbf{K}}_k$, which is typically the sample covariance matrix (which we assume here).

The small-sample problem for QDA can be appreciated by considering the spectral decompositions of the covariance matrices,

$$\mathbf{K}_k = \sum_{j=1}^{d} \lambda_{kj} \mathbf{v}_{kj} \mathbf{v}_{kj}^t, \tag{5.1}$$

where $\lambda_{k1}, \lambda_{k2}, \ldots, \lambda_{kd}$ are the eigenvalues of $\mathbf{K}_k$ in decreasing order and $\mathbf{v}_{kj}$ is the eigenvector corresponding to $\lambda_{kj}$. Then,

$$\mathbf{K}_k^{-1} = \sum_{j=1}^{d} \frac{\mathbf{v}_{kj}\mathbf{v}_{kj}^t}{\lambda_{kj}} \tag{5.2}$$

and the quadratic discriminant of equation (4.73) takes the form

$$d_k(\mathbf{x}) = -\sum_{j=1}^{d} \frac{[\mathbf{v}_{kj}^t(\mathbf{x} - \mathbf{u}_k)]^2}{\lambda_{kj}} - \sum_{j=1}^{d} \log \lambda_{kj} + 2 \log f(k). \tag{5.3}$$

It is clear that the discriminant is strongly influenced by the smallest eigenvalues. This creates a difficulty because the large eigenvalues of the sample covariance matrix are biased high and the small eigenvalues are biased low—and this phenomenon is accentuated for small samples. This problem also affects least-squares linear discriminants because they involve inversion of the sample autocorrelation matrix. It has long been addressed by trying to improve estimation of the eigenvalues (Haff, 1980; Dey and Srmivasan, 1985).

Relative to quadratic discriminant analysis, a simple method of regularization is to apply LDA even though the covariance matrices are not equal. This means estimating a single covariance matrix by pooling the data. This reduces the number of parameters to be estimated and increases the sample size relative to the smaller set of parameters. In effect, it reduces the VC dimension of the classification rule. Generally, regularization reduces variance at the cost of bias, and the goal is substantial variance reduction with negligible bias.

Figure 5.1 illustrates QDA regularization by applying LDA instead of QDA. The data come from two-dimensional spherical Gaussian class conditional distributions, the one on the left possessing a smaller variance. Figure 5.1a and 5.1b show the LDA and QDA classifiers for a sample of size 30, respectively, with the quadratic curve depicting the optimal QDA classifier for the full distributions. The LDA classifier has a smaller error than the QDA classifier, illustrating the advantage of regularization. The situation is different in figure 5.1c and 5.1d, which show the LDA and QDA classifiers for a sample of size 90, respectively. Here the large sample size has facilitated accurate design of the QDA classifier, and it is quite close to optimal, thereby yielding a smaller error than that of the LDA classifier.

A softer approach than strictly going from QDA to LDA is to shrink the individual covariance estimates in the direction of the pooled estimate. This can be accomplished by introducing a parameter $\alpha$ between 0 and 1 and using the estimates

$$\hat{\mathbf{K}}_k(\alpha) = \frac{n_k(1-\alpha)\hat{\mathbf{K}}_k + n\alpha\hat{\mathbf{K}}}{n_k(1-\alpha) + \alpha n}, \tag{5.4}$$

where $n_k$ is the number of points corresponding to $Y = k$, $n$ is the total number of points, and $\hat{\mathbf{K}}$ is the pooled estimate of the covariance matrix. QDA results from

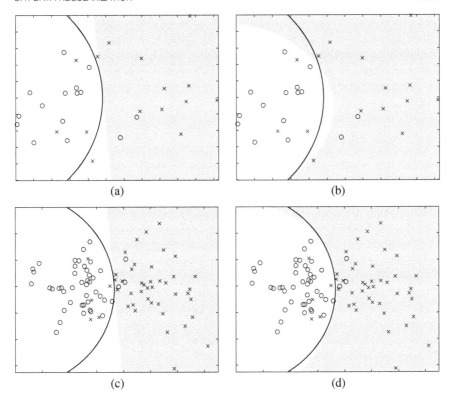

Figure 5.1. QDA regularization by using LDA applied to two unequal-variance circular Gaussian class conditional distributions, with a quadratic line showing the optimal classifier. (a) LDA applied to a 30-point sample, error 0.0998; (b) QDA applied to a 30-point sample, error 0.1012; (c) LDA applied to a 90-point sample, error 0.1036; (d) QDA applied to a 90-point sample, error 0.0876.

$\alpha = 0$ and LDA from $\alpha = 1$, with different amounts of shrinkage occurring for $0 < \alpha < 1$ (Titterington, 1985). While reducing variance, one must be prudent in choosing $\alpha$, especially when the covariance matrices are very different.

To obtain more regularization while producing little bias, one can shrink the regularized sample covariance matrix $\hat{\mathbf{K}}_k(\alpha)$ toward the identity multiplied by the average eigenvalue of $\hat{\mathbf{K}}_k(\alpha)$. This has the effect of decreasing large eigenvalues and increasing small eigenvalues, thereby offsetting the biasing effect seen in equation (5.3) (Friedman, 1989). Thus, we consider the estimate

$$\hat{\mathbf{K}}_k(\alpha, \beta) = (1 - \beta)\hat{\mathbf{K}}_k(\alpha) + \frac{\beta}{n} \, \mathrm{tr}[\hat{\mathbf{K}}_k(\alpha)]\mathbf{I}, \qquad (5.5)$$

where $\mathrm{tr}[\hat{\mathbf{K}}_k(\alpha)]$ is the trace of $\hat{\mathbf{K}}_k(\alpha)$, $\mathbf{I}$ is the identity, and $0 \le \beta \le 1$. To apply this *regularized discriminant analysis* using $\hat{\mathbf{K}}_k(\alpha, \beta)$ requires selecting two model parameters. Model selection is critical to advantageous regularization and typically is problematic. Cross-validation error estimation has been suggested for selecting

$\alpha$ and $\beta$ in equation (5.5). As we have seen, cross-validation estimation tends to be unreliable with small samples, precisely the situation in which we want to apply regularization; nevertheless, simulation results for Gaussian conditional distributions indicate a benefit of regularization for various covariance models and very little increase in error even in models where it does not appear to help (Friedman, 1989).

### 5.1.2 Noise Injection

Rather than regularizing the estimated covariance matrix, one can regularize the data itself by *noise injection*. This can be done by "spreading" the sample data by generating synthetic data about each sample point, thereby creating a large synthetic sample from which to design the classifier while at the same time making the designed classifier less dependent on the specific points in the small data set. For instance, one can place a circular Gaussian distribution at each sample point, randomly generate points from each such distribution, and then apply classifier design. This approach is not limited to any particular classification rule. However, it can be posed analytically in terms of matrix operations for linear classification, and this is critical to situations in which a large number of feature sets must be examined. We discuss analytic noise injection for least-squares linear classification as proposed for microarray-based classification, where a vast number of potential feature sets are involved (Kim et al., 2002a).

Let $(\mathbf{U}, V)$ be the random vector having equally likely outcomes $(\mathbf{x}_1, y_1)$, $(\mathbf{x}_2, y_2), \dots, (\mathbf{x}_n, y_n)$ among the sample data. Let $\mathbf{Z}$ be a zero-mean random $d$-vector independent of $(\mathbf{U},V)$ and consider the random vector $(\mathbf{U}+\mathbf{Z}, V)$. By the way in which the distribution of $(\mathbf{U},V)$ is defined across the data, the autocorrelation matrix and cross-correlation vector for $\mathbf{U}^{(1)} = (1,\mathbf{U}^t)^t$ are given by

$$\mathbf{R}_{\mathbf{U}^{(1)}} = \hat{\mathbf{R}}_{\mathbf{X}^{(1)}}, \tag{5.6}$$

$$E[\mathbf{U}^{(1)}V] = \hat{E}[\mathbf{X}^{(1)}Y], \tag{5.7}$$

where $\mathbf{X}^{(1)} = (1, \mathbf{X}^t)^t$. Since

$$(\mathbf{U}+\mathbf{Z})^{(1)} = \mathbf{U}^{(1)} + \mathbf{Z}^{(0)}, \tag{5.8}$$

where $\mathbf{Z}^{(0)} = (0, \mathbf{Z}^t)^t$, by independence,

$$\mathbf{R}_{(\mathbf{U}+\mathbf{Z})^{(1)}} = \mathbf{R}_{\mathbf{U}^{(1)}} + \mathbf{R}_{\mathbf{Z}^{(0)}} = \hat{\mathbf{R}}_{\mathbf{X}^{(1)}} + \mathbf{R}_{\mathbf{Z}^{(0)}}, \tag{5.9}$$

$$E[(\mathbf{U}+\mathbf{Z})^{(1)}V] = E[\mathbf{U}^{(1)}V] = \hat{E}[\mathbf{X}^{(1)}Y]. \tag{5.10}$$

The optimal nonhomogeneous linear estimator of $V$ in terms of $\mathbf{U}+\mathbf{Z}$ has the weight vector

$$\mathbf{a} = \mathbf{R}_{(\mathbf{U}+\mathbf{Z})^{(1)}}^{-1} E[(\mathbf{U}+\mathbf{Z})^{(1)}V] = (\hat{\mathbf{R}}_{\mathbf{X}^{(1)}} + \mathbf{R}_{\mathbf{Z}^{(0)}})^{-1} \hat{E}[\mathbf{X}^{(1)}Y]. \tag{5.11}$$

It defines an *analytic noise injection regularization*. We refer to the resulting classifier as the σ-*classifier*.

If we wish to spread the data isotropically, then we can make **Z** possess a spherical uniform distribution or a spherical Gaussian distribution with $\mathbf{R_Z} = \sigma^2 \mathbf{I}$. In the case of a Gaussian spread, equation (5.9) takes the form

$$\mathbf{R}_{(U+Z)^{(1)}}(\sigma) = \hat{\mathbf{R}}_{\mathbf{X}^{(1)}} + \sigma^2 \begin{pmatrix} 0 & \mathbf{0}' \\ \mathbf{0} & \mathbf{I} \end{pmatrix}. \tag{5.12}$$

One might also consider a more general case in which the spread is isotropic but points have different dispersions associated with them—for instance, owing to a quality metric associated with each point. In this case, the variance vector for **Z** would not consist of equal variances.

When estimating the error of a linear classifier designed by noise injection, one needs to resort to error estimation using the training data. In the present setting, a natural error estimate is the measure of the mass of the $n$ distributions on the wrong side of the estimated hyperplane. This error mass can be found analytically, which is important when considering large numbers of feature sets; indeed, it is simply a special case of bolstered error estimation.

There is a close relation to equation (5.5). Ignoring the augmentation, which would not be there if we were considering only homogeneous optimization, equation (5.12) becomes

$$\mathbf{R}_{U+Z}(\sigma) = \hat{\mathbf{R}}_{\mathbf{X}} + \sigma^2 \mathbf{I}. \tag{5.13}$$

Setting $\alpha = 1$, equation (5.5) becomes

$$\hat{\mathbf{K}}(\beta) = (1 - \beta)\hat{\mathbf{K}} + \frac{\beta}{n} \operatorname{tr}[\hat{\mathbf{K}}]\mathbf{I}. \tag{5.14}$$

In each method, the estimate used to measure correlation, either the autocorrelation or covariance matrix, is offset by the scaled identity. In one case the scaling is motivated by eigenvalue estimation, and in the other, by the geometry of noise injection.

We illustrate application of the σ-classifier with a study using expression data from microarrays for 597 genes to identify gene combinations for use as glioma classifiers (Kim et al., 2002b). Gliomas are the most common malignant primary brain tumors. These tumors are derived from neuroepithelial cells and can be divided into two principal lineages: astrocytomas and oligodendrogliomas. Using a test set of 25 patients, single genes and two- to three-gene combinations are identified in distinguishing four types of gliomas: oligodendroglioma (OL), anaplastic oligodendroglioma (AO), anaplastic astrocytoma (AA), and glioblastoma multiforme (GM). Isotropic Gaussian σ-classifiers are designed for the classification of OL from the others, AO from the others, AA from the others, and GM from the others.

When using the σ-classifier, it is reasonable (although certainly not required) to employ bolstered error estimation with the same value of σ because the σ-classifier

Table 5.1 Errors and Increments for Discriminating AO from Other Gliomas

| Gene 1 | Gene 2 | Gene 3 | Error | Gain |
|--------|--------|--------|-------|------|
| DNase X | | | 0.1556 | |
| TNFSF5 | | | 0.1658 | |
| RAD50 | | | 0.1659 | |
| HBEGF | | | 0.1670 | |
| NF45 | | | 0.1731 | |
| DNase X | TNFSF5 | | 0.0750 | 0.0806 |
| DNase X | PTGER4 | | 0.0784 | 0.0772 |
| TNFSF5 | GNA13 | | 0.0826 | 0.0832 |
| DNase X | HGF | | 0.0892 | 0.0664 |
| TNFRSF5 | PTGER4 | | 0.0907 | 0.0947 |
| TNFSF5 | RAB5A | | 0.0909 | 0.0749 |
| TNFSF5 | SNF2L4 | | 0.0950 | 0.0708 |
| erbB4 | PTGER4 | | 0.1012 | 0.0841 |
| DNase X | β-PPT | | 0.1013 | 0.0544 |
| TNFSF5 | MERLIN | | 0.1020 | 0.0638 |
| DNase X | TNFSF5 | RAB5A | 0.0441 | 0.0309 |
| DNase X | TNFSRF5 | PTGER4 | 0.0454 | 0.0330 |
| TNFSF5 | RAB5A | GNA13 | 0.0464 | 0.0362 |
| DNase X | PTGER4 | SAP97 | 0.0476 | 0.0308 |
| TNFSF5 | GNA13 | HGF | 0.0526 | 0.0300 |
| TNFSF5 | β-PPT | PKA C-α | 0.0529 | 0.0549 |
| DNase X | β-PPT | R κ B | 0.0534 | 0.0479 |
| TNFSF5 | PKA C-α | LIG4 | 0.0591 | 0.0488 |
| TNFSF5 | LIG4 | HBGF-1 | 0.0616 | 0.0474 |
| TNFSF5 | β-PPT | SMARCA4 | 0.0625 | 0.0325 |

is designed in accordance with a Gaussian-spread distribution, and error estimation using Gaussian bolstered resubstitution with $\sigma_1 = \sigma_2 = \cdots = \sigma_n = \sigma$ is relative to the same distribution (Kim et al., 2002a). The spread $\sigma$ has been determined in the following manner. For feature vector $\mathbf{X} = (X_1, X_2, \ldots, X_d)$, let $\sigma_{k,j}$ be the standard deviation of $X_j$ on the class $k$, where $\sigma_{k,j}$ is estimated from the data, and let $\sigma_{max}$ be the maximum value of $\sigma_{k,j}$ for $k = 0, 1$ and $j = 1, 2, \ldots, d$. A normalized spread $\sigma_{nor}$, between 0 and 1, has been determined according to a model-based method, and the spread for the sample data is taken to be $\sigma = \sigma_{nor}\sigma_{max}$. Using $\sigma_{max}$ for normalization provides a conservative estimate of the misclassification error. For the glioma data, we focus on $\sigma_{nor} = 0.6$ because it provides conservative error estimation.

Table 5.1 gives the best 5 single-gene sets and the best 10 two-gene sets. It also gives the three-gene sets among the top 50 three-gene sets for which the error of the three-gene set is at least 0.03 less than the error of its best two-gene subset. The purpose of placing this requirement on the *marginal gain* of a three-gene set over its two-gene subsets is to avoid redundancy caused by adjoining features to already strong performing feature sets (see section 5.3.4 for a discussion of this kind of redundancy). For each feature set table 5.1 gives the error and the marginal gain. The top 3 three-gene sets on the full list (not correcting for redundancy) are the top 3 in the table; however, the fourth-best performer, {PTGER4, TNFSF5, DNase1L1}, does not appear in the table because its marginal gain is only 0.0275.

One could take a Monte Carlo approach to the vector **Z** in equations (5.9)–(5.11) and thereby not depend upon an analytic solution. Such an approach has been examined relative to linear discriminant analysis (Skurichina et al., 2000). An immediate advantage is that noise injection can be used in strictly data-driven iterative classifier designs. Moreover, aside from independence from the uniform distribution over the data points and zero mean, there are no distributional assumptions on **Z**. A spherical distribution need not be employed. Indeed, it has been demonstrated that it can be advantageous to base noise injection at a sample point on the nearest neighbors of the point (Skurichina et al., 2000). In (Holmström and Koistinen, 1992), the distribution of noise to be injected is determined through Gaussian kernel density estimation, and in this way it is shown that noise injection–based design can be asymptotically consistent as the sample size goes to infinity provided that the noise is chosen correctly. Noise injection can be applied to any classification rule and has been extensively examined relative to the rule and the amount of noise injected (Hua et al., 2006). Note that noise injection can take a different form in which the sample data points themselves are perturbed by additive noise instead of new synthetic points being generated. This approach has been used in designing neural networks, in particular, where owing to a small sample, the same data points are used repeatedly (Sietsma and Dow, 1988; Matsuoka, 1992; Bishop, 1995).

## 5.2 COMPLEXITY REGULARIZATION

As remarked previously, VC classes can perform arbitrarily poorly. Since performance depends on the distribution to which a class is applied, this indicates the risk involved in a poor choice of constraint. Our true interest is in finding a classifier whose performance is as close as possible to the Bayes error given the sample data. When designing a classifier from sample $S_n$ our interest is in the error increment between the designed classifier in the class $\mathcal{C}$ and the Bayes error,

$$\Delta_n^{\mathcal{C}} = \varepsilon_{n,\mathcal{C}} - \varepsilon_d. \tag{5.15}$$

Rewriting equation (4.36) in terms of this total cost yields

$$E[\Delta_n^{\mathcal{C}}] = \Delta_{\mathcal{C}} + E[\Delta_{n,\mathcal{C}}], \tag{5.16}$$

which reveals the balance we wish to achieve between constraint error and estimation error.

### 5.2.1 Regularization of the Error

Rather than consider a single class, one can consider a sequence of VC classes $\mathcal{C}^1, \mathcal{C}^2, \ldots$, find the best classifier in each class according to the data, and then choose from among these according to which class is of appropriate VC dimension for the sample size. For instance, one might assume a nested sequence $\mathcal{C}^1 \subset \mathcal{C}^2 \subset \ldots$; however, we do not make this assumption. We would then have a system of equations like equation (5.16), one for each class. We could then choose $k$ such that

$$E[\Delta_n^{\mathcal{C}^k}] = \inf_{j=1,2,\ldots} \{\Delta_{\mathcal{C}^j} + E[\Delta_{n,\mathcal{C}^j}]\}. \tag{5.17}$$

This would yield a result superior to equation (5.16), in which only a single class is considered. Of course, all of this must be achieved with sample data and apply to a particular estimated classifier in $\mathcal{C}^k$. Moreover, we do not know the expectations $E[\Delta_{n,\mathcal{C}^j}]$. However, suppose that we have bounds on these expectations, say, $E[\Delta_{n,\mathcal{C}^j}] \leq \lambda_{n,j}$, and these bounds are not too loose. Then, we could try to find some classifier $\phi$ such that

$$\varepsilon[\phi] - \varepsilon_d \approx \inf_{j=1,2,\ldots} \{\lambda_{n,j} + \Delta_{\mathcal{C}^j}\}. \tag{5.18}$$

The idea is to define a new error measurement that takes into account both the error estimate of a designed classifier and the complexity of the class from which it has been chosen—the more complex the class, the greater the penalty. Applying theorem 4.7 in terms of $\mathcal{C}$ and in conjunction with equation (4.53) yields

$$P(\hat{\Delta}_n(\mathcal{C}) > \tau) \leq 8n^{V_\mathcal{C}} e^{-n\tau^2/8}. \tag{5.19}$$

Arguing as we did following theorem 4.7, we conclude that

$$E[\hat{\Delta}_n(\mathcal{C})] \leq 8\sqrt{\frac{V_\mathcal{C} \log(8en)}{2n}}. \tag{5.20}$$

Applying equation (4.44) in conjunction with the definition of $\hat{\Delta}_n(\mathcal{C})$ yields

$$E[|\hat{\varepsilon}_n[\hat{\psi}_{n,\mathcal{C}}] - \varepsilon[\hat{\psi}_{n,\mathcal{C}}]|] \leq E[\sup_{\psi \in \mathcal{C}} |\hat{\varepsilon}_n[\psi] - \varepsilon[\psi]|] \leq 8\sqrt{\frac{V_\mathcal{C} \log(8en)}{2n}} \tag{5.21}$$

(Devroye et al., 1996). Hence, on average,

$$\varepsilon[\hat{\psi}_{n,\mathcal{C}}] \leq \hat{\varepsilon}_n[\hat{\psi}_{n,\mathcal{C}}] + 8\sqrt{\frac{V_\mathcal{C} \log(8en)}{2n}}. \tag{5.22}$$

The square root term acts like a penalty added to the empirical error relative to the true error of the empirical error classifier. The more complex a class, the greater the risk of underestimating the true error with the empirical error, but adding the penalty ensures that the true error will, on average, be below the penalized empirical error.

In light of the preceding considerations, we define a new *penalized error* that is a sum of the empirical error and a complexity penalty $\rho(n)$:

$$\tilde{\varepsilon}_n[\psi] = \hat{\varepsilon}_n[\psi] + \rho(n). \tag{5.23}$$

Recognizing that the square root term in equation (5.22) provides an upper bound, we define the complexity penalty by

$$\rho(n) = \sqrt{\frac{8V_C \log(en)}{n}},$$  (5.24)

which provides a penalty less than the bound and for which the reason for the exact form will become apparent in the next theorem.

A VC-dimension complexity penalty of the kind given in equation (5.24) is a choice motivated by our preceding discussion; however, equation (5.23) is quite general, and $\rho$ can take different forms. As written, equations (5.23) and (5.24) define a penalized error in terms of classifier error and classifier complexity as measured by the VC dimension. More generally, one can consider a function of the error and some other measure of complexity. The central idea is to obtain a measurement involving both error and classifier complexity to achieve *complexity regularization* relative to the classifier class. Put another way, from a collection of classifier models we want to choose a model most suited for the amount of data. For instance, *minimum-description-length* complexity regularization tries to balance increased error and increased model complexity by replacing error minimization by minimization of a sum of entropies, one relative to encoding the error and the other relative to encoding the classifier description (Rissanen, 1978, 1986; Barron and Cover, 1989). We will proceed in the context of the VC dimension, noting that comparison of these kinds of model selection paradigms appears not to be easy (Kearns et al., 1997).

### 5.2.2 Structural Risk Minimization

Relative to a sequence $\mathcal{K} = \{C^1, C^2, \dots\}$ of classes, *structural risk minimization* proceeds by selecting the classifier in each class that minimizes the empirical error over the sample and then choosing from among these the one possessing minimal penalized error, where in each case the penalty is relative to the class containing the classifier (Vapnik and Chervonenkis, 1974; Vapnik, 1982). The *structural-risk-minimization classifier* is given by $\tilde{\psi}_{n,\mathcal{K}} = \hat{\psi}_{n,C^k}$, where

$$k = \arg\min_{j=1,2,\dots} \hat{\varepsilon}_n[\hat{\psi}_{n,C^j}] + \rho_j(n)$$  (5.25)

and $\rho_j(n)$ is the complexity penalty for $C^j$. The total cost, estimation plus constraint, is given by

$$\tilde{\Delta}_n^{\mathcal{K}} = \varepsilon[\tilde{\psi}_{n,\mathcal{K}}] - \varepsilon_d.$$  (5.26)

The next theorem asserts strong universal consistency and employs the constant

$$\gamma_{\mathcal{K}} = \sum_{j=1}^{\infty} e^{-V_{C^j}}$$  (5.27)

associated with $\mathcal{K}$. Here $\gamma_{\mathcal{K}}$ is assumed to be finite, which is assured if $V_{\mathcal{C}^{j+1}} > V_{\mathcal{C}^j}$.

**THEOREM 5.1 (Lugosi and Zeger, 1996)** *If* $\mathcal{K} = \{\mathcal{C}^1, \mathcal{C}^2, \ldots\}$ *is a sequence of VC classes such that* $\gamma_{\mathcal{K}}$ *is finite and, for any feature-label distribution,* $\varepsilon_{\mathcal{C}^j} \to \varepsilon_d$ *as* $j \to \infty$, *then the structural-risk-minimization classifier is strongly universally consistent.*

*Proof.* The proof consists of showing the convergence to zero with probability 1 for the two parts of the decomposition

$$\tilde{\Delta}_n^{\mathcal{K}} = (\varepsilon[\tilde{\psi}_{n,\mathcal{K}}] - \tilde{\varepsilon}_n[\tilde{\psi}_{n,\mathcal{K}}]) + (\tilde{\varepsilon}_n[\tilde{\psi}_{n,\mathcal{K}}]) - \varepsilon_d), \tag{5.28}$$

where we have denoted the empirical error classifier on $\mathcal{C}^j$ by $\hat{\psi}_{n,j}$. Applying the Vapnik-Chervonenkis theorem (to obtain the fourth inequality) yields

$$P(\varepsilon[\tilde{\psi}_{n,\mathcal{K}}] - \tilde{\varepsilon}_n[\tilde{\psi}_{n,\mathcal{K}}] > \tau) \le P(\sup_j(\varepsilon[\hat{\psi}_{n,j}] - \tilde{\varepsilon}_n[\hat{\psi}_{n,j}]) > \tau)$$

$$\le P(\sup_j(\varepsilon[\hat{\psi}_{n,j}] - \hat{\varepsilon}_n[\hat{\psi}_{n,j}]) > \tau + \rho_j(n))$$

$$= P\left(\bigcup_{j=1}^{\infty}\left[\varepsilon[\hat{\psi}_{n,j}] - \hat{\varepsilon}_n[\hat{\psi}_{n,j}] > \tau + \rho_j(n)\right]\right)$$

$$\le \sum_{j=1}^{\infty} P(|\varepsilon[\hat{\psi}_{n,j}] - \hat{\varepsilon}_n[\hat{\psi}_{n,j}]| > \tau + \rho_j(n))$$

$$\le \sum_{j=1}^{\infty} 8n^{V_{\mathcal{C}^j}} e^{-n(\tau+\rho_j(n))^2/8}$$

$$\le 8e^{-n\tau^2/8} \sum_{j=1}^{\infty} n^{V_{\mathcal{C}^j}} e^{-n\rho_j^2(n)/8}$$

$$= 8\gamma_{\mathcal{K}} e^{-n\tau^2/8}, \tag{5.29}$$

where the last equality follows from the definition of the complexity penalty. By the Borel-Cantelli lemma, $\varepsilon[\tilde{\psi}_{n,\mathcal{K}}] - \tilde{\varepsilon}_n[\tilde{\psi}_{n,\mathcal{K}}] \to 0$ with probability 1 as $n \to \infty$. Regarding the second term in the decomposition, for any $\tau > 0$ there exists $k$ such that $\varepsilon_{\mathcal{C}^k} - \varepsilon_d < \tau$. Let $n$ be sufficiently large that $\rho_k(n) < \tau/2$. The theorem will be proved by showing that

$$\limsup_{n \to \infty} \tilde{\varepsilon}_n[\tilde{\psi}_{n,\mathcal{K}}] - \varepsilon_{\mathcal{C}^k} \le 0 \tag{5.30}$$

with probability 1. Again using the Vapnik-Chervonenkis theorem,

$$P(\tilde{\varepsilon}_n[\hat{\psi}_{n,\mathcal{K}}] - \varepsilon_{\mathcal{C}^k} > \tau) \leq P(\tilde{\varepsilon}_n[\hat{\psi}_{n,k}] - \varepsilon_{\mathcal{C}^k} > \tau)$$

$$= P(\hat{\varepsilon}_n[\hat{\psi}_{n,k}] + \rho_k(n) - \varepsilon_{\mathcal{C}^k} > \tau)$$

$$\leq P(\hat{\varepsilon}_n[\hat{\psi}_{n,k}] - \varepsilon_{\mathcal{C}^k} > \tau/2)$$

$$\leq P(\sup_{\psi \in \mathcal{C}^k} |\hat{\varepsilon}_n[\psi] - \varepsilon[\psi]| > \tau/2)$$

$$\leq 8n^{V_{\mathcal{C}^k}} e^{-n\tau^2/32}, \tag{5.31}$$

which goes to 0 as $n \to \infty$, thereby completing the proof. $\qquad\square$

Whereas the preceding theorem shows that structural risk minimization is universally consistent, a slight modification of the proof provides bounds on the rate of convergence. Suppose the Bayes classifier lies in one of the VC classes and let $k$ be the smallest integer such that $\mathcal{C}^k$ contains the Bayes classifier. Then, in place of equation (5.28) we have the decomposition

$$\tilde{\Delta}_n^{\mathcal{K}} = (\varepsilon[\tilde{\psi}_{n,\mathcal{K}}] - \tilde{\varepsilon}_n[\tilde{\psi}_{n,\mathcal{K}}]) + (\tilde{\varepsilon}_n[\tilde{\psi}_{n,\mathcal{K}}]) - \varepsilon_{\mathcal{C}^k}). \tag{5.32}$$

Since

$$P(\tilde{\Delta}_n^{\mathcal{K}} > \tau) \leq P(\varepsilon[\tilde{\psi}_{n,\mathcal{K}}] - \tilde{\varepsilon}[\tilde{\psi}_{n,\mathcal{K}}] > \tau/2) + P(\tilde{\varepsilon}[\tilde{\psi}_{n,\mathcal{K}}] - \varepsilon_{\mathcal{C}^k} > \tau/2), \tag{5.33}$$

we can bound the left-hand side by bounding the two probabilities on the right-hand side in the same manner as in the previous theorem. The only difference is that in this case the probabilities involve $\tau/2$ instead of $\tau$, which will affect the resulting constants and mean that in bounding the second summand we must have $\rho_k(n) < \tau/4$ instead of $\rho_k(n) < \tau/2$. We obtain the next theorem.

**THEOREM 5.2 (Lugosi and Zeger, 1996)** *Let* $\mathcal{K} = \{\mathcal{C}^1, \mathcal{C}^2, \dots\}$ *be a sequence of VC classes such that* $\gamma_{\mathcal{K}}$ *is finite, the Bayes classifier is contained in one of the classes, $k$ is the smallest integer for which the Bayes classifier is in $\mathcal{C}^k$, and $\tau > 0$. If* $V_{\mathcal{C}^j} \log en < n\tau^2/128$, *then*

$$P(\tilde{\Delta}_n^{\mathcal{K}} > \tau) \leq 8\gamma_{\mathcal{K}} e^{-n\tau^2/32} + 8n^{V_{\mathcal{C}^k}} e^{-n\tau^2/128}. \tag{5.34}$$

Again reasoning as we did following theorem 4.7, there exists a constant $\alpha$ independent of $\mathcal{K}$ and an integer $k$ dependent on $\mathcal{K}$ such that

$$E[\tilde{\Delta}_n^{\mathcal{K}}] \leq \alpha \sqrt{\frac{V_{\mathcal{C}^k} \log n}{n}}. \tag{5.35}$$

Hence, $E[\tilde{\Delta}_n^{\mathcal{K}}] = O(\sqrt{\log n/n})$. An immediate consequence of equation (5.35) is that

$$E[\tilde{\Delta}_n^{\mathcal{K}}] \leq \inf_{j=1,2,\dots} \left( \alpha \sqrt{\frac{V_{\mathcal{C}^j} \log n}{n}} + \Delta_{\mathcal{C}^j} \right). \tag{5.36}$$

Recalling our desire to find a classifier satisfying equation (5.18), $\tilde{\psi}_{n,\mathcal{K}}$ is a candidate as long as the bound in equation (5.36) is not too loose.

The latter proviso points to the shortcomings of the method. The distribution-free character of the VC dimension means that the penalty term may unduly penalize well-behaved distributions while simultaneously contributing to the looseness of the bound in equation (5.36). Moreover, while the need to know the VC dimensions of the classes is mitigated by the fact that one can use upper bounds on the VC dimensions, replacing the VC dimensions in the penalty term by their upper bounds and adjusting the bounds accordingly, this will further add to the penalty and the bounds.

### 5.2.3  Empirical Complexity

To get around using upper bounds on the VC dimension and mitigate the distribution-free character of the VC dimension, one can try to replace the VC-based penalty term with a term based on empirical complexity—that is, a complexity penalty term estimated from the data. We briefly describe a theoretical approach to empirical complexity, called *adaptive model selection using empirical complexity* (AMSEC), for which there are impediments to practical implementation and whose full description we leave to the literature (Lugosi and Nobel, 1999).

The sample $S_n$ is split into two equal subsets $S_{n,1}$ and $S_{n,2}$. Based on $S_{n,1}$, the key part of the construction is to associate with each class $\mathcal{C}^j$ an *empirical complexity* $\kappa_{n,j}$ and a finite subset $\hat{\mathcal{C}}^j$ of $\mathcal{C}^j$ that serves as an approximation of $\mathcal{C}^j$ relative to the errors on the sample committed by classifiers in $\mathcal{C}^j$ (in a way that needs to be rigorously defined). Once this has been accomplished, let $\hat{\psi}^{n,j}$ be the empirical error classifier in $\hat{\mathcal{C}}^j$ relative to $S_{n,2}$, meaning that $\hat{\psi}^{n,j}$ has minimal empirical error over the second half of the sample among all classifiers in $\hat{\mathcal{C}}^j$. Letting $\hat{\varepsilon}_{n,2}$ denote empirical error over $S_{n,2}$, the AMSEC classifier is defined by $\hat{\psi}_{n,\mathcal{K}} = \hat{\psi}^{n,k}$, where

$$k = \arg\min_{j=1,2,\dots} \frac{2}{n}\hat{\varepsilon}_{n,2}[\hat{\psi}^{n,j}] + \kappa_{n,j} \tag{5.37}$$

and $2\hat{\varepsilon}_{n,2}[\hat{\psi}^{n,j}]/n$ is the relative empirical error, which serves as an estimate of the error. Note the similarity between equations (5.25) and (5.37). By analogy with equation (5.26), define

$$\hat{\Delta}_n^{\mathcal{K}} = \varepsilon[\hat{\psi}_{n,\mathcal{K}}] - \varepsilon_d. \tag{5.38}$$

The next theorem corresponds to equation (5.36).

**THEOREM 5.3 (Lugosi and Nobel, 1999)** *For a sequence* $\mathcal{K} = \{\mathcal{C}^1, \mathcal{C}^2, \dots\}$ *of classes,*

$$E[\hat{\Delta}_n^{\mathcal{K}}] \le \inf_{j=1,2,\dots}\left(55\sqrt{\frac{V_{\mathcal{C}^j}\log n}{n}} + 5.2\sqrt{\frac{\log j}{n}} + \Delta_{\mathcal{C}^j}\right). \tag{5.39}$$

Two important points can be made with regard to equation (5.39): first, ignoring the $5.2\sqrt{\log j/n}$ term, which is usually small in comparison to the first term, the bound is of the same order as the bound in equation (5.36); and second, the bound in equation (5.39) comes close to satisfying equation (5.18) (Lugosi and Nobel, 1999).

## 5.3 FEATURE SELECTION

High-throughput technologies such as gene expression microarrays offer the ability to simultaneously measure vast numbers of biological variables relatively fast and relatively cheaply, thereby providing enormous amounts of multivariate data with which to model biological processes. Yet there are obstacles inherent in dealing with extremely large numbers of interacting variables. In particular, large data sets may have the perverse effect of limiting the amount of scientific information that can be extracted because, as we have seen with overfitting, the ability to build models with scientific validity is negatively impacted by classification rules that are too complex for the number of available training data. Thus far we have focused on complexity owing to the structure of the classification rule; however, complexity also results from the number of variables. This can be seen in the VC dimension, for instance, of a linear classification rule whose VC dimension is $d + 1$, where $d$ is the number of variables. This dimensionality problem motivates feature (variable) selection when designing classifiers.

When there is feature selection, the feature selection algorithm is part of the classification rule. Feature selection may occur integrally with the classification rule, the *wrapper* method, it may occur prior to the application of a standard classification rule applied to a subset of features, the *filter* method, or it may be a combination of both. In any case, feature selection is part of the overall classification rule and, relative to this rule, the number of variables is the number in the data measurements, not the final number used in the designed classifier. Feature selection results in a subfamily of the original family of classifiers and thereby constitutes a form of constraint. For instance, if there are $d$ features available and linear discriminant analysis is used directly, then the classifier family consists of all hyperplanes in $d$-dimensional space, but if a feature selection algorithm reduces the number of variables to $m < d$ prior to the application of LDA, then the classifier family consists of all hyperplanes in $d$-dimensional space that are confined to $m$-dimensional subspaces. Feature selection yields classifier constraint, not a reduction in the dimensionality of the feature space relative to design. Since its role is constraint, assessing the worth of feature selection involves the standard dilemma: increasing constraint (greater feature selection) reduces design error at the cost of optimality. And we must not forget that the benefit of feature selection depends on the feature selection method and how it interacts with the rest of the classification rule.

### 5.3.1 Peaking Phenomenon

Perhaps the key issue in feature selection concerns error monotonicity. The Bayes error is monotone: if $A$ and $B$ are feature sets for which $A \subset B$, then $\varepsilon_B \leq \varepsilon_A$, where

$\varepsilon_A$ and $\varepsilon_B$ are the Bayes errors corresponding to $A$ and $B$, respectively. However, if $\varepsilon_{A,n}$ and $\varepsilon_{B,n}$ are the corresponding errors resulting from designed classifiers on a sample of size $n$, then it cannot be asserted that $\varepsilon_{A,n} \geq \varepsilon_{B,n}$. It may even be that $E[\varepsilon_{B,n}] > E[\varepsilon_{A,n}]$. Indeed, it is typical for the expected design error to decrease and then increase for increasingly large feature sets. Thus, monotonicity does not apply to designed classifiers. Moreover, even if $\varepsilon_{A,n} \geq \varepsilon_{B,n}$, this relation can be reversed when using estimates of the errors.

To more closely examine the lack of monotonicity for designed classifiers, consider a collection of classifier families $C_1 \subset C_2 \subset C_3 \subset \cdots$ for a fixed sample size $n$. A typical situation might be that, while the smaller families extensively reduce design cost, their reduction in optimality is too great. In such a case we might expect that the expected errors of the designed classifiers at first decrease as we utilize increasingly large families but then begin to increase when the increasing design cost outweighs the decreasing cost of constraint. If we now apply this reasoning to feature selection and consider a sequence $x_1, x_2, \ldots, x_d, \ldots$ of features, what will result is at first a decrease in the expected error as $d$ increases and then, after some point, an increase in the error for increasing $d$. While our description is idealized and the situation can be more complex, it serves to describe what is commonly called the *peaking phenomenon* (Hughes, 1968; Jain and Chandrasekaran, 1982).

In light of the peaking phenomenon, given a set of features, what is the optimal number of features? The answer depends on the classification rule, feature-label distribution, and sample size. To illustrate some of the possibilities we consider two models (Hua et al., 2005). For the *linear model*, the class conditional distributions are Gaussian with identical covariance matrix $\mathbf{K}$, the classes are equally likely, and the Bayes classifier results from linear discriminant analysis. In the *nonlinear model*, the class conditional distributions are Gaussian, with different covariance matrices $\mathbf{K}_0$ and $\mathbf{K}_1$, the classes are equally likely, and the Bayes classifier results from quadratic discriminant analysis. The maximum dimension is $D = 30$. Hence, the number of features available is less than or equal to 30, and the peaking phenomenon shows up only in graphs for which peaking occurs with less than 30 features. We assume the basic blocked covariance structure

$$\mathbf{K} = \sigma^2 \begin{pmatrix} 1 & \rho & \rho & & & & & \\ \rho & 1 & \rho & & & 0 & & \\ \rho & \rho & 1 & & & & & \\ & & & \ddots & & & & \\ & & & & 1 & \rho & \rho \\ & 0 & & & \rho & 1 & \rho \\ & & & & \rho & \rho & 1 \end{pmatrix}. \tag{5.40}$$

Features within the same block are correlated with correlation coefficient $\rho$, and features within different blocks are uncorrelated. There are $m$ groups, with $m$ being a divisor of 30. We denote a particular feature with the label $x_{ij}$, where $i$, $1 \leq i \leq g$, denotes the group to which the feature belongs and $j$, $1 \leq j \leq r$, denotes its position in the group. If we list the feature sets in the order $x_{11}, x_{21}, \ldots, x_{m1}, x_{12}, \ldots, x_{mr}$, then the first $d$ features are selected. In the nonlinear model, $2\mathbf{K}_0 = \mathbf{K}_1 = \mathbf{K}$.

Figure 5.2 shows the effect of correlation for LDA with the linear model. Note that the sample size must exceed the number of features to avoid degeneracy. The variance $\sigma^2$ is set to give a Bayes error of 0.05. In figure 5.2a there are $m = 5$ groups and $\rho = 0.125$. Peaking occurs with very few features for sample sizes 30 and below but then exceeds 30 features for sample sizes above 90. Matters are different in figure 5.2b, where the the features are highly correlated with $\rho = 0.5$. Here, even with a sample size of 200, the optimal number of features is only 8. Similar results have been observed for the nonlinear model. The behavior in figure 5.2 corresponds to the usual understanding of the peaking phenomenon; however, the situation can be far more complicated. In the uncorrelated case (not shown), the optimal feature size is about $n - 1$, which matches well with previously reported LDA results (Jain and Waller, 1978).

Figure 5.3a and 5.3b show results for the 3NN classifier on the linear and nonlinear models, respectively. In both cases the variance $\sigma^2$ is set to give a Bayes error of 0.05, and there is a single group with $\rho = 0.25$. In the linear model, there is no peaking up to 30 features. Peaking has been observed in some cases at up to 50 features (ignoring the wobble at sample size 10 owing to Monte Carlo simulation). An interesting phenomenon is seen in figure 5.3b. The optimal-feature-number curve is not increasing as a function of sample size. The optimal feature size is larger at very small sample sizes, rapidly decreases, and then stabilizes as the sample size increases. To check this stabilization, the 3NN classifier has been tested on the nonlinear model case in figure 5.3b for sample sizes up to 5000. The result shows that the optimal feature size increases very slowly with sample size. In particular, for $n = 100$ and $n = 5000$, the optimal feature sizes are 9 and 10, respectively.

It is often assumed that the peaking phenomenon results in a concave error curve. While this might be typical, it is not always the case. Figure 5.4a shows the error surface for the Rosenblatt perceptron in the uncorrelated linear model with Bayes error 0.05. For some sample sizes the error decreases, increases, and then decreases again as the feature set size grows, thereby forming a ridge across the error surface. The design cost equation, equation (4.11), can be rewritten as

$$E[\varepsilon_{n,d}] = \varepsilon_d + E[\Delta_{n,d}] \qquad (5.41)$$

to include the feature set size. Corresponding to the situation observed in figure 5.4a is the following phenomenon: there are feature sizes $d_0 < d_1$ such that for $d < d_0$, $\varepsilon_d$ is falling faster than $E[\Delta_{n,d}]$ is rising; for $d_0 < d < d_1$, $\varepsilon_d$ is falling slower than $E[\Delta_{n,d}]$ is rising; and for $d > d_1$, $\varepsilon_d$ is falling faster than $E[\Delta_{n,d}]$ is rising. For sample size $n = 10$, simulations have been run with up to 400 features and $\varepsilon_d$ still falls no slower than $E[\Delta_{n,d}]$ rises. This nonconcave phenomenon is not that unusual. In the study being discussed, it occurred in all cases for the perceptron and in the correlated cases for the linear SVM. Figure 5.4b shows the error surface for the linear SVM in the correlated nonlinear model with a single block, $\rho = 0.25$, and Bayes error 0.05.

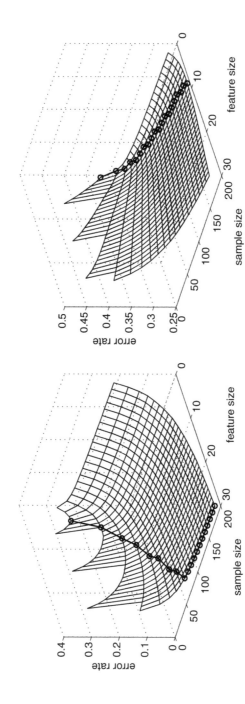

Figure 5.2. Optimal number of features for LDA in a linear model. (a) Slightly correlated features; (b) highly correlated features. (Hua et al., 2005.)

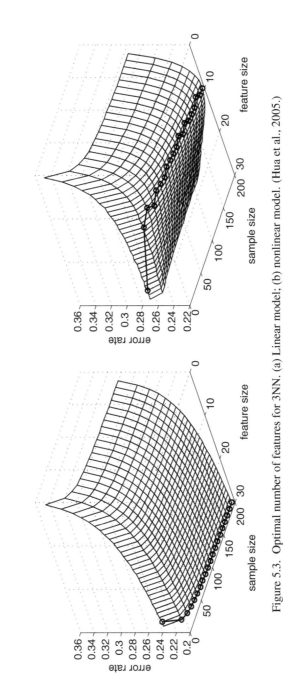

Figure 5.3. Optimal number of features for 3NN. (a) Linear model; (b) nonlinear model. (Hua et al., 2005.)

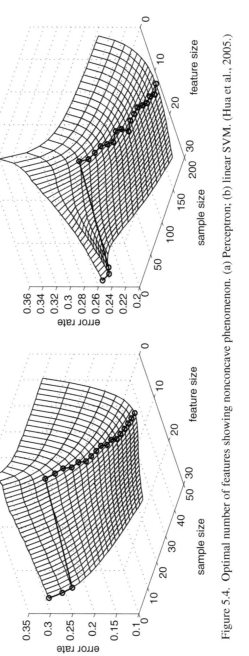

Figure 5.4. Optimal number of features showing nonconcave phenomenon. (a) Perceptron; (b) linear SVM. (Hua et al., 2005.)

### 5.3.2 Feature Selection Algorithms

The examples given in figures 5.2–5.4 utilize a model in which the features are ordered. For applications, this is not known. In fact, we are confronted by a fundamental limiting principle: to select a subset of $k$ features from a set of $d$ features and be assured that it provides an optimal classifier with minimum error among all optimal classifiers for subsets of size $k$, all $k$-element subsets must be checked unless there is distributional knowledge that mitigates the search requirement. This principle is formalized in the following theorem, in which $\varepsilon[A]$ denotes the Bayes error corresponding to feature set $A$.

**THEOREM 5.4 (Cover and Van Campenhout, 1977)** *If $\{U_1, U_2, \ldots, U_r\}$ is the family of all possible feature sets formed from the set $\{X_1, X_2, \ldots, X_k\}$ of random variables under the assumption that $i < j$ if $U_i \subset U_j$, then there exists a distribution of the random variables $X_1, X_2, \ldots, X_k, Y$, where $Y$ is binary, such that $\varepsilon[U_1] > \varepsilon[U_2] > \cdots > \varepsilon[U_r]$. The requirement that $i < j$ if $U_i \subset U_j$ is necessary because the subset condition implies that $\varepsilon[U_i] \geq \varepsilon[U_j]$.*

The situation remains difficult even if $X_1, X_2, \ldots, X_k$ are conditionally independent given $Y$. For instance, the individually best two random variables may not form the best pair of random variables for classifying $Y$. Specifically, there exits binary-valued random variables $X_1, X_2, X_3$, and $Y$ such that $X_1, X_2$, and $X_3$ are conditionally independent given $Y$, $\varepsilon[X_1] < \varepsilon[X_2] < \varepsilon[X_3]$, and $\varepsilon[X_1, X_2] > \varepsilon[X_1, X_3] > \varepsilon[X_2, X_3]$ (Toussaint, 1971).

An exhaustive search can be avoided by using a branch and bound feature selection algorithm that takes advantage of the monotonicity property (Narendra and Fukunaga, 1977). If $A \subset B$ and $C$ is a feature set for which $\varepsilon[C] \geq \varepsilon[A]$, then $\varepsilon[C] \geq \varepsilon[B]$. In principle, this approach yields an optimal solution; however, it suffers from two problems. First, a worst-case performance can be exponentially complex, making it problematic for large feature sets; and second, monotonicity does not hold for designed classifiers. Thus, we require suboptimal search algorithms.

A common approach to suboptimal feature selection is sequential selection, either forward or backward, and their variants. *Sequential forward selection* (SFS) begins with a small set of features, perhaps one, and iteratively builds the feature set. When there are $k$ features $x_1, x_2, \ldots, x_k$ in the growing feature set, all feature sets of the form $\{x_1, x_2, \ldots, x_k, w\}$ are compared and the best one is chosen to form the feature set of size $k + 1$. A problem with SFS is that there is no way to delete a feature adjoined early in the iteration that may not perform as well in combination as other features. The SFS look-back algorithm aims to mitigate this problem by allowing deletion. For it, when there are $k$ features $x_1, x_2, \ldots, x_k$ in the growing feature set, all feature sets of the form $\{x_1, x_2, \ldots, x_k, w, z\}$ are compared and the best one is chosen. Then, all $(k + 1)$-element subsets are checked to allow for the possibility that one of the features chosen earlier can be deleted, the result being the $(k + 1)$ features that will form the basis for the next stage of the algorithm. Flexibility can be added by considering *sequential forward floating selection* (SFFS), where the number of features to be adjoined and deleted is not fixed (Pudhil et al., 1994). For a large number of potential features, feature selection is problematic and the best method depends on the circumstances. The evaluation of methods is generally comparative and based on simulations, and in practice SFFS appears to perform well (Jain and Zongker, 1997; Kudo and Skansky, 2000).

It should be recognized that feature selection is part of the classification rule and as such is a form of complexity regularization. For instance, in the case of gene selection, a normalized maximum-likelihood method has been proposed for feature selection in Boolean models by restating the classification problem as a modeling problem involving a class of parametric models (Tabus et al., 2003). The method is related to the MDL principle and has immediate application in model selection for gene regulatory networks. Since feature selection is part of the classification rule, if one has $D$ features and chooses $d$ of them, then the classification rule is being applied to $D$-dimensional space and this must be taken into account when treating the complexity of classifier design. Moreover, if an error estimation method relates to the classification rule, then one must incorporate feature selection into the error estimation procedure. This is why when cross-validation is used for error estimation, the cross-validation procedure must be applied to the overall classification rule, including feature selection.

### 5.3.3 Impact of Error Estimation on Feature Selection

When selecting features via an algorithm like SFFS that employs error estimation within it, one should expect the choice of error estimator to impact feature selection, the degree depending on the classification rule and on the feature-label distribution (Sima et al., 2005). To illustrate the issue, we consider two 20-dimensional unit-variance spherical Gaussian class conditional distributions with means at $\delta \mathbf{a}$ and $-\delta \mathbf{a}$, where $\mathbf{a} = (a_1, a_2, \ldots, a_n)$, $\|\mathbf{a}\| = 1$, $a_i > 0$, and $\delta > 0$ is a separation parameter. The Bayes classifier is a hyperplane perpendicular to the axis joining the means. The best feature set of size $k$ corresponds to the $k$ largest parameters among $\{a_1, a_2, \ldots, a_n\}$.

We consider SFS and SFFS feature selection, and the LDA and 3NN rules, and select 4 features from samples of size 30. Table 5.2 gives the average true errors of the feature sets found by SFS, SFFS, and an exhaustive search using various error estimators. The top row gives the average true error when the true error is used in feature selection. This is for comparison purposes only because in practice one cannot use the true error during feature selection. Note that both SFS and SFFS perform close to an exhaustive search when the true error is used. Of key interest is that the choice of error estimator can make a greater difference than the manner of feature selection. For instance, for LDA an exhaustive search using leave-one-out results in an average true error of 0.2224, whereas SFFS using bolstered resubstitution yields an average true error of only 0.1918. SFFS using semibolstered resubstitution (0.2016) or .632 bootstrap (0.2129) is also superior to an exhaustive search using leave-one-out, although not as good as bolstered resubstitution. In the case of 3NN, once again SFFS with bolstered resubstitution, semibolstered resubstitution, or bootstrap outperforms a full search using leave-one-out.

Table 5.2 Error Rates for Feature Selection Using Various Error Estimators

| Error Estimator | LDA | | | 3NN | | |
|---|---|---|---|---|---|---|
| | Exhaustive | SFS | SFFS | Exhaustive | SFS | SFFS |
| true | 0.1440 | 0.1508 | 0.1494 | 0.1525 | 0.1559 | 0.1549 |
| resub | 0.2256 | 0.2387 | 0.2345 | 0.2620 | 0.2667 | 0.2543 |
| loo | 0.2224 | 0.2403 | 0.2294 | 0.2301 | 0.2351 | 0.2364 |
| cv5 | 0.2289 | 0.2367 | 0.2304 | 0.2298 | 0.2314 | 0.2375 |
| b632 | 0.2190 | 0.2235 | 0.2129 | 0.2216 | 0.2192 | 0.2201 |
| bresub | 0.1923 | 0.2053 | 0.1918 | 0.2140 | 0.2241 | 0.2270 |
| sresub | 0.1955 | 0.2151 | 0.2016 | 0.2195 | 0.2228 | 0.2230 |

It is evident from the preceding example that finding a good feature set is a challenging task when there is a large number of potential features and a small number of samples. Two basic questions arise: (1) Can one expect feature selection to yield a feature set whose error is close to that of an optimal feature set? (2) If a good feature set is not found, should it be expected that good feature sets do not exist? These questions translate quantitatively into questions concerning conditional expectation: (1) Given the error of an optimal feature set, what is the conditionally expected error of the selected feature set? (2) Given the error of the selected feature set, what is the conditionally expected error of the optimal feature set? In a study addressing these questions by considering LDA, linear support vector machines, and $k$NN classification using SFFS and t-test feature selection, it was demonstrated that under the conditions of the study, one should not expect to find a feature set whose error is close to optimal, nor should the inability to find a good feature set lead to the conclusion that good feature sets do not exist (Sima and Dougherty, 2006). The situation is exacerbated for counting-based estimators like cross-validation on account of a multitude of ties among the feature set errors (Zhou and Mao, 2006).

### 5.3.4 Redundancy

The search problem is exacerbated by the twin issues of redundancy and multivariate prediction. Given two feature sets $G$ and $H$, we define the *incremental gain* owing to adjoining $H$ to $G$ by

$$\Delta_\bullet(G \oplus H) = \varepsilon[G] - \varepsilon[G \cup H], \tag{5.42}$$

where $\varepsilon[G]$ is the Bayes error for $G$. The notation is not commutative since $\Delta_\bullet(G \oplus H)$ is usually not equal to $\Delta_\bullet(H \oplus G)$. Moreover, note that $\Delta_\bullet(G \oplus H) \geq 0$. $\Delta_\bullet(G \oplus H)$ gives the reduction in Bayes error from using $G \cup H$ instead of just $G$. We define the *joint incremental gain* by

$$\Delta_\bullet(G, H) = \min\{\varepsilon[G], \varepsilon[H]\} - \varepsilon[G \cup H]$$

$$= \min\{\Delta_\bullet(G \oplus H), \Delta_\bullet(H \oplus G)\}, \tag{5.43}$$

which is the decrease in error from using the union of the two feature sets relative to the best of the feature sets used individually. In the extreme case, $\Delta_\bullet(G, H) = 0$, meaning that at least one of the feature sets does as well by itself as does the union of the feature sets. If $\Delta_\bullet(G \oplus H) = 0$, then $H$ is totally redundant relative to $G$ since adjoining $H$ gives no benefit.

Now consider two features $x_1$ and $x_2$, where for notational convenience we write $\varepsilon[x_j]$ instead of $\varepsilon[\{x_j\}]$. Since $\Delta_\bullet(x_1 \oplus x_2) \geq 0$, there can in theory be no loss in classification accuracy from using both features instead of just $x_1$; however, if $\Delta_\bullet(x_1 \oplus x_2)$ is very small, then overfitting in classifier design may well result in a worse classifier, so that practically it may not be beneficial to adjoin $x_2$ to $x_1$. This is just the peaking phenomenon. In a sense, what the peaking phenomenon tells us is that given a feature set $G$, it may not be beneficial to adjoin a feature $x_k$, especially if $\Delta_\bullet(G \oplus x_k)$ is very small, that is, if $x_k$ is strongly redundant relative to $G$.

Redundancy comes into play if one wishes to take a naive filter approach to feature selection by selecting a set of features with the highest correlations and then applying a standard classification rule to these features, the reasoning being that these are the best predictors of the class. The problem of redundancy arises because the top-performing features might themselves be highly correlated, say, by the fact that they share a similar regulatory pathway, and using more than one or two of them may provide little added benefit. Thus, even for very large samples, redundancy can result in a poor choice of features, and for small samples the results can be far worse owing to extreme overfitting. Even if one takes a hybrid approach by choosing a set of features based on single-feature prediction and then applies an algorithm like SFFS to this preliminary set, redundancy can result in applying SFFS to a collection of features that has already been stripped of the features needed for good classification. The situation can be even more pernicious in that the first cut has produced many features that are not correlated but which do not work well in combination. This multivariate prediction issue can arise because top-performing single features may not be significantly more beneficial when used in combination with other features. It also applies in the opposite direction, where features that perform very poorly when used alone may provide outstanding classification when used in combination. In an extreme case, we might have $\varepsilon[x_1] = \varepsilon[x_2] = 1$, so that both features are omitted from further consideration, but $\Delta(x_1, x_2) = 1$, meaning that $\varepsilon[x_1, x_2] = 0$.

Given a feature set $G = \{x_1, x_2, \ldots, x_d\}$, a natural question is whether it can be reduced in size without unduly affecting classification accuracy. To treat this problem, we define the *marginal gain* of $G$ by

$$\kappa(G) = \min_{k=1,2,\ldots,d} \Delta_\bullet(G - \{x_k\} \oplus x_k). \tag{5.44}$$

$\kappa(G)$ is the decrease in error obtained by using $G$ in place of the best subset of $G$ of size $d - 1$, and we refer to the subset for which the minimum is achieved as the *most discriminatory subset* of $G$. If $\kappa(G)$ is small, then the most discriminatory subset of $G$ has a performance close to $G$ and there is little purpose in using $G$ instead of just using this subset.

Thus far we have been considering Bayes classifiers for feature sets. With a bit of prudence, the basic definitions extend to classification rules. Relative to the classification rule $\Psi$, the *incremental gain* owing to adjoining $H$ to $G$ is defined by

$$\Delta_\Psi(G \oplus H) = E[\varepsilon[G] - \varepsilon[G \cup H]], \tag{5.45}$$

where the expectation is taken relative to the random sample and the errors are the errors of the designed classifiers on the feature sets. The joint incremental gain relative to $\Psi$ is given by

$$\Delta_\Psi(G, H) = \min\{\Delta_\Psi(G \oplus H), \Delta_\Psi(H \oplus G)\}. \tag{5.46}$$

Owing to design error, the incremental gain may be negative. Therefore, it can be useful to adjust the incremental gain and joint incremental gain by defining them to be 0 should they be negative. We denote these adjusted gains by $\Delta_\Psi^+$. Finally, we define the marginal gain relative to $\Psi$ by

$$\kappa_\Psi(G) = \min_{k=1,2,\ldots,d} \Delta_\Psi^+(G - \{x_k\} \oplus x_k), \tag{5.47}$$

where by definition $\kappa_\Psi(G) \geq 0$.

In practice, there are two difficulties with the preceding definitions. When working with a single sample $S$, the expectation in equation (5.45) must be replaced by the error of the designed classifiers on the sample, and this error must be replaced by an error estimate. Hence, equation (5.45) must be replaced by the estimate

$$\hat{\Delta}_{\Psi,S}(G \oplus H) = \hat{\varepsilon}_{\Psi,S}[G] - \hat{\varepsilon}_{\Psi,S}[G \cup H]. \tag{5.48}$$

Corresponding changes must be made in equations (5.46) and (5.47) to give the estimates $\hat{\Delta}_{\Psi,S}(G \oplus H)$ and $\hat{\kappa}_{\Psi,S}(G)$, respectively.

The results reported in table 5.1 for discriminating anaplastic oligodendroglioma from the other gliomas take redundancy into account, in particular, the marginal gain. The table shows the best 5 single-gene sets, the best 10 two-gene sets, and the three-gene sets among the top 50 three-gene sets for which $\hat{\kappa}_{\Psi,S}(G) \geq 0.03$. Since there were only 587 genes in the study, and these were selected based on prior knowledge, not according to the data, all possible one-, two-, and three-gene feature sets have been evaluated. The total number of possible feature sets of sizes two and three are $C(587,2) = 171,991$ and $C(587,3) = 22,538,244$, respectively, vastly smaller numbers than would have been found if thousands of genes or large feature sets had been allowed. We now closely examine table 5.1 with respect to redundancy and multivariate prediction.

The effects of redundancy and multivariate prediction can be seen in the manner in which the top single-performing features can lose their advantage as the number of features increases. The top five single-gene performers in the table are DNase1L1, TNFSF5, RAD50, HBEGR, and NF45, with classification errors 0.1556, 0.1658, 0.1659, 0.1670, and $= 0.1731$, respectively. DNase1L1 and TNFSF5 provide very good classification results when used with other genes, the best three two-gene feature sets both in the table and on the full two-gene list being {DNase1L1, TNFSF5}, {DNase1L1, PTGER4}, and {GNA13, TNFSF5}, with errors 0.0750, 0.0784, and 0.0826, respectively. On the other hand, RAD50, HBEGR, and NF45 do not do as well in combination. RAD50 appears first on the two-gene list in combination with TNFSF5 at rank 150 with error 0.1378 and joint incremental gain 0.0300. HGEGF appears first on the two-gene list in combination with GNB1 at rank 198 with error 0.1412 and joint incremental gain 0.0258. NS45 appears first on the two-gene list with DNase1L1 at rank 265 with error 0.1451 and joint incremental gain 0.0105. When taken together, two of the top five genes barely do better than either individually, a clear sign of two-gene redundancy.

In the opposite direction, the best three-gene feature set consists of DNase1L1, TNSF5A, and RAB5A, with error 0.0411. Since RAB5A is ranked only 32 on the single-gene list with error 0.1946, it might well have been missed in a single-gene cut. HBGF-1 appears first on the single-gene list at rank 415 with error 0.2080, but it appears in table 5.1 in the set {TNFSF5, LIG4, HBGF-1} with error 0.0616 and a strong marginal gain 0.0474. Continuing in this vein, {PKA C-α, RK B, β-PPT} has the largest three-gene marginal gain, 0.0722. This is a large leap over its subsets. The point is that poor-performing singletons and pairs may perform better when used in large combinations. This is why the SFFS algorithm allows features to be deleted. With a growing number of features, the features that predict well in combination tend to dominate.

There is a case in which the incremental gain is easy to describe. Suppose the class conditional distributions are Gaussian, possess an identical covariance matrix, and have uncorrelated variables. Let $\mu_{01}, \mu_{02}, \ldots, \mu_{0m}$ and $\mu_{11}, \mu_{12}, \ldots, \mu_{1m}$ denote the means of the variables for classes 0 and 1, respectively, $\sigma_1^2, \sigma_2^2, \ldots, \sigma_m^2$ denote their common variances, and $G$ denote the set of variables. The Bayes classifier is determined by LDA, and the Bayes error can be shown to be

$$\varepsilon[G] = \frac{1}{\sqrt{2\pi}} \int_{\Delta(G)/2}^{\infty} e^{-u^2/2} \, du, \tag{5.49}$$

where for each variable $x_j$ we define

$$\delta^2(x_j) = \left( \frac{\mu_{0j} - \mu_{1j}}{\sigma_j} \right)^2 \tag{5.50}$$

and

$$\Delta(G)^2 = \sum_{j=1}^{m} \delta^2(x_j). \tag{5.51}$$

Each term in the sum gives the contribution of the feature in reducing the error, and these contributions are additive. The contribution is increased for greater separation of the means and smaller variance. The error tends to 0 as $\Delta(G) \to \infty$. If $G$ and $H$ are two feature sets, then the incremental gain owing to adjoining $H$ to $G$ is given by

$$
\Delta_{\bullet}(G \oplus H) = \frac{1}{\sqrt{2\pi}} \left[ \int_{\Delta(G)/2}^{\infty} e^{-u^2/2} \, du - \int_{\Delta(G \cup H)/2}^{\infty} e^{-u^2/2} \, du \right]
$$

$$
= \int_{\Delta(G)/2}^{\Delta(G \cup H)/2} e^{-u^2/2} \, du. \tag{5.52}
$$

We close this section by noting that the notion of incremental gain is immediately applicable to the coefficient of determination by simply reversing the order. For the CoD we have

$$
\Delta_\theta(G \oplus H) = \theta[G \cup H] - \theta[G], \tag{5.53}
$$

where $\theta[G]$ is the CoD for $G$. The joint incremental gain is given by

$$
\Delta_\theta(G, H) = \min \{ \Delta_\theta(G \oplus H), \Delta_\theta(H \oplus G) \}. \tag{5.54}
$$

### 5.3.5 Parallel Incremental Feature Selection

Where possible, it is beneficial to take an analytic approach to the evaluation of feature selection and provide metrics to assess the risk and gain posed by feature selection. We will consider a feature selection method based on a marginal-increment-type approach. The method, which we call *parallel incremental feature selection* (PIFS), is akin to forward selection, but it differs in that it requires some minimum improvement to adjoin a feature and keeps track of a large number of growing feature sets, all those that are successfully grown based on a minimum increment. This means it is most profitably used in a high-performance computing environment and for situations in which only small feature sets are considered. The latter requirement is appropriate for small samples and where features represent network predictors and network connectivity is limited. Rather than use the error directly, we explain the algorithm in terms of the coefficient of determination, as was the case for the microarray application to which the analysis was originally applied (Hashimoto et al., 2003).

A suboptimal approach for the optimal two-feature set is to skip the computation of the CoD for two variables $X_i$ and $X_j$ relative to a target variable $Y$ if the individual CoDs of $X_i$ and $X_j$ relative to $Y$ are both small. Let $\theta_i$, $\theta_{ij}$, and $\theta_{ijl}$ denote the CoDs for $\{X_i\}$, $\{X_i, X_j\}$, and $\{X_i, X_j, X_l\}$, respectively. A suboptimal algorithm is defined by ignoring $\theta_{ij}$ if both $\theta_i$ and $\theta_j$ are very small. To exploit this behavior, rather than consider each CoD a deterministic quantity associated with a fixed feature-label distribution, we can consider the feature-label distribution random, thereby making the CoDs random and allowing us to discuss the probability that $\theta_{ij}$ is small given that $\theta_i$ and $\theta_j$ are small. Specifically, we can consider the probability that $\theta_{ij} > \delta$ given that $\theta_i < \lambda$ and $\theta_j < \lambda$. If we are concerned only with feature sets for which the CoD exceeds $\delta$, then (on average) little may be lost by not computing the joint CoD if both individual CoDs are less than $\lambda$ and $\lambda$ is sufficiently smaller than $\delta$. Skip the computation of $\theta_{ij}$ if and only if $\max\{\theta_i, \theta_j\} < \lambda$.

The requirement that $\max\{\theta_i, \theta_j\} \geq \lambda$ for the calculation of $\theta_{ij}$ under the supposition that $\theta_{ij}$ must exceed $\delta$ to be significant can be viewed as a condition on the incremental determinations for the set $\{X_i, X_j\}$ relative to the sets $\{X_i\}$ and $\{X_j\}$, namely, $\theta_{ij} - \theta_i$ and $\theta_{ij} - \theta_j$. The condition can be interpreted as the assumption that the increments cannot exceed $\delta - \lambda$. Under this assumption, $\max\{\theta_i, \theta_j\} < \lambda$ implies $\theta_{ij} \leq \delta$, and therefore there is no point in computing $\theta_{ij}$.

A number of issues are relevant to the condition that $\max\{\theta_i, \theta_j\} \geq \lambda$ for computing $\theta_{ij}$. A measure of *computational saving* is given by the probability of not computing $\theta_{ij}$:

$$\gamma_{ij}(\lambda) = P(\max\{\theta_i, \theta_j\} < \lambda). \tag{5.55}$$

If we are interested in CoDs above a threshold level, then we say that $\theta_{ij}$ is *significant* at level $\delta$ if $\theta_{ij} > \delta$. A measure of the *risk of losing significant CoDs* is provided by the probability

$$\rho_{ij}(\lambda, \delta) = P(\theta_{ij} > \delta | \max\{\theta_i, \theta_j\} < \lambda). \tag{5.56}$$

It gives the probability of a pairwise CoD being significant at level $\delta$ given that it is not computed (at level $\lambda$). There is a conflict between the risk of losing significant CoDs and computational saving: as $\lambda$ increases, so too does $\rho_{ij}(\lambda, \delta)$.

Expanding the conditional probability defining $\rho_{ij}(\lambda, \delta)$ and dividing both sides of the equation by $P(\theta_{ij} > \delta)$ yields

$$\frac{\rho_{ij}(\lambda, \delta)}{P(\theta_{ij} > \delta)} = \frac{\kappa_{ij}(\lambda, \delta)}{P(\max\{\theta_i, \theta_j\} < \lambda)}, \tag{5.57}$$

where

$$\kappa_{ij}(\lambda, \delta) = P(\max\{\theta_i, \theta_j\} < \lambda | \theta_{ij} > \delta) \tag{5.58}$$

provides a measure of *loss of significant CoDs*. It gives the probability of not computing a significant CoD. As with $\rho_{ij}(\lambda, \delta)$, we would like $\kappa_{ij}(\lambda, \delta)$ to be small. Also, as with $\rho_{ij}(\lambda, \delta)$, increasing $\lambda$ to increase computational saving also increases $\kappa_{ij}(\lambda, \delta)$. The effect on $\kappa_{ij}(\lambda, \delta)$ of increasing $\lambda$ for fixed $\delta$ is seen by expanding $\kappa_{ij}(\lambda, \delta)$. If $\lambda_1 < \lambda_2$, then $\kappa_{ij}(\lambda_1, \delta) \leq \kappa_{ij}(\lambda_2, \delta)$.

$\kappa_{ij}(\lambda, \delta)$ is important because it quantifies the noncomputation of significant pairwise CoDs; however, if we take the view that we wish to discover only whether or not $\theta_{ij}$ is significant, without necessarily finding its value, then the problem can be looked at slightly differently. If $\max\{\theta_i, \theta_j\} > \delta$, then $\theta_{ij} > \delta$ and there is no need to compute $\theta_{ij}$. Omitting such computations means that the computational saving is enhanced, with $\gamma_{ij}(\lambda)$ being replaced by

$$\eta_{ij}(\lambda, \delta) = P(\max\{\theta_i, \theta_j\} < \lambda) + P(\max\{\theta_i, \theta_j\} > \delta). \tag{5.59}$$

We say that $\theta_{ij}$ is *nonredundantly significant* at level $\delta$ if $\max\{\theta_i, \theta_j\} \leq \delta$ and $\theta_{ij} > \delta$. Having computed the single-variable CoDs, we need find only those that are nonredundantly significant. We can adjust our notion of loss, replacing $\kappa_{ij}(\lambda, \delta)$ by the probability

$$\nu_{ij}(\lambda, \delta) = P(\max\{\theta_j, \theta_j\} < \lambda | \theta_{ij} > \delta, \max\{\theta_j, \theta_j\} \leq \delta). \tag{5.60}$$

of not computing nonredundantly significant CoDs. Assuming $\lambda < \delta$, $\kappa_{ij}(\lambda, \delta) \leq \nu_{ij}(\lambda, \delta)$.

The analysis can be extended to three or more random variables. The criterion for not computing the CoD can be just an extension of the criterion for two variables: $\max\{\theta_i, \theta_j, \theta_l\} < \lambda$. It is straightforward to show that if $\max\{\theta_{ij}, \theta_{jl}, \theta_{il}\} < \lambda$, then $\max\{\theta_i, \theta_j, \theta_l\} < \lambda$. Hence, the condition $\max\{\theta_i, \theta_j, \theta_l\} < \lambda$ is more restrictive than the condition $\max\{\theta_{ij}, \theta_{jl}, \theta_{il}\} < \lambda$, and the computational saving is greater using the condition $\max\{\theta_i, \theta_j, \theta_l\} < \lambda$ than using the condition $\max\{\theta_{ij}, \theta_{jl}, \theta_{il}\} < \lambda$.

Under the condition $\max\{\theta_i, \theta_j, \theta_l\} < \lambda$, $\rho_{ij}(\lambda, \delta)$ becomes $\rho_{ijl}(\lambda, \delta)$. The form of $\rho_{ij}(\lambda, \delta)$ remains the same except that $\max\{\theta_i, \theta_j\}$ is replaced by $\max\{\theta_i, \theta_j, \theta_l\}$. Analogous comments apply to $\gamma_{ij}(\lambda)$, $\kappa_{ij}(\lambda, \delta)$, $\eta_{ij}(\lambda, \delta)$, and $\nu_{ij}(\lambda, \delta)$. Similar considerations apply to using more than three random variables.

To illustrate the various measures, we consider variable selection in the context of a binary model for the joint probability density $P(X_1, X_2, \ldots, X_d, Y)$. We use the model discussed in section 4.8, in particular, the Dirichlet model in equations (4.170) and (4.178). Here we choose $q$ randomly from a uniform distribution. Moreover, to model the practical situation in which it is unlikely to have a constant error contribution $\rho$ across the configurations, Gaussian noise with mean 0 and standard deviation $\sigma_\rho$ is added to the conditional probabilities $p_i$. Given a realization $r_i$ and $p_i$ taken from the model $(\kappa, \rho, \sigma_\rho)$, we can directly compute the CoDs $\theta_i$ and $\theta_{ij}$.

Figure 5.5 provides graphical results for various measures as a function of $\lambda$. (The results are based on $d = 5$ variables, $\kappa = 1.5$, $\rho = 0.2$, and $\sigma_\rho = 0.2$. As an illustration of how to interpret these graphs, consider $\rho_{ij}(\lambda, \delta)$ in figure 5.5b. Once $\lambda$ is selected, the graph can be used to check if the risk of losing CoDs $\theta_{ij} > \delta$ is smaller than a desired tolerance. For instance, if the tolerance for the risk pertaining to coefficients $\theta_{ij} > 0.6$ is 10 percent, then for $\lambda = 0.4$ this tolerance is satisfied since the risk is less than 1 percent. One should keep in mind that these performance curves depend on the number of features and the model parameters.

### 5.3.6 Bayesian Variable Selection

Here we describe a model-based approach for feature selection using Bayesian methods (Lee et al., 2003) and refer to (Bae and Mallick, 2004) for a two-level hierarchical Bayesian model. We assume a *probit regression model* in which the probabilities $P(Y_k = 1)$ for the sample $S_n = \{(\mathbf{X}_1, Y_1), (\mathbf{X}_2, Y_2), \ldots, (\mathbf{X}_n, Y_n)\}$ are modeled as

$$P(Y_k = 1) = \Phi(\mathbf{X}_k^t \boldsymbol{\beta}), \tag{5.61}$$

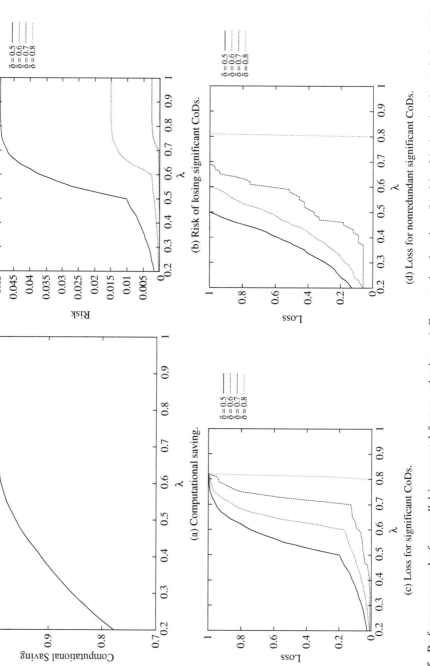

Figure 5.5. Performance graphs for parallel incremental feature selection. a) Computational saving; (b) risk of losing significant CoDs; (c) loss for significant CoDs; (d) loss for nonredundant significant CoDs. (Hashimoto et al., 2003.)

where $\Phi$ is the standard Gaussian cumulative distribution function and $\boldsymbol{\beta} = (\beta_1, \beta_2, \ldots, \beta_n)$ is an unknown parameter vector. Latent variables $Z_1, Z_2, \ldots, Z_n$ are introduced, with $Z_k \sim N(\mathbf{X}_k^t \boldsymbol{\beta}, 1)$, and where $Y_k = 1$ if $Z_k > 0$ and $Y_k = 0$ if $Z_k \leq 0$ (Albert and Chib, 1993).

For variable selection, the indicator vector $\boldsymbol{\gamma} = (\gamma_1, \gamma_2, \ldots, \gamma_n)$ is introduced, where $\gamma_k = 0$ if $\beta_k = 0$ and $\gamma_k = 1$ if $\beta_k \neq 0$. Given $\boldsymbol{\gamma}$, let $\boldsymbol{\beta}_\gamma$ be the vector of all nonzero elements of $\boldsymbol{\beta}$ and let $\mathbf{X}_\gamma$ denote the matrix whose columns are formed by the sample data corresponding to those variables for which $\gamma_k = 1$. If there are $q$ such variables for which $\gamma_k = 1$, then $\mathbf{X}_\gamma$ is an $n \times q$ matrix. Given $\boldsymbol{\gamma}$, we assume that $\boldsymbol{\beta}_\gamma$ possesses a zero-mean Gaussian prior distribution with covariance matrix $c(\mathbf{X}_\gamma^t \mathbf{X}_\gamma)^{-1}$, where $c$ is a positive scale factor suggested to be between 10 and 100 (Smith and Khon, 1996). Here $c = 100$, which means that the prior of $\boldsymbol{\beta}_\gamma$ contains little information. Finally, we assume that the probabilities $\pi_k = P(Y_k = 1)$ of the independent random variables $Y_1, Y_2, \ldots, Y_n$ are small to keep down the number of features in the model.

Gibbs sampling is used to generate the parameters from the posterior distributions. The unknown parameters are $\mathbf{Z} = (Z_1, Z_2, \ldots, Z_n)^t$, $\boldsymbol{\beta}$, and $\boldsymbol{\gamma}$. Straightforward implementation of Gibbs sampling would require simulation from the complete conditional distributions; however, the algorithm can be modified by drawing $\boldsymbol{\gamma}$ from the marginal conditional distribution $\boldsymbol{\gamma} | \mathbf{Z}$ after integrating out $\boldsymbol{\beta}$, thereby speeding up the computation while leaving the target posterior distribution invariant. Indeed, integrating out $\boldsymbol{\beta}$ yields

$$\pi(\mathbf{Z}|\boldsymbol{\gamma}) \propto \exp\left[ -\frac{1}{2}\left( \mathbf{Z}^t \mathbf{Z} - \frac{c}{1-c} \mathbf{Z}^t \mathbf{X}_\gamma (\mathbf{X}_\gamma^t \mathbf{X}_\gamma)^{-1} \mathbf{X}_\gamma^t \mathbf{Z} \right) \right] \qquad (5.62)$$

(Lee et al., 2003). We let $\theta(\mathbf{Z}, \mathbf{X}_\gamma, c)$ denote the exponential. The conditional distribution for $\boldsymbol{\gamma}$ is

$$\pi(\boldsymbol{\gamma}|\mathbf{Z}) \propto \pi(\mathbf{Z}|\boldsymbol{\gamma})\pi(\boldsymbol{\gamma}) \propto \theta(\mathbf{Z}, \mathbf{X}_\gamma, c) \prod_{k=1}^{n} \pi_k^{\gamma_k}(1-\pi_k)^{1-\gamma_k}. \qquad (5.63)$$

Rather than draw $\boldsymbol{\gamma}$ as a vector, we draw it componentwise from

$$\pi(\gamma_k|\mathbf{Z}, \gamma_{j\neq k}) \propto \pi(\mathbf{Z}|\boldsymbol{\gamma})\pi(\gamma_k) \propto \theta(\mathbf{Z}, \mathbf{X}_\gamma, c)\pi_k^{\gamma_k}(1-\pi_k)^{1-\gamma_k}. \qquad (5.64)$$

Continuing, $\boldsymbol{\beta}$ is drawn from

$$\pi(\boldsymbol{\beta}|\boldsymbol{\gamma}, \mathbf{Z}) \sim N\left( \frac{c}{1-c}(\mathbf{X}_\gamma^t \mathbf{X}_\gamma)^{-1}\mathbf{X}_\gamma^t \mathbf{Z}_\gamma, \frac{c}{1-c}(\mathbf{X}_\gamma^t \mathbf{X}_\gamma)^{-1} \right), \qquad (5.65)$$

where $\mathbf{Z}_\gamma$ is $\mathbf{Z}$ restricted to the components $\gamma_k = 1$. Finally, the full conditional distribution of $Z_k$ is given by $Z_k|\boldsymbol{\beta}, Y_k = 1 \propto N(\mathbf{X}_k^t \boldsymbol{\beta}, 1)$ truncated at the left by 0 and by $Z_k|\boldsymbol{\beta}, Y_k = 0 \propto N(\mathbf{X}_k^t \boldsymbol{\beta}, 1)$ truncated at the right by 0. The samples $(\boldsymbol{\beta}^{(r)}, \mathbf{Z}^{(r)}, \boldsymbol{\gamma}^{(r)})$ are used for posterior inference and prediction and are obtained by the following algorithm: initialize $(\boldsymbol{\beta}^{(0)}, \mathbf{Z}^{(0)}, \boldsymbol{\gamma}^{(0)})$ and at step $r$ draw $\boldsymbol{\gamma}^{(r)}$ from $\pi(\boldsymbol{\gamma}|\mathbf{Z}^{(r-1)})$, draw $\mathbf{Z}^{(r)}$ from $\pi(\mathbf{Z}|\boldsymbol{\beta}^{(r-1)}, \boldsymbol{\gamma}^{(r)})$, and draw $\boldsymbol{\beta}^{(r)}$ from $\pi(\boldsymbol{\beta}|\mathbf{Z}^{(r)}, \boldsymbol{\gamma}^{(r)})$.

A measure of the relative classification importance of a variable is given by its posterior probability of inclusion. An estimate of this probability is given by the relative number of times it appears in the sample, which means the number of times the corresponding value of $\gamma$ is 1.

The Bayesian feature selection method has been applied to the hereditary breast cancer data (Hedenfalk et al., 2001) discussed in section 4.2 (Lee et al., 2003). A two-sample $t$-statistic was used to find five significant genes for initialization, and the Gibbs sampling was run with $\pi_k = 0.005$ for $k = 1, 2, \ldots, n$. The two most significant genes, those with the largest frequencies, found by the Bayesian feature selection method are KRT8 (keratin 8) and TOB1. KRT8 is a member of the cytokeratin family of genes. Cytokeratins are frequently used to identify breast cancer metastases by immunohistochemistry, and cytokeratin 8 abundance has been shown to correlate well with node-positive disease. TOB1, second on the list, interacts with the oncogene receptor ERBB2 and is found to be more highly expressed in BRCA2 and sporadic cancers, which are likewise more likely to harbor ERBB2 gene amplifications. TOB1 has an antiproliferative activity that is apparently antagonized by ERBB2. We note that the gene for the receptor was not in the arrays, so that the gene selection algorithm was blinded to its input.

## 5.4 FEATURE EXTRACTION

Rather than reduce dimensionality by selecting from the original features, one might take the approach of *feature extraction*, where a transform is applied to the original features to map them into a lower-dimensional space. Since the new features involve a transform of the original features, the original features remain (although some may be eliminated by compression) and are still part of the classification process. A disadvantage of feature extraction is that the new features lack the physical meaning of the original features—for instance, gene expression levels. A potential advantage of feature extraction is that, given the same number of reduced features, the transform features may provide better classification than the selected original features. Perhaps the most common form of feature extraction is principal component analysis, which is described in this section.

Principal component analysis depends on a fundamental representation for random functions. Owing to the importance of the representation theorem, we state it fairly generally for random functions and then specialize to the case of a random vector, which is our main concern. We refer to the literature for a more complete discussion of the representation and its relation to data compression (Dougherty, 1999). Note its relation to Mercer's theorem.

**THEOREM 5.5 (Karhunen-Loeve)** *Suppose $X(t)$ is a zero-mean random function on the real interval $T$ and $w(t) \geq 0$ is a weight function such that*

$$\int_T \int_T |K_X(t, s)|^2 w(t) w(s)\, dt\, ds < \infty, \tag{5.66}$$

*where $K_X(t, s)$ is the covariance function for $X(t)$. Then, for the integral equation*

$$\int_T K_X(t, s)u(s)w(s)\, ds = \lambda u(t), \tag{5.67}$$

*there exists a discrete (finite or infinite) set of eigenvalues $\lambda_1 \geq \lambda_2 \geq \lambda_3 \geq \cdots 0$ with the corresponding eigenfunctions $u_1(t), u_2(t), u_3(t), \ldots$ such that*

$$\int_T K_X(t, s)u_k(s)w(s)\, ds = \lambda_k u_k(t) \tag{5.68}$$

*for all $k$ and such that $\{u_k(t)\}$ is a deterministic orthonormal system on $T$ relative to the weight function $w(t)$, meaning that*

$$\int_T u_k(t)\overline{u_j(t)}w(t)\, dt = \begin{cases} 1 & \text{if } k = j, \\ 0 & \text{if } k \neq j. \end{cases} \tag{5.69}$$

*$X(t)$ is represented in the mean square sense by the expansion*

$$X(t) = \sum_{k=1}^{\infty} Z_k u_k(t), \tag{5.70}$$

*where $Z_1, Z_2, \ldots$ are the generalized Fourier coefficients of $X(t)$ relative to the eigenfunctions,*

$$Z_k = \int_T X(t)\overline{u_k(t)}w(t)\, dt. \tag{5.71}$$

*$Z_1, Z_2, \ldots$ are uncorrelated, and $\text{Var}[Z_k] = \lambda_k$.*

In the discrete setting over finitely many points $1, 2, \ldots, n$, the random function reduces to the zero-mean random vector $\mathbf{X} = (X_1, X_2, \ldots, X_n)^t$. If the weight function is identically 1, then the integral equation in equation 5.67 reduces to the finite system

$$\mathbf{Ku} = \lambda\mathbf{u}, \tag{5.72}$$

where $\mathbf{K}$ is the covariance matrix (autocorrelation matrix, since $\mu_X = 0$) and $\mathbf{u} = (u_1, u_2, \ldots, u_n)^t$. The required eigenvalues and eigenfunctions are the eigenvalues and eigenvectors of the covariance matrix. Since $\mathbf{K}$ is symmetric, if $\lambda_1$ and $\lambda_2$ are distinct eigenvalues, then their respective eigenvectors will be orthogonal and the desired orthonormal eigenvectors can be found by dividing each by its own magnitude. On the other hand, if an eigenvalue has repeated eigenvectors, then these will be linearly independent and an algebraically equivalent set can be found by the Gram-Schmidt orthogonalization procedure.

According to the Karhunen-Loeve theorem, if the vectors $\mathbf{u}_1, \mathbf{u}_2, ..., \mathbf{u}_n$ are the orthonormalized eigenvectors of $\mathbf{K}$ corresponding to the eigenvalues $\lambda_1 \geq \lambda_2 \geq \cdots \geq \lambda_n$, then

$$\mathbf{X} = \sum_{i=1}^{n} Z_i \mathbf{u}_i, \tag{5.73}$$

where $Z_1, Z_2, \ldots, Z_n$ are uncorrelated and are the generalized Fourier coefficients of $\mathbf{X}$ with respect to $\mathbf{u}_1, \mathbf{u}_2, \ldots, \mathbf{u}_n$,

$$Z_i = \mathbf{X}^t \bar{\mathbf{u}}_i = \sum_{j=1}^{n} X_j \bar{u}_{ij}, \tag{5.74}$$

where $\mathbf{u}_i = (u_{i1}, u_{i2}, \ldots, u_{in})^t$. $Z_1, Z_2, \ldots, Z_n$ are called the *principal components* for $\mathbf{X}$. For $m < n$, data compression is achieved by approximating $\mathbf{X}$ by

$$\mathbf{X}_m = \sum_{i=1}^{m} Z_i \mathbf{u}_i. \tag{5.75}$$

We define the *mean square error* between $\mathbf{X}$ and $\mathbf{X}_m$ by

$$E[\mathbf{X}, \mathbf{X}_m] = \sum_{k=1}^{n} E[|X_k - X_{m,k}|^2], \tag{5.76}$$

where the components of $\mathbf{X}_m$ are $X_{m,1}, X_{m,2}, \ldots, X_{m,n}$. It can be shown that

$$E[\mathbf{X}, \mathbf{X}_m] = \sum_{k=m+1}^{n} \lambda_k. \tag{5.77}$$

Since the eigenvalues are decreasing with increasing $k$, the error is minimized when keeping only $m$ terms by keeping the first $m$ terms.

For a matrix transform perspective, let $\mathbf{Z} = (Z_1, Z_2, \ldots, Z_n)^t$. In terms of the matrix

$$\mathbf{U} = \begin{pmatrix} \mathbf{u}_1^t \\ \mathbf{u}_2^t \\ \vdots \\ \mathbf{u}_n^t \end{pmatrix} = \begin{pmatrix} u_{11} & u_{12} & \cdots & u_{1n} \\ u_{21} & u_{22} & \vdots & u_{2n} \\ \vdots & \vdots & \ddots & \vdots \\ u_{n1} & u_{n2} & \cdots & u_{nn} \end{pmatrix}, \tag{5.78}$$

whose rows are the orthonormal covariance matrix eigenvectors, $\mathbf{Z} = \bar{\mathbf{U}}\mathbf{X}$, which according to equation (5.75) forms a transform pair with $\mathbf{X} = \mathbf{U}^t\mathbf{Z}$. Since the rows of $\mathbf{U}$ form an orthonormal system, $\mathbf{U}$ is a unitary matrix, meaning that its inverse is equal to its transpose $\mathbf{U}^{-1} = \mathbf{U}^t$. Since $\mathbf{U}$ is unitary, $|\det[\mathbf{U}]| = 1$. The covariance matrix of $\mathbf{Z}$ is diagonal, with the eigenvalues of $\mathbf{K}$ running down the diagonal,

$$\mathbf{K_Z} = \begin{pmatrix} \lambda_1 & 0 & \cdots & 0 \\ 0 & \lambda_2 & \cdots & 0 \\ 0 & 0 & \ddots & \vdots \\ 0 & 0 & \cdots & \lambda_n \end{pmatrix}. \tag{5.79}$$

This output covariance matrix shows the decorrelation effect of the transform as well as the manner in which variances of decreasing size are strung down the diagonal.

To apply the transform for the purpose of compression, some number $r = n - m$ is chosen and only the first $m$ terms of $\mathbf{Z}$ are retained; the rest are set to zero. In terms of the transform pair, instead of exactly reconstructing $\mathbf{X}$ by $\mathbf{X} = \mathbf{U}^t \mathbf{Z}$, the inverse transform is applied to the truncated vector $\mathbf{Z}_m = (Z_1, Z_2, \ldots, Z_m, 0, \ldots, 0)^t$. Applying the inverse transform $\mathbf{U}^t$ to $\mathbf{Z}_m$ yields the approximation of equation (5.75):

$$\mathbf{X}_m = \mathbf{U}^t \mathbf{Z}_m. \tag{5.80}$$

To illustrate the transform, let $\mathbf{X} = (X_1, X_2)^t$, where $X_1$ and $X_2$ are jointly normal zero-mean random variables having common variance $\sigma^2$ and correlation coefficient $\rho > 0$. Then,

$$\mathbf{K} = \begin{pmatrix} \sigma^2 & \rho\sigma^2 \\ \rho\sigma^2 & \sigma^2 \end{pmatrix}. \tag{5.81}$$

The eigenvalues of $\mathbf{K}$ are found by solving $\det[\mathbf{K} - \lambda\mathbf{I}] = 0$. The two distinct solutions are given by $\lambda_1 = (1 + \rho)\sigma^2$ and $\lambda_2 = (1 - \rho)\sigma^2$. The corresponding orthonormal eigenvectors are $\mathbf{u}_1 = 2^{-1/2}(1, 1)^t$ and $\mathbf{u}_2 = 2^{-1/2}(1, -1)^t$. The generalized Fourier coefficients are

$$Z_1 = \mathbf{X}^t \mathbf{u}_1 = \frac{1}{\sqrt{2}}(X_1 + X_2),$$

$$Z_2 = \mathbf{X}^t \mathbf{u}_2 = \frac{1}{\sqrt{2}}(X_1 - X_2). \tag{5.82}$$

As it must, $\mathbf{X} = Z_1\mathbf{u}_1 + Z_2\mathbf{u}_2$. According to the Karhunen-Loeve theorem,

$$\text{Var}[Z_1] = \lambda_1 = (1 + \rho)\sigma^2,$$

$$\text{Var}[Z_2] = \lambda_2 = (1 - \rho)\sigma^2. \tag{5.83}$$

Hence, if the variables are highly correlated with $\rho \approx 1$, then $\text{Var}[Z_2] \approx 0$ and $Z_2\mathbf{u}_2$ can be dropped from the expression for $\mathbf{X}$ with little effect on $\mathbf{X}$.

In practical compression settings, the covariance matrix must usually be estimated from observations, and the eigenvalues must be found by numerical computation. Thus, various transforms are employed for compression in which the transform matrix is fixed, independent of the statistics of the signal—for instance, the discrete cosine transform. Instead of the transform $\mathbf{Z} = \overline{\mathbf{U}}\mathbf{X}$ of equation 5.74, an orthonormal transform $\mathbf{Y} = \overline{\mathbf{V}}\mathbf{X}$ of a similar form is applied. The decreasing order of the eigenvalues and full decorrelation of the coefficients make Karhunen-Loeve compression very appealing. However, might there be some other orthonormal set of vectors besides the eigenvectors of the covariance matrix that yields less error when employing an equivalent amount of compression by suppressing higher-order components? In fact, the Karhunen-Loeve transform is optimal relative to the mean-square-error expression in equation (5.76): if any other orthonormal matrix transform is applied to $\mathbf{X}$ and $\mathbf{X}_m$ is reconstructed by the inverse transform, then $E[\mathbf{X}, \mathbf{X}_m]$ is minimal for the Karhunen-Loeve transform. Although we will not go into detail, the line of the argument is to first show that for the matrix transform $\mathbf{Y} = \overline{\mathbf{V}}\mathbf{X}$,

$$E[\mathbf{X}, \mathbf{X}_m] = \sum_{k=m+1}^{n} \text{Var}[Y_k], \qquad (5.84)$$

and then to show that minimization of the preceding variance sum under the orthonormal constraint yields the sum of the eigenvalues of the covariance matrix from $m + 1$ to $n$, which is precisely the error for the Karhunen-Loeve transform. Hence, the latter is optimal.

# Bibliography

Albert J, Chib S. (1993) Bayesian analysis of binary and polychotomous response data. *J Am Stat Assoc* 88: 669–79.

Bae K, Mallick BK. (2004) Gene selection using a two-level hierarchical Bayesian model. *Bioinformatics* 20: 3423–30.

Barron AR, Cover TM. (1991) Minimum complexity density estimation. *IEEE Trans Inform Theory* 37(4): 1034–54.

Bishop CM. (1995) *Neural Networks for Pattern Recognition*, Oxford University Press, New York.

Cover T, Van Campenhout J. (1977) On the possible orderings in the measurement selection problem. *IEEE Trans Systems Man Cybern* 7: 657–61.

Dey DK, Srmivasan C. (1985) Estimation of a covariance matrix under Stein's loss. *Ann Stat* 13: 1581–91.

Dougherty ER. (1999) *Random Processes for Image and Signal Processing*, IEEE and SPIE Presses, Bellingham, WA.

Friedman JH. (1989) Regularized discriminant analysis. *J Am Stat Assoc* 84(405): 165–75.

Haff LR. (1980) Empirical Bayes estimation of the multivariate normal covariance matrix. *Ann Stat* 8: 586–97.

Hashimoto R, Dougherty ER, Brun M, Zhou Z, Bittner ML Trent JM. (2003). Efficient selection of feature sets possessing high coefficients of determination based on incremental determinations. *Signal Process* 83(4): 695–712.

Hedenfalk I, Duggan D, Chen Y, Radmacher M, Bittner M, Simon R, Meltzer P, Gusterson B, Esteller M, Raffeld M, Yakhini Z, Ben-Dor A, Dougherty E, Kononen J, Bubendorf L, Fehrle W, Pittaluga S, Gruvverger S, Loman N, Johannsson O, Olsson H, Wifond B, Sauter G, Kallioniemi OP, Borg A, Trent J. (2001) Gene expression profiles in hereditary breast cancer. *N Engl J Med* 344(8): 539–48.

Holmström L, Koistinen P. (1992) Using additive noise in back-propagation training. *IEEE Trans Neural Netw* 3: 24–38.

Hua J, Lowey J, Xiong Z, Dougherty ER. (2006) Noise-injected neural networks show promise for use on small-sample expression data. *BMC Bioinformatics* 7: 274.

Hua J, Xiong Z, Lowey J, Suh E, Dougherty ER. (2005) Optimal number of features as a function of sample size for various classification rules. *Bioinformatics* 21(8): 1509–15.

Hughes GF. (1968) On the mean accuracy of statistical pattern recognizers. *IEEE Trans Inform Theory* 14: 55–63.

Jain AK, Chandrasekaran B. (1982) Dimensionality and sample size considerations in pattern recognition practice. In *Classification, Pattern Recognition and Reduction of Dimensionality*, Krishnaiah PR, Kanal LN, eds., vol. 2 of *Handbook of Statistics*, North Holland, Amsterdam, pp. 835–56.

Jain AK, Waller WG. (1978) On the optimal number of features in the classification of multivariate Gaussian data. *Pattern Recognition* 10: 365-74.

Jain AK, Zongker D. (1997) Feature selection: evaluation, application, and small sample performance. *IEEE Trans Pattern Anal Mach Intell* 19: 153–8.

Kearns M, Mansour Y, Ng AY, Ron D. (1997) An experimental and theoretical comparison of model selection methods. *Mach Learning* 27: 7–50.

Kim S, Dougherty ER, Barrera J, Chen Y, Bittner M, Trent JM. (2002a) Strong feature sets from small samples. *Comput Biol* 9(1): 127–46.

Kim S, Dougherty ER, Shmulevich I, Hess KR, Hamilton SR, Trent JM, Fuller GN, Zhang W. (2002b) Identification of combination gene sets for glioma classification. *Mol Cancer Ther* 1(13): 1229–36.

Kudo M. Sklansky J. (2000) Comparison of algorithms that select features for pattern classifiers. *Pattern Recognition* 33: 25–41.

Lee KE, Sha N, Dougherty ER, Vannucci M, Mallick BK. (2003) Gene selection: a Bayesian variable selection approach. *Bioinformatics* 19: 90–7.

Lugosi G, Nobel A. (1999) Adaptive model selection using empirical complexities. *Ann Stat* 27(6): 1830–64.

Lugosi G, Zeger K. (1996) Concept learning using complexity regularization. *IEEE Trans Inform Theory* 42: 48–54.

Matsuoka K. (1992) Noise injection into inputs in back-propagation learning. *IEEE Trans Systems Man Cybern* 22(3): 436–40.

Narendra PM, Fukunaga K. (1977) A branch and bound algorithm for feature subset selection. *IEEE Trans Comput* 26(9): 917–22.

Pudil P, Novovicova J, Kittler J. (1994) Floating search methods in feature selection. *Pattern Recognition Lett* 15: 1119–25.

Rissanen J. (1978) Modeling by shortest data description. *Automatica* 14: 465–71.

Rissanen J. (1986) Stochastic complexity and modeling. *Ann Stat* 14: 1080–1100.

Sietsma J, Dow RJF. (1988) Neural network pruning—why and how. *Proc IEEE Intern Conf Neural Netw* I: 325–33.

Sima C, Attoor S, Braga-Neto U, Lowey J, Suh E, Dougherty ER. (2005b) Impact of error estimation on feature-selection algorithms. *Pattern Recognition* 38: 2472–82.

Sima C, Dougherty ER. (2006) What should be expected from feature selection in small-sample settings. *Bioinformatics* 22(19): 2430–6.

Skurichina M, Duin RPW, Raudys S. (2000) $K$-nearest neighbours noise injection in multilayer perceptron training. *IEEE Trans Neural Netw* 11(2): 504–11.

Smith M, Kohn R. (1997) Nonparametric regression using Bayesian variable selection. *Econometrics* 75: 317–44.

Tabus I, Rissanen J, Astola J. (2003) Classification and feature gene selection using the normalized maximum likelihood model for discrete regression. *Signal Process* 83(4): 713–27.

Titterington DM. (1985) Common structure of smoothing techniques in statistics. *Int Stat Rev* 53: 141–70.

Toussaint G. (1971) Note on optimal selection of independent binary-valued features for pattern recognition. *IEEE Trans Inform Theory* 17: 618.

Vapnik VN. (1982) *Estimation of Dependencies Based on Empirical Data*, Springer-Verlag, New York.

Vapnik VN. (1998) *Statistical Learning Theory*, John Wiley & Sons, New York.

Zhou X, Mao KZ. (2006) The ties problem resulting from counting-based error estimators and its impact on gene selection algorithms. *Bioinformatics* 22(19):2507–15.

# Chapter Six

## Clustering

A classification operator takes a single data point and outputs a class label; a cluster operator takes a set of data points and partitions the points into clusters (subsets). Clustering has become a popular data analysis technique in genomic studies using gene expression microarrays (Ben-Dor et al., 1999). Time series clustering groups together genes whose expression levels exhibit similar behavior through time. Similarity indicates possible coregulation. Another way to use expression data is to take expression profiles over various tissue samples and then cluster these samples based on the expression levels for each sample. This approach is used to indicate the potential for discriminating pathologies based on their differential patterns of gene expression.

Classification exhibits two fundamental characteristics: (1) classifier error can be estimated under the assumption that the sample data come from a feature-label distribution; and (2) given a family of classifiers, sample data can be used to learn a classifier in the family. Once designed, the classifier represents a mathematical model that provides a decision mechanism relative to real-world measurements. The model represents scientific knowledge to the extent that it has predictive capability. Testing (error estimation) quantifies the worth of the model.

Clustering has generally lacked both fundamental characteristics of classification. In particular, lacking inference in the context of a probability model, it has remained essentially a subjective visualization tool. Jain et al. write, "Clustering is a subjective process; the same set of data items often needs to be partitioned differently for different applications. This subjectivity makes the process of clustering difficult" (Jain et al., 1999). This criticism raises the question as to whether clustering can be used for scientific knowledge, because to the extent that clustering is subjective, it lacks scientific content. The epistemological issue is raised specifically in the context of gene expression microarrays by Kerr and Churchill when they write, "How does one make statistical inferences based on the results of clustering?" (Kerr and Churchill, 2001) Unless clustering leads to predictions that can be tested with physical data, it has no scientific content because, as Richard Feynman states, "It is whether or not the theory gives predictions that agree with experiment. It is not a question of whether a theory is philosophically delightful, or easy to understand, or perfectly reasonable from the point of view of common sense" (Feynman, 1985). Subjective appreciations are certainly useful in the formulation of hypotheses, but these are constitutive of scientific knowledge only if they are set in a predictive framework.

Many validation techniques have been proposed for evaluating clustering results. These are generally based on the degree to which clusters derived from a set of sample data satisfy certain heuristic criteria. This is significantly different from classification, where the error of a classifier is given by the probability of an erroneous decision.

No doubt clustering can be used as a visualization tool to see how points can be partitioned and therefore assist the scientist in the formation of hypotheses. From this perspective clustering provides grist for the scientific imagination and can serve a useful purpose, but as such it is at the lowest level of the path from observation to model formation to model validation via predictive experiments. Indeed, the whole notion of "validation" as used in clustering is a bit confusing because it does not necessarily refer to scientific validation, as does error estimation in classification (Dougherty and Braga-Neto, 2006).

For clustering, error estimation must assume that clusters resulting from a cluster algorithm can be compared to the correct partition of the data set in the context of a probability distribution, thereby providing an error measure. The key to a general probabilistic theory of clustering, including both error estimation and learning, is to recognize that classification theory is based on operators on random variables and that a probabilistic theory of clustering needs to be based on operators on random sets. Implicit in this recognition is that clustering is theoretically more difficult than classification. Whereas issues concerning random variables can be phrased in terms of the probability distribution function of the random variable, in general, random sets are characterized via their capacity functionals (Matheron, 1975; Stoyan et al., 1987). Modeling via capacity functionals is extremely difficult (Cressie and Lasslett, 1987; Dougherty, 1999).

Regarding learning, clustering has sometimes been referred to as "unsupervised learning." Even if we do not argue that such a term is an oxymoron, there is certainly no learning going on when clustering is used in the typical manner. With clustering, a heuristically chosen operator is applied to a point set to yield a partition of the point set. If there is no randomization within the algorithm, then the operator is simply a function on point sets; if there is randomization within the algorithm, then the operator is a random function on point sets. In either event, there is no learning.

In this chapter we will introduce some clustering algorithms and validation procedures. Since our interest is in cluster operators as models that can carry scientific content, our focus will be on a probabilistic understanding of clustering as it relates to cluster operators on random point sets—in terms of both prediction and learning.

## 6.1 EXAMPLES OF CLUSTERING ALGORITHMS

A host of clustering algorithms has been proposed in the literature. We will discuss several in this section. Our intent is not to be exhaustive but to present algorithms that illustrate different generic approaches. Many are based on criteria related to Euclidean distance clustering, and we begin there.

### 6.1.1 Euclidean Distance Clustering

If we envision clusters formed by points $\mathbf{x}$ generated by a random sample $S$ from a mixture of $m$ circular Gaussian conditional distributions, then the points $\mathbf{a}_0^S, \mathbf{a}_1^S, \ldots,$ $\mathbf{a}_{m-1}^S$ that minimize

$$\rho_S(\mathbf{a}_0, \mathbf{a}_1, \ldots, \mathbf{a}_{m-1}) = \frac{1}{|S|} \sum_{\mathbf{x} \in S} \min_{0 \le j \le jm-1} ||\mathbf{x} - \mathbf{a}_j||^2 \tag{6.1}$$

are a reasonable choice for the centroids of the $m$ subsamples arising from the conditional distributions. Let $\mathcal{V} = \{V_0, V_1, \ldots, V_{m-1}\}$ be the Voronoi partition of $\mathfrak{R}^d$ induced by $\mathbf{a}_0^S, \mathbf{a}_1^S, \ldots, \mathbf{a}_{m-1}^S$: a point lies in $V_k$ if its distance to $\mathbf{a}_k^S$ is no more than its distance to any of the other points $\mathbf{a}_0^S, \mathbf{a}_1^S, \ldots, \mathbf{a}_{m-1}^S$. For *Euclidean distance clustering*, the sample points are clustered according to how they fall into the Voronoi partition.

While perhaps motivated by a mixture of Gaussians, the quantity $\rho_S$ can be computed for any set $S$ of points. The measure $\rho_S(\mathbf{a}_0, \mathbf{a}_1, \ldots, \mathbf{a}_{m-1})$ provides an empirical distance error for the points $\mathbf{a}_0, \mathbf{a}_1, \ldots, \mathbf{a}_{m-1}$, and $\mathbf{a}_0^S, \mathbf{a}_1^S, \ldots, \mathbf{a}_{m-1}^S$ minimize this distance. $\rho_S(\mathbf{a}_0, \mathbf{a}_1, \ldots, \mathbf{a}_{m-1})$ does not provide a true error measure because it is based on a specific random sample. The true error is given in terms of the distribution of the random set $\Xi$ whose realizations are the random samples. For the minimizing points $\mathbf{a}_0^S, \mathbf{a}_1^S, \ldots, \mathbf{a}_{m-1}^S$, the true distance error is an expectation conditioned relative to the sample, namely,

$$\rho_\Xi(\mathbf{a}_0, \mathbf{a}_1, \ldots, \mathbf{a}_{m-1}) = E_\Xi \left[ \min_{0 \le j \le m-1} ||\mathbf{X} - \mathbf{a}_j||^2 \, | \, S \right], \tag{6.2}$$

where the expectation is taken with respect to the distribution of the random set $\Xi$. For a realization $S$ of $\Xi$, $\rho_S$ estimates $\rho_\Xi$. The empirical and true distance errors may differ significantly for small samples; however, if there exists a compact set $K$ such that $P(\Xi \subset K) = 1$, then their difference converges to zero with probability 1 as the number of points in $\Xi$ tends to $\infty$ (Linder et al., 1994).

Direct implementation of Euclidean distance clustering is computationally prohibitive. A classical iterative approximation is given by the *k-means algorithm*, where $k$ refers to the number of clusters provided by the algorithm. Each sample point is placed into a unique cluster during each iteration, and the means are updated based on the classified samples. Given a sample $S$ with $n$ points to be placed into $k$ clusters, initialize the algorithm with $k$ means $\mathbf{m}_1, \mathbf{m}_2, \ldots, \mathbf{m}_k$ among the points; for each point $\mathbf{x} \in S$, calculate the distance $||\mathbf{x} - \mathbf{m}_i||$ for $i = 1, 2, \ldots, k$; form clusters $C_1, C_1, \ldots, C_k$ by placing $\mathbf{x}$ into $C_i$ if $||\mathbf{x} - \mathbf{m}_i|| \le ||\mathbf{x} - \mathbf{m}_j||$ for $j = 1, 2, \ldots, k$; update $\mathbf{m}_1, \mathbf{m}_2, \ldots, \mathbf{m}_k$ as the means of $C_1, C_1, \ldots, C_k$, respectively; and repeat until the means do not change. At each stage of the algorithm, the clusters are determined by the Voronoi diagram associated with $\mathbf{m}_1, \mathbf{m}_2, \ldots, \mathbf{m}_k$. Two evident problems with the $k$-means algorithm are the prior assumption on the number of means and the choice of means to seed the algorithm.

Equation (6.1) can be rewritten in the form

$$\rho_S(\mathbf{a}_0, \mathbf{a}_1, \ldots, \mathbf{a}_{m-1}) = \frac{1}{|S|} \sum_{i=1}^{n} \sum_{j=0}^{m-1} P(C_j|\mathbf{x}_i)^b \left\| \mathbf{x}_i - \mathbf{a}_j \right\|^2, \tag{6.3}$$

where $b = 1$ and $P(C_j|\mathbf{x}_i)$ is the probability that $\mathbf{x}_i \in C_j$, which is either 0 or 1 and is 1 only for the minimizing $j$ in equation (6.1). A fuzzy approach results from letting the conditional probabilities reflect uncertainty, so that cluster inclusion is not crisp, and letting $b > 0$ be a parameter affecting the degree to which a point can belong to more than a single cluster. The conditional probabilities are constrained by the requirement that their sum is 1 for any fixed $\mathbf{x}_i$,

$$\sum_{j=0}^{m-1} P(C_j|\mathbf{x}_i) = 1. \tag{6.4}$$

Let $p_j$ denote the prior probability of $C_j$. Since the conditional probabilities $P(C_j|\mathbf{x}_i)$ are not estimable and are heuristically set, we view them as fuzzy membership functions. In this case, for the minimizing values of $\rho_S(\mathbf{a}_0, \mathbf{a}_1, \ldots, \mathbf{a}_{m-1})$, the partial derivatives with respect to $\mathbf{a}_j$ and $p_i$ satisify $\partial\rho_S/\partial\mathbf{a}_j = 0$ and $\partial\rho_S/\partial p_j = 0$. These partial-derivative identities yield

$$\mathbf{m}_j = \frac{\sum_{i=1}^{n} P(C_j|\mathbf{x}_i)^b \mathbf{x}_i}{\sum_{i=1}^{n} P(C_j|\mathbf{x}_i)^b}, \tag{6.5}$$

$$P(C_j|\mathbf{x}_i) = \frac{\|\mathbf{x}_i - \mathbf{m}_j\|^{-1/(b-1)}}{\sum_{l=1}^{k} \|\mathbf{x}_i - \mathbf{m}_l\|^{-1/(b-1)}}. \tag{6.6}$$

These lead to the *fuzzy k-means* iterative algorithm (Dunn, 1973; Bezdek, 1981). Initialize the algorithm with $b$, $k$ means $\mathbf{m}_1, \mathbf{m}_2, \ldots \mathbf{m}_k$, and the membership functions $P(C_j|\mathbf{x}_i)$ for $j = 1, 2, \ldots, k$ and $i = 1, 2, \ldots, n$, where the membership functions must be normalized so that their sum is 1 for any fixed $\mathbf{x}_i$; recompute $\mathbf{m}_j$ and $P(C_j|\mathbf{x}_i)$ by equations (6.5) and (6.6); and repeat until there are only small prespecified changes in the means and membership functions. The intent of fuzzifying the $k$-means algorithm is to keep the means from getting "stuck" during the iterative procedure.

### 6.1.2 Self-Organizing Maps

*Self-organizing maps* provide a different extension of the $k$-means concept (Kohonen, 1982, 1995). The idea is to map high-dimensional vectors in Euclidean space to a low-dimensional grid in a neighborhood-preserving manner, whereby we mean that vectors that are close in the high-dimensional space have close representations in the low-dimensional grid. To describe self-organizing maps, we begin with an ordered lattice, typically a one- or two-dimensional grid, that we index by $I = \{1, 2, \ldots, k\}$. Associated with $I$ is a *neighborhood function* $\eta_t$, defined on $I \times I$ and parameterized by $t = 0, 1, \ldots$, satisfying three properties: (1) $\eta_t$ depends only on the distance between points in $I$, meaning that $\eta_t(i, j) = \eta_t(\|i - j\|)$; (2) $\eta_t$ is nonincreasing relative to distance, meaning $\eta_t(\|i - j\|) \leq \eta_t(\|u - v\|)$ if $\|u - v\| \leq \|i - j\|$; and (3) the domain of $\eta_t$ is nonincreasing relative to $t$, meaning that for each $t$ there exists a nonnegative integer parameter $\alpha(t)$ such that

$\eta_t(||i-j||) = 0$ if $||i-j|| > \alpha(t)$ and $\alpha(t+1) \leq \alpha(t)$. As will become evident, the properties of the neighborhood function have been chosen so that points nearby a chosen point in $I$ will be updated in conjunction with the chosen point, with less adjustment for points further away from the chosen point, and this updated neighborhood will decrease in size as the algorithm proceeds through its iterations, usually with the neighborhood reduced to a single point $[\alpha(t) = 0]$ during the last steps of the algorithm. It is assumed that the input vectors to the algorithm are derived from a random sample and that they lie in some bounded convex set in $\mathfrak{R}^d$.

Each iteration of a self-organizing map algorithm is characterized by a state vector

$$\mathbf{m}(t) = (\mathbf{m}_1(t), \mathbf{m}_2(t), \ldots, \mathbf{m}_k(t)), \tag{6.7}$$

where $\mathbf{m}_i(t) \in \mathfrak{R}^d$ for $i = 1, 2, \ldots, k$. The algorithm is initialized at $\mathbf{m}(0)$. The algorithm proceeds in the following recursive fashion. Given an input vector $\mathbf{x}$ at time $t$, the index $\iota_t$ of the component state closest to $\mathbf{x}$ is selected, namely,

$$\iota_t = \arg \min_{i \in I} ||\mathbf{x} - \mathbf{m}_i(t)||. \tag{6.8}$$

The vector $\mathbf{m}(t)$ is then updated according to

$$\mathbf{m}_i(t+1) = \mathbf{m}_i(t) + \beta_t \eta(\iota_t, i)[\mathbf{m}_i(t) - \mathbf{x}] \tag{6.9}$$

for $i = 1, 2, \ldots, k$, where $\beta_t > 0$ is a parameter that can be lowered over time to lessen the adjustment of $\mathbf{m}(t)$. A critical role is played by the neighborhood function in achieving the preservation of neighborhood relations. Many theoretical questions remain regarding the organizational nature of the algorithm and convergence (Cottrell et al., 1998). Clustering is achieved in the same manner as the $k$-means algorithm, with the clusters determined by the Voronoi diagram associated with $\mathbf{m}_1(t), \mathbf{m}_2(t) \ldots, \mathbf{m}_k(t)$ (Flexer, 2001).

### 6.1.3 Hierarchical Clustering

Both $k$-means and fuzzy $k$-means are based on Euclidean distance clustering. Another approach (albeit one also often related to Euclidean distance) is to iteratively join clusters according to a similarity measure. The general *hierarchical clustering* algorithm is given by the following procedure: initialize the clusters by $C_i = \{\mathbf{x}_i\}$ for $i = 1, 2, \ldots, n$ and by a desired final number $k$ of clusters; then proceed to iteratively merge the nearest clusters according to the similarity measure until there are only $k$ clusters. An alternative is to continue to merge until the similarity measure satisfies some criterion. The merging process can be pictorially represented in the form of a dendrogram, where joined arcs represent merging. As stated, the hierarchical clustering is *agglomerative* in the sense that points are agglomerated into growing clusters. One can also consider *divisive* clustering, in which, beginning with a single cluster, the algorithm proceeds to iteratively split clusters. Various similarity measures have been proposed. Three popular ones are the minimum, maximum, and average measures given by

$$d_{\min}(C_i, C_j) = \min_{\mathbf{x} \in C_i, \mathbf{x}' \in C_j} \|\mathbf{x} - \mathbf{x}'\|, \tag{6.10}$$

$$d_{\max}(C_i, C_j) = \max_{\mathbf{x} \in C_i, \mathbf{x}' \in C_j} \|\mathbf{x} - \mathbf{x}'\|, \tag{6.11}$$

$$d_{\mathrm{av}}(C_i, C_j) = \frac{1}{|C_i| \cdot |C_j|} \sum_{\mathbf{x} \in C_i, \mathbf{x}' \in C_j} \|\mathbf{x} - \mathbf{x}'\|. \tag{6.12}$$

Hierarchical clustering using the minimum distance is called *nearest-neighbor* clustering. If it halts when the distance between the nearest clusters exceeds a prespecified threshold, then it is called *single-linkage clustering*. Given a set of clusters at any stage of the algorithm, it merges the clusters possessing the nearest points. If we view the points as the nodes of a graph, then when two clusters are merged, an edge is placed between the nearest nodes in the two clusters. Hence, no closed loops are created and the resulting graph is a tree. If the algorithm is not stopped until there is a single cluster, the result is a spanning tree, and the spanning tree is minimal. While the algorithm may be intuitively pleasing owing to the manner in which it generates a hierarchical subcluster structure based on nearest neighbors, it is extremely sensitive to noise and can produce strange results, such as elongated clusters. It is also very sensitive to early mergings since, once joined, points cannot be separated.

*Farthest-neighbor* clustering results from using the maximum distance. If it halts when the distance between the nearest clusters exceeds a prespecified threshold, then it is called *complete-linkage clustering*. Given a set of clusters at any stage of the algorithm, it merges the clusters for which the greatest distance between points in the two clusters is minimized. This approach counteracts the tendency toward elongation from which nearest-neighbor clustering suffers. Finally, if the algorithm halts when the average distance between the nearest clusters exceeds a prespecified threshold, then it is called *average-linkage clustering*.

A relative of hierarchical clustering is an iterative cluster-growing algorithm called the *cluster affinity search technique* (CAST) (Ben-Dor et al., 1999). For a point set $S = \{\mathbf{x}_1, \mathbf{x}_2, \ldots, \mathbf{x}_n\}$, let $s(i, j)$ denote the similarity measure between $\mathbf{x}_i$ and $\mathbf{x}_j$. For any subset $C \subset S$ and point $\mathbf{x}_i$, define the affinity of $\mathbf{x}_i$ with $C$ by

$$A(\mathbf{x}_i; C) = \sum_{\mathbf{x}_j \in C} s(i, j). \tag{6.13}$$

The algorithm has as input an *affinity threshold* $t$. It has at each stage of iteration a growing cluster $C_{\mathrm{open}}$, and a set $U$ of points not yet contained in $C_{\mathrm{open}}$ or a previously completed cluster. It is initialized by $C_{\mathrm{open}} \leftarrow \emptyset$ and proceeds by joining high-affinity points to the adapting cluster $C_{\mathrm{open}}$, followed by deleting low-affinity points from $C_{\text{open}}$ and then repeating the JOIN-DELETE cycle until changes cease to occur, at which point the current $C_{\text{open}}$ is set as a cluster and a new growing cluster is initialized, $C_{\text{open}} \leftarrow \emptyset$, as long as there remain points outside finished clusters.

The JOIN part of the algorithm proceeds in the following manner. If there exists a point $\mathbf{x}_i \in U$ such that $A(\mathbf{x}_i; C_{\text{open}}) \geq t|C_{\text{open}}|$, then select a point $\mathbf{x}_k \in U$ having maximum affinity with $C_{\text{open}}$, update $C_{\text{open}}$ by adjoining $\mathbf{x}_k (C_{\text{open}} \leftarrow C_{\text{open}} \cup \{\mathbf{x}_k\})$, update $U$ by removing $\mathbf{x}_k (U \leftarrow U - \{\mathbf{x}_k\})$, and update $A(\mathbf{x}_i; C_{\text{open}})$ for all points $\mathbf{x}_i \in U \cup C_{\text{open}}$. Repeat this procedure as long as there exists a point $\mathbf{x}_i \in U$ such that $A(\mathbf{x}_i; C_{\text{open}}) \geq t|C_{\text{open}}|$. If no such point exists, go to DELETE.

The DELETE part of the algorithm proceeds in the following manner. If there exists a point $\mathbf{x}_i \in C_{\text{open}}$ such that $A(\mathbf{x}_i; C_{\text{open}}) < t|C_{\text{open}}|$, then select a point $\mathbf{x}_l \in C_{\text{open}}$ having minimum affinity with $C_{\text{open}}$, update $C_{\text{open}}$ by deleting $\mathbf{x}_l (C_{\text{open}} \rightarrow C_{\text{open}} - \{\mathbf{x}_l\})$, update $U$ by joining $\mathbf{x}_l (U \rightarrow U \cup \{\mathbf{x}_l\})$, and update $A(\mathbf{x}_i; C_{\text{open}})$ for all points $\mathbf{x}_i \in U \cup C_{\text{open}}$. Repeat this procedure as long as there is a point $\mathbf{x}_i \in C_{\text{open}}$ such that $A(\mathbf{x}_i; C_{\text{open}}) < t|C_{\text{open}}|$. If no such point exists, go to JOIN. And if no point satisfies the condition for joining, the current cluster is complete and a new one is initialized.

When the preceding iteration is finished, for any point $\mathbf{x}_i$ it must be that $A(\mathbf{x}_i; C(\mathbf{x}_i)) \geq t|C(\mathbf{x}_i)|$, where $C(\mathbf{x}_i)$ is the cluster containing $\mathbf{x}_i$, but we are not assured that $A(\mathbf{x}_i; C(\mathbf{x}_i)) \geq A(\mathbf{x}_i; C)$ for every cluster $C$. The algorithm has the following postprocessing. If there exists a point $\mathbf{x}_i$ and a $C \neq C(\mathbf{x}_i)$ such that $A(\mathbf{x}_i; C(\mathbf{x}_i)) < A(\mathbf{x}_i; C)$, then move $\mathbf{x}_i$ to the cluster for which it has the highest affinity. Repeat this process until there no longer exists such a point or until some maximum number of iterations. The clustering is called *stable* if the process halts because no such points exist.

### 6.1.4 Model-Based Cluster Operators

If random point sets are generated by taking random samples from a mixture of Gaussian distributions, then a natural way to approach clustering is to model the data by a mixture of Gaussian distributions: try to estimate the conditional distributions from the data and cluster the points based upon these (Banfield and Raftery, 1993; Celeux and Govaert, 1993; Fraley and Raftery, 1998; Dasgupta and Raftery, 1998). The estimation will be on the data arising from a single realization $S$ of $\Xi$. In fact, the conditional distributions need not be Gaussian.

In general, suppose there are $l$ conditional distributions possessing the parameterized densities $f_0(\mathbf{x}|\boldsymbol{\theta}_0), f_1(\mathbf{x}|\boldsymbol{\theta}_1), \ldots, f_{l-1}(\mathbf{x}|\boldsymbol{\theta}_{l-1})$ corresponding to the labels $0, 1, \ldots, l-1$. For the sample data $\mathbf{x}_1, \mathbf{x}_2, \ldots, \mathbf{x}_n$, the likelihood function for this mixture model is given by

$$L(\boldsymbol{\theta}_0, \boldsymbol{\theta}_1, \ldots, \boldsymbol{\theta}_{l-1}|\mathbf{x}) = \prod_{i=1}^{n} \sum_{k=0}^{l-1} p_k f_k(\mathbf{x}_i|\boldsymbol{\theta}_k), \qquad (6.14)$$

where $p_k$ is the probability that a sample point is drawn from $f_k(\mathbf{x}|\boldsymbol{\theta}_k)$. For a Gaussian mixture, the parameters take the form $\boldsymbol{\theta}_k = (\boldsymbol{\mu}_k, \boldsymbol{\Sigma}_k)$. Several standard models for Gaussian mixtures have been proposed according to the form of the covariance matrices, in particular, with regard to their unitary similarity transformations

$$\Sigma_k = \mathbf{U}_k \mathbf{L}_k \mathbf{U}_k^t, \tag{6.15}$$

where $\mathbf{L}$ is the diagonal matrix whose diagonal is composed, in order, of the eigenvalues of $\Sigma_k$ and $\mathbf{U}$ is the unitary matrix whose columns are, in order, the orthonormalized eigenvectors corresponding to the eigenvalues (Banfield and Raftery, 1993). A special case is the spherical model $\Sigma_k = a_k \mathbf{I}$. The less constrained the model, the better it can fit the data, but constrained models can have far fewer parameters.

The clustering algorithm proceeds by first specifying the number of clusters and then using the EM algorithm to estimate the remaining parameters. In the E step, the probability of each data point belonging to each cluster is estimated conditionally on the model parameters for the current step; in the M step, the model parameters are estimated given the current membership probabilities. Upon convergence of the EM algorithm, each point is assigned to the cluster for which it has the maximum conditional probability.

A key issue for model-based clustering is comparing models of different complexities relative to the parameters, a particular case being deciding upon the number of clusters. We will visit this issue in section 6.4.4.

## 6.2 CLUSTER OPERATORS

As noted at the outset of this chapter, testing a clustering algorithm has historically been considered different from testing a classification algorithm. For the latter, the error is defined as the probability of misclassification or the expected number of misclassifications. This error is with respect to the labeling of an individual point by a classifier and depends on the existence of a feature-label distribution. A clustering algorithm operates on the point set as a whole, and individual points can be labeled differently by the same algorithm (even if there is no internal randomization within the algorithm) because generally the cluster into which a point is placed depends on the entire observed point set. Thus, an inferential theory of clustering must address the role of cluster operators as operators on random sets—and we address this role in the framework of labeled point processes (Dougherty and Brun, 2004).

### 6.2.1 Algorithm Structure

If a set $S$ is to be partitioned into two clusters (subsets), we can consider the partitioning to be a label function $\phi_S : S \to \{0, 1\}$, the clusters being defined by $\phi_S^{-1}(0)$ and $\phi_S^{-1}(1)$. If $S$ is selected randomly, then we can consider it to be a realization of a random point set $\Xi$. Let $\lambda$ be the mapping on $\Xi$ defined by $\lambda(S) = \phi_S$. Here $\lambda$ is a label operator in the sense that it assigns a label function to each set $S$. We assume the sets $S$ are finite, $S \subset \Re^d$, and that there exists a probability distribution on the product space $\Re^d \times \{0, 1\}$. A deterministic clustering algorithm corresponds to a label operator, where by deterministic we mean that there is a unique output $\phi_S$ of the algorithm for any input set $S$. Clustering algorithms need not be deterministic, but we will consider this issue subsequently.

Consider a distribution $\Omega$ on $\Re^d \times \{0, 1\}$, where the conditional distributions $\mathbf{X}|0$ and $\mathbf{X}|1$ are Gaussian with $\mathbf{X}|0 \sim N(\mu_0, \Sigma_0)$ and $\mathbf{X}|1 \sim N(\mu_1, \Sigma_1)$. A random set $\Xi$ is determined by random samples from $\Omega$. A realization $S$ of $\Xi$ corresponds to a set of sample points drawn from $\Omega$. If $S$ is a realization of $\Xi$ and $\mathbf{x} \in S$, then $\Omega$ induces a label $\phi_S(\mathbf{x})$ by $\phi_S(\mathbf{x}) = 0$ or $\phi_S(\mathbf{x}) = 1$, depending on whether $\mathbf{x}$ has been drawn from $N(\mu_0, \Sigma_0)$ or from $N(\mu_1, \Sigma_1)$, respectively, meaning $\mathbf{x}$ corresponds to a point $(\mathbf{x}, 0)$ or a point $(\mathbf{x}, 1)$, respectively, drawn from $\Omega$. The label of $\mathbf{x}$ depends on the sample and is therefore random. Letting $\Phi_S(\mathbf{x})$ denote the random label of $\mathbf{x}$, a label $\phi_S(\mathbf{x})$ is a sample point for the random variable $\Phi_S(\mathbf{x})$. For the realization $S$ of $\Xi$, $\phi_S$ is a realization of the random function $\Phi_S$ on $S$. In choosing a clustering algorithm, we would like the algorithm to accurately cluster the points, meaning that we would like a label operator $\lambda$ such that $\lambda(S)$ is close to the underlying random label function $\Phi_S$. The main point is that when $S$ is observed, a label operator (deterministic clustering algorithm) $\lambda$ labels the points in a fixed way and the accuracy of the operator must be measured in terms of the closeness (on average) of its labeling to the random labeling of $S$ arising from $\Phi_S$. The matter is complicated by the fact that both $\lambda$ and $\Phi_S$ must be viewed as operators on the random set $\Xi$. In this sense, $\lambda$ is a deterministic operator defined on the random set $\Xi$, whereas $\Phi_S$ is a random operator defined on the random set $\Xi$.

Suppose we want to design an algorithm to cluster realizations of $\Xi$ for the present Gaussian example. For each realization $S$ of $\Xi$, let $\mathbf{a}_0^S$ and $\mathbf{a}_1^S$ be points that minimize $\rho_S(\mathbf{a}_0, \mathbf{a}_1)$ in equation (6.1). The points $\mathbf{a}_0^S$ and $\mathbf{a}_1^S$ determine a hyperplane perpendicular to the chord connecting them that splits the space into two regions $R_0^S$ and $R_1^S$, where the region labels 0 and 1 have been randomly assigned. Define a label function by $\phi_{\rho,S}(\mathbf{x}) = 0$ if $\mathbf{x} \in R_0^S$ and by $\phi_{\rho,S}(\mathbf{x}) = 1$ if $\mathbf{x} \in R_1^S$. The label operator $\lambda_\rho$ is defined by $\lambda_\rho(S) = \phi_{\rho,S}$. A seemingly obvious way to measure the closeness of $\lambda_\rho$ as an estimator of $\Phi_S$ is to define the error by $P(\phi_{\rho,S}(\mathbf{x}) \neq \Phi_S(\mathbf{x}))$ (where for now we are not concerned with a rigorous mathematical interpretation of the probability expression). The problem with this approach is that on account of the random labeling of the regions, $P(\phi_{\rho,S}(\mathbf{x}) \neq \Phi_S(\mathbf{x}))$ will be large even if the hyperplane does a good job of splitting the points. But the problem is not substantive because we are interested in clusters of the sample points, not in their labels per se. Hence, both

$$P(\phi_{\rho,S}(\mathbf{x}) \neq \Phi_S(\mathbf{x})) = 0 \tag{6.16}$$

and

$$P(\phi_{\rho,S}(\mathbf{x}) \neq \Phi_S(\mathbf{x})) = 1 \tag{6.17}$$

indicate perfect results. In the latter case, we need only replace $\phi_{\rho,S}$ by $1 - \phi_{\rho,S}$ (flip the labels) and the labeling will be perfect. Hence, a more appropriate error measure is

$$\varepsilon[\lambda_\rho] = \min\{P(\phi_{\rho,S}(\mathbf{x}) \neq \Phi_S(\mathbf{x})), P(1 - \phi_{\rho,S}(\mathbf{x}) \neq \Phi_S(\mathbf{x}))\}. \tag{6.18}$$

Let us consider a different label function for $\Omega$. In this case there is a given hyperplane $H$ that partitions the space into regions $R_0^S$ and $R_1^S$, independent of $S$, and we define a label function by $\phi_{H,S}(\mathbf{x}) = 0$ if $\mathbf{x} \in R_0^S$ and $\phi_{H,S}(\mathbf{x}) = 1$ if $\mathbf{x} \in R_1^S$. The label operator $\lambda_H$ is defined by

$$\lambda_H(S) = \phi_{H,S}. \tag{6.19}$$

There is a critical difference between $\lambda_\rho$ and $\lambda_H$: the decision boundary for $\lambda_\rho$ depends on the data, whereas the decision boundary for $\lambda_H$ is independent of the data. In fact, $\lambda_H$ is a classifier. No matter the realization of $\Xi$, it simply provides binary classification of each sample point irrespective of the other sample points. The hyperplane forms the decision boundary of a linear classifier, and $\varepsilon[\lambda_\rho]$ is the minimum of the classifier errors for $\lambda_H$ and $1 - \lambda_H$.

Although a clustering algorithm can be defined via a classifier, classification is much more restrictive because the decision on labeling a point is independent of the points among which it is observed. Classification rules are useful when $\Phi_S(\mathbf{x})$ is independent of the set $S$ to which $\mathbf{x}$ belongs. The impact of this strong restriction can be seen by slightly modifying the random set $\Xi$. Instead of inducing $\Xi$ by random sampling from two fixed Gaussians, let the means of the Gaussians be random in such a way that the means are far apart relative to the variances of the distributions. In this situation, it may be impossible to find a fixed hyperplane independent of the realization of $\Xi$ that produces a decent classifier and in turn a good clustering algorithm. On the other hand, the label operator $\lambda_\rho$ can still produce a good clustering algorithm. For instance, consider the spherical Gaussians $N(\mathbf{0}, \Sigma_0)$ and $N(\mathbf{a}, \Sigma_0)$. Treating these Gaussians as class conditional densities, the Bayes classifier is the hyperplane perpendicular to the line segment between $\mathbf{0}$ and $\mathbf{a}$ that is also equidistant from $\mathbf{0}$ and $\mathbf{a}$. Letting $H$ denote this hyperplane, $\lambda_H$ is a reasonable label function for which the quality of clustering depends on the norm of $\mathbf{a}$. Now if $\mathbf{a}$ is a random vector with norm $|\mathbf{a}| = \delta$, no single hyperplane will result in a good performance for $\lambda_H$; however, $\lambda_\rho$ will perform well for sufficiently large values of $\delta$.

### 6.2.2 Label Operators

We now formulate matters rigorously in the context of random sets. A point process $\Xi$ is a measurable mapping of a probability space $(\Omega, \mathcal{A}, \xi)$ into $(\mathbf{N}, \mathcal{N})$, where $\mathbf{N}$ is the family of locally finite, simple (no repetition) sequences in $\mathfrak{R}^d$ and $\mathcal{N}$ is the smallest $\sigma$-algebra on $\mathbf{N}$ such that all mappings $S \to |S \cap B|$ are measurable, where $S \in \mathbf{N}$, $B$ is a Borel set, and $|S \cap B|$ is the number of points in $S \cap B$ (Stoyan et al., 1987). The distribution $\nu$ of $\Xi$ is determined by the probabilities $\nu(Y)$ for $Y \in \mathcal{N}$. Here we restrict $\mathbf{N}$ to consist of finite sequences.

For $S \in \mathbf{N}$, a *label function* is a mapping $\phi_S : \{S \to 0, 1, \ldots, l - 1\}$, the set of labels. $L^S$, the family of functions, is finite, and therefore we can assign a probability density (mass function) $P_S$ over $L^S$. A *random label function* $\Phi_S$ is a random selection of the function $\phi_S$ under the distribution $P_S$. This means that if

$L^S = \{\phi_S^1, \phi_S^2, \ldots, \phi_S^{\tau(S)}\}$, then

$$P(\Phi_S = \phi_S^k) = P_S(\phi_S^k). \tag{6.20}$$

A random labeling on $\mathbf{N}$ is a family $\Lambda = \{\Phi_S : S \in \mathbf{N}\}$ of random label functions for the sets in $\mathbf{N}$. A *labeled point process* is a pair $(\Xi, \Lambda)$, where $\Xi$ is a point process on $(\mathbf{N}, \mathcal{N})$ and $\Lambda$ is a random labeling on $\mathbf{N}$. For a labeled point process $(\Xi, \Lambda)$, the random labeling $\Lambda$ can be approximated by a deterministic labeling. A *label operator* is a mapping $\lambda \to \cup_{S \in \mathbf{N}} L^S$ such that $\lambda(S) \in L^S$ for any $S \in \mathbf{N}$. A label operator $\lambda$ assigns a label function $\phi_{S,\lambda}$ to each set $S \in \mathbf{N}$.

We need to measure the probability that a label operator $\lambda$ correctly labels a point $\mathbf{x}$ in a set $S$ relative to a given random label function $\Phi_S$. Let $P(\phi_{S,\lambda}(\mathbf{x}) \neq \Phi_S(\mathbf{x}))$ be the probability that the point $\mathbf{x} \in S$ is wrongly labeled. This probability is computed directly from the distribution $P_S$: if $\phi_{S,\lambda} = \phi_S^{k_0} \in \text{in} L^S$, then

$$P(\phi_{S,\lambda}(\mathbf{x}) \neq \Phi_S(\mathbf{x})) = \sum_{k \neq k_0} I_{\phi_{S,\lambda}(\mathbf{x}) \neq \phi_S^k(\mathbf{x})} P_S(\phi_S^k). \tag{6.21}$$

The error probability of $\lambda$ for the set $S$ is defined by

$$\varepsilon_\lambda(S) = \sum_{\mathbf{x} \in S} \frac{1}{|S|} P(\phi_{S,\lambda}(\mathbf{x}) \neq \Phi_S(\mathbf{x})). \tag{6.22}$$

This can be written as

$$\varepsilon_\lambda(S) = \sum_{\mathbf{x} \in S} h_\lambda(\mathbf{x}, S), \tag{6.23}$$

where

$$h_\lambda(\mathbf{x}, S) = |S|^{-1} P(\phi_{S,\lambda}(\mathbf{x}) \neq \Phi_S(\mathbf{x})) \tag{6.24}$$

if $\mathbf{x} \in S$ and $h_\lambda(\mathbf{x}, S) = 0$ if $\mathbf{x} \notin S$. Here $h_\lambda$ is a function over $\Re^d \times \mathbf{N}$. The error of the label operator $\lambda$ with respect to the random labeling $\Lambda$ is defined by the expected value of $\varepsilon_\lambda(S)$,

$$\varepsilon(\lambda) = E[\varepsilon_\lambda] = \int \sum_{\mathbf{x} \in S} h_\lambda(\mathbf{x}, S) \, \nu(dS) \tag{6.25}$$

with respect to the distribution of $\Xi$. This error can be expressed via the Campbell measure (Dougherty and Brun, 2004).

A number of classical clustering algorithms are internally random, meaning that $\lambda(S)$ is random for fixed $S$. For instance, the algorithm may have a random seed. In this case, the preceding error expression is simply altered by taking the expected value of $\varepsilon(\lambda)$ relative to the internal randomness of the algorithm.

Label functions serve as the basis for a theoretical clustering framework: given a labeled point process, a cluster operator should partition the points in such a way as to estimate the partition created by the labeling. A label function $\phi_S$ induces a partition $\mathcal{P}_{\phi_S}$ of $S$ according to the equivalence relation $\mathbf{x} \sim \mathbf{y}$ if $\phi_S(\mathbf{x}) = \phi_S(\mathbf{y})$. Conversely, if $\mathcal{P} = \{S_1, S_2, \ldots, S_m\}$ is a partition of $S$, $(S_{j1}, S_{j2}, \ldots, S_{jm})$ is a permutation of the subsets of $\mathcal{P}$, and we define $\phi_S^{\mathcal{P},j}(\mathbf{x}) = k$ if $\mathbf{x} \in S_{jk}$, then $\phi_S^{\mathcal{P},1}$, $\phi_S^{\mathcal{P},2}, \ldots, \phi_S^{\mathcal{P},m}$ induce $\mathcal{P}$. In general, we define an equivalence relation between label functions by $\phi_S^i \sim \phi_S^j$ if $\mathcal{P}_{\phi_S^i} = \mathcal{P}_{\phi_S^j}$. If we index all the partitions of $S$ according to the scheme $\{\mathcal{P}_S^1, \mathcal{P}_S^2, \ldots, \mathcal{P}_S^{\omega(S)}\}$, then the equivalence relation between label functions induces a surjection

$$\pi_S : L^S \to \{\mathcal{P}_S^1, \mathcal{P}_S^2, \ldots, \mathcal{P}_S^{\omega(S)}\} \tag{6.26}$$

by

$$\pi_S(\phi_S^k) = \mathcal{P}_{\phi_S^k}. \tag{6.27}$$

A *cluster operator* is a mapping $\zeta$ that assigns a partition $\zeta(S)$ to each set $S \in \mathbf{N}$. A label operator $\lambda$ induces a cluster operator $\zeta_\lambda$: if $\lambda(S) = \phi_{S,\lambda}$, then $\zeta_\lambda(S) = \mathcal{P}_{\phi_{S,\lambda}}$. In the opposite direction, a cluster operator $\zeta$ has associated with it a family $F_\zeta$ of induced label operators that themselves induce the same partition: $\lambda \in F_\zeta$ if and only if $\pi_S(\phi_{S,\lambda}) = \zeta(S)$ for all $S \in \mathbf{N}$.

To measure the degree to which a cluster operator partitions a labeled point process $(\Xi, \Lambda)$ in accordance with the partitions induced by $\Lambda$, we compare the label operators induced by the cluster operator with the random label function itself. Hence, the *error of a cluster operator* $\zeta$ as an estimator of $(\Xi, \Lambda)$ is given by

$$\varepsilon(\zeta) = \min_{\lambda \in F_\zeta} \varepsilon(\lambda). \tag{6.28}$$

If $\lambda$ is a minimizing label operator for the preceding equation, then we say that $\lambda$ is the *matching* label operator for $\zeta$.

### 6.2.3 Bayes Clusterer

Given a labeled point process $(\Xi, \Lambda)$, the *Bayes label operator* $\lambda^*$ is the label operator possessing minimal error $\varepsilon(\lambda^*)$. Minimization of $\varepsilon(\lambda)$ can be obtained by minimization of $h_\lambda$: if $h_{\lambda^*}(\mathbf{x}, S) \leq h_\lambda(\mathbf{x}, S)$ for any label operator $\lambda$, set $S \in \mathbf{N}$, and $\mathbf{x} \in S$, then equation (6.25) shows that $\varepsilon(\lambda^*) \leq \varepsilon(\lambda)$.

For $L = \{0, 1\}$, the Bayes label operator $\lambda^*$ is defined by $\phi_{S,\lambda^*}(\mathbf{x}) = 1$ if $P(\Phi_S(\mathbf{x}) = 1) \geq 0.5$ and by $\phi_{S,\lambda^*}(\mathbf{x}) = 0$ if $P(\Phi_S(\mathbf{x}) = 1) < 0.5$ for each $S \in \mathbf{N}$ and $\mathbf{x} \in S$. It is straightforward to show that under this definition,

$$P(\phi_{S,\lambda^*}(\mathbf{x}) \neq \Phi_S(\mathbf{x})) \leq P(\phi_{S,\lambda}(\mathbf{x}) \neq \Phi_S(\mathbf{x})) \tag{6.29}$$

for any label operator $\lambda$, $S \in \mathbf{N}$, and $\mathbf{x} \in S$. Consequently, $\varepsilon(\lambda^*) \leq \varepsilon(\lambda)$. More generally, for $L = \{0, 1, \ldots, l-1\}$,

$$\phi_{S,\lambda^*}(\mathbf{x}) = \arg\max_{j=0,1,\ldots,l-1} P(\Phi_S(\mathbf{x}) = j). \tag{6.30}$$

If $\zeta^*$ is the cluster operator induced by $\lambda^*$, then $\varepsilon(\zeta^*) = \varepsilon(\lambda^*)$ because $\lambda^* \in F_{\zeta^*}$ and $\varepsilon(\lambda^*)$ is minimal among all the label operators. Hence, $\zeta^*$ has minimal error among the family of clustering operators. $\zeta^*$ is called the *Bayes clusterer*.

As noted previously, every classifier induces a label function by partitioning $\Re^d$, which in turn induces a cluster operator defined independently of $S \in \mathbf{N}$. When can such a classifier-induced cluster operator perform well?

Consider the special case of $l$ conditional distributions $f(\mathbf{x}|0)$, $f(\mathbf{x}|1)$, ..., $f(\mathbf{x}|l-1)$ with

$$f(\mathbf{x}) = \sum_{j=0}^{l-1} f(\mathbf{x}|j) P(Y=j), \qquad (6.31)$$

where $Y$ is the class label. If the labeled point process $(\Xi, \Lambda)$ is generated from the conditional distributions via random sampling, then we say that $(\Xi, \Lambda)$ is *conditional-distribution-induced* by these distributions. In this case, the distributions of the random label functions $\Phi_S$ are given by

$$P(\Phi_S(\mathbf{x}) = j) = P(Y = j|\mathbf{x}) \qquad (6.32)$$

and are independent of the set $S$, the Bayes label operator reduces to the Bayes classifier relative to the feature-label distribution for $(\mathbf{X}, Y)$, and realizations of $(\Xi, \Lambda)$ are samples from $(\mathbf{X}, Y)$.

In practice, we approximate the Bayes classifier by a designed classifier, and the Bayes clusterer by a clustering algorithm. For conditional-distribution-induced labeled point processes, we can design a cluster operator by designing a classifier and clustering via the label operator induced by the classifier.

Rather than going from classifier to cluster operator, it is possible to go the opposite way. Consider a conditional-distribution-induced labeled point process $(\Xi, \Lambda)$, a cluster operator $\zeta$, and a label operator $\lambda \in F_\zeta$ induced by $\zeta$. Since realizations of $(\Xi, \Lambda)$ are samples from $(\mathbf{X}, Y)$, given a realization $(S, \phi_S)$ of $(\Xi, \Lambda)$, we can apply $\lambda$ to $S$ to obtain the label function $\lambda(S) = \phi_{S,\lambda}$. We can use the labeled points $(\mathbf{x}, \phi_{S,\lambda}(\mathbf{x}))$ to design a classifier. If $\lambda$ is the matching label operator for $\zeta$, then we can look for the resulting classifier to perform well if $\zeta$ performs well and a suitable classification rule is applied to the labeled points $(\mathbf{x}, \phi_{S,\lambda}(\mathbf{x}))$. Depending on the classification rule, we say that the resulting classifier has been *induced* by $\zeta$ with respect to the classification rule. As an example, the *nearest-neighbor $\zeta$-classification rule* defines a classifier $\psi$ in the following manner: for $\mathbf{z} \in \Re^d$, $\psi(\mathbf{z})$ is the label, under the matching label operator, of the point nearest to $\mathbf{z}$ among the points of $S$.

### 6.2.4 Distributional Testing of Cluster Operators

Equations (6.25) and (6.28) provide a distribution-based measure of performance for a clustering algorithm in the framework of operators on labeled point processes. A key issue for evaluating the performance of cluster operators is to obtain independent test data.

We are helped in this regard by a common property of cluster operators. We typically would like a cluster operator to partition a point set in the same way no matter where the point set is located in space, which is not the case for a classifier. To formalize matters, we say that a labeled point process $(\Xi, \Lambda)$ is a *random translation* of the labeled point process $(\Xi_0, \Lambda_0)$ if there exists a random translation vector $\mathbf{Z}$ such that $\Xi = \Xi_0 + \mathbf{Z}$ and, for any $\Phi_S \in \Lambda$, $\Phi_S(\mathbf{x}) = \Phi_{S_0}(\mathbf{x} - \mathbf{Z})$, where $S = S_0 + \mathbf{Z}$. A cluster operator $\zeta$ is said to be *translation-invariant* if $\zeta(\Xi + \mathbf{z}) = \zeta(\Xi) + \mathbf{z}$ for any random point process $\Xi$ and point $\mathbf{z} \in \Re^d$. If $(\Xi, \Lambda)$ is a random translation of the labeled point process $(\Xi_0, \Lambda_0)$ and $\zeta$ is translation-invariant, then the error of $\zeta$ as an estimator of $(\Xi, \Lambda)$ is equal to the error of $\zeta$ as an estimator of $(\Xi_0, \Lambda_0)$. This equivalence has practical consequences in testing. Under it, we need generate only realizations of $(\Xi_0, \Lambda_0)$ to test $\zeta$. Thus, if we consider conditional-distribution-induced labeled point processes and translation-invariant cluster operators, then the results apply to processes resulting from random translations of the considered labeled point processes.

If we assume a probability model for a conditional-distribution-induced labeled point process, then we can generate independent synthetic data to test the performance of a cluster operator. For instance, we can test clustering using a distribution composed of $m$ independent conditional Gaussian random vectors $\mathbf{X}|1, \mathbf{X}|2, \ldots, \mathbf{X}|m$ with mean vectors $\mathbf{u}_1, \mathbf{u}_2, \ldots, \mathbf{u}_m$, respectively, with $\mathbf{X}|k$ having covariance matrix $\mathbf{C}_k$ and $r_k$ as the number of sample points drawn from $\mathbf{X}|k$. (Dougherty et al., 2002). If we assume that the number of clusters is known beforehand, which is not necessary, then the clustering algorithm is preset to have $m$ clusters $C_1, C_2, \ldots, C_m$. An obvious simplification is assuming that the component random variables of $\mathbf{X}|k$ are uncorrelated so that $\mathbf{C}_k$ is diagonal with variance vector $\boldsymbol{\sigma}_k^2 = (\sigma_{k1}^2, \sigma_{k2}^2, \ldots, \sigma_{kn}^2)$ or, going further, that the component variables possess equal variance.

For real data the problem becomes more difficult because estimating the error of a cluster operator $\zeta$ requires knowledge of the true partition. Since this is unknown in practice, one way to proceed is to estimate a model from the data and apply the operator on synthetic data generated from the model to estimate the error. The difficulty is that one needs to estimate the conditional distributions for the labels. One way to estimate them is to apply a clustering algorithm to the data and then estimate the conditional distributions based on the resulting clusters; however, this makes the estimation dependent on the chosen clustering. One way to approach this problem is to take a pooled approach: cluster operators $\zeta_1, \zeta_2, \ldots, \zeta_q$, called *seed operators*, are applied to the data along with $\zeta$, the error of $\zeta$ is computed relative to outputs of $\zeta_1, \zeta_2, \ldots, \zeta_q$, and these errors are averaged to obtain an estimate of $\varepsilon(\zeta)$. Here $\zeta$ may or may not be among the family of operators $\zeta_1, \zeta_2, \ldots, \zeta_q$.

Specifically, suppose there are $q$ seed cluster operators $\zeta_1, \zeta_2, \ldots, \zeta_q$. For $k = 1, 2, \ldots, q$, $\zeta_k$ is applied to form $m$ clusters $U_{k1}, U_{k2}, \ldots, U_{km}$. A model $\mathcal{M}_k$ is fit by estimating its parameters from $U_{k1}, U_{k2}, \ldots, U_{km}$. Error estimation is achieved for a cluster operator $\zeta$ by applying it to $\mathcal{M}_1, \mathcal{M}_2, \ldots, \mathcal{M}_q$. Here $\zeta$ will produce expected error rates of $\varepsilon_{m,1}(\zeta), \varepsilon_{m,2}(\zeta), \ldots, \varepsilon_{m,q}(\zeta)$ corresponding to seeding by $\zeta_1, \zeta_2, \ldots, \zeta_q$, respectively. Bread-basket error estimation for $\zeta$ is given by

$$\varepsilon_m(\zeta) = \frac{1}{q} \sum_{k=1}^{q} \varepsilon_{m,k}(\zeta_k). \tag{6.33}$$

Not only will the result depend on the seed operators, but it will also depend on the models. The usual tradeoff occurs: simple models favor good estimation, whereas complex models can fit (perhaps overfit) the data better. For instance, the model $\mathcal{M}_k$ might consist of $m$ Gaussians fit to the $m$ clusters $U_{k1}, U_{k2}, \ldots, U_{km}$, and these may be made less or more general, putting more or fewer restrictions on the form of the covariance matrices.

## 6.3  CLUSTER VALIDATION

The error of a cluster operator can be defined in the context of random point sets in a manner analogous to classification error. One might loosely define a "valid" cluster operator as one possessing small error—or even more loosely as one that produces "good" clusters. Of course, such a definition is vacuous unless it is supported by a definition of goodness. What if one tries to evaluate a cluster operator in the absence of a probability distribution (labeled point process)? Then, measures of validity (goodness) need to be defined. They will not apply to the cluster operator as an operator on random sets but will depend on heuristic criteria. Their relation to future application of the operator will not be understood, for the whole notion of prediction depends on the existence of a random process. Nevertheless, a heuristic criterion may serve to give some indication of how the cluster operator is performing on the data at hand relative to some criterion. We will consider some methods that address the validity of a clustering output, or compare clustering outputs, based on heuristic criteria. These can be roughly divided into two categories. *Internal validation* methods evaluate the resulting clusters based solely on the data. *External validation* methods evaluate the resulting clusters based on prespecified information.

### 6.3.1  External Validation

For external validation we consider criteria based on the comparison of two label functions for a specific point set $S$. Suppose we are given $S$ and a cluster operator $\zeta$ and wish to compare the partitioning of $S$ produced by $\zeta$ with a partition chosen by some other means, say, one produced by the investigator's understanding of the data. Let $\phi_S$ and $\phi_{S,\lambda}$ be label functions corresponding the heuristic and the $\zeta$ partition, respectively. We cannot compute the error $\varepsilon(\lambda) = E[\varepsilon_\lambda]$ for $\lambda$ without a random set to generate $S$. Indeed, we cannot compute $\varepsilon_\lambda(S)$ because it depends on $P(\phi_{S,\lambda}(\mathbf{x}) \neq \Phi_S(\mathbf{x}))$, and there is no random label function $\Phi_S$. Nonetheless, we can compute an empirical value $e[\phi_S, \phi_{S,\lambda}]$ as the fraction of incorrect labels made by $\phi_{S,\lambda}$ relative to $\phi_S$. This can be done for every $\lambda \in F_\zeta$, and then equation (6.28) can be mimicked to produce the empirical error

$$e(\zeta; \phi_S) = \min_{\lambda \in F_\zeta} e[\phi_S, \phi_{S,\lambda}].  \qquad (6.34)$$

Actually, if we randomly generated $S$ and its labels according to a labeled random process, then $e(\zeta; \phi_S)$ could serve as an estimator for $\varepsilon(\zeta)$, but in the present situation there is no random process. Finally, note that it need not be the case that one of the partitions is postulated a priori. Both can arise from cluster operators, the result being that an empirical misclassification error is computed for the given point set relative to one of the cluster operators.

Rather than compute an empirical error directly, we can consider how pairs of points are commonly and uncommonly clustered by a cluster operator and a heuristic partitioning. Suppose that $\mathcal{P}_S$ and $\mathcal{P}_{S,\lambda}$ are the heuristic and $\zeta$ partitions, respectively. Define four quantities: $a$ is the number of pairs of points in $S$ such that the pair belongs to the same class in $\mathcal{P}_S$ and the same class in $\mathcal{P}_{S,\lambda}$; $b$ is the number of pairs such that the pair belongs to the same class in $\mathcal{P}_S$ and different classes in $\mathcal{P}_{S,\lambda}$; $c$ is the number of pairs such that the pair belongs to different classes in $\mathcal{P}_S$ and the same class in $\mathcal{P}_{S,\lambda}$; and $d$ is the number of pairs $S$ such that the pair belongs to different classes in $\mathcal{P}_S$ and different classes in $\mathcal{P}_{S,\lambda}$. If the partitions match exactly, then all the pairs are either in the $a$ or the $d$ class. The *Rand index* is defined by

$$R = \frac{a+d}{a+b+c+d} \tag{6.35}$$

(Rand, 1971). The Rand index lies between 0 and 1. Two other indices utilizing the quantities $a$, $b$, $c$, and $d$ are the *Jaccard coefficient*

$$J = \frac{a}{a+b+c} \tag{6.36}$$

and the *Folkes and Mallows index*

$$FM = \sqrt{\frac{a}{a+b}\frac{a}{a+c}}. \tag{6.37}$$

Numerous external validation procedures have been proposed (Milligan and Cooper, 1986). It may seem vacuous to apply an empirical measure to compare a cluster operator to a heuristic partition because why would one partition by a clustering algorithm if the partition is already known? However, agreement between the heuristic and $\zeta$ partitions indicates that the scientific understanding behind the heuristic partition is being reflected in the measurements, thereby providing supporting evidence for the measurement process.

### 6.3.2 Internal Validation

Internal validation methods evaluate the clusters based solely on the data, without external information. Typically, a heuristic measure is defined to indicate the goodness of the clustering. It is important to keep in mind that the measure applies only to the data at hand and therefore is not predictive of the worth of a clustering algorithm—even with respect to the measure itself.

A common heuristic for spatial clustering is that if the algorithm produces tight clusters and cleanly separated clusters, then it has done a good job clustering. We consider two indices based on this heuristic. Let $\mathcal{P} = \{S_1, S_2, \ldots, S_k\}$ be a partition of $S$, $\delta(S_i, S_j)$ be a between-cluster distance, and $\sigma(S_i)$ be a measure of cluster dispersion. The *Davies-Bouldin index* is defined by

$$\alpha(\mathcal{P}) = \frac{1}{k} \sum_{i=1}^{k} \max_{j \neq i} \left\{ \frac{\sigma(S_i) + \sigma(S_j)}{\delta(S_i, S_j)} \right\} \tag{6.38}$$

(Davies and Bouldin, 1979), and the *Dunn index* is defined by

$$\beta(\mathcal{P}) = \min_{i} \min_{j \neq i} \frac{\delta(S_i, S_j)}{\max_{l} \sigma(S_l)} \tag{6.39}$$

(Dunn, 1973). Low and high values are favorable for $\alpha(\mathcal{P})$ and $\beta(\mathcal{P})$, respectively. As defined, the indices leave open the distance and dispersion measures, and different ones have been employed. If we want the classes far apart, an obvious choice for $\delta$ is

$$\delta(S_i, S_j) = \min_{\mathbf{x} \in S_i, \mathbf{z} \in S_j} \|\mathbf{x} - \mathbf{z}\| . \tag{6.40}$$

For tight classes, an evident choice is the diameter of the class

$$\sigma(S_i) = \max_{\mathbf{x}, \mathbf{z} \in S_i} \|\mathbf{x} - \mathbf{z}\| . \tag{6.41}$$

Since these kinds of measures do not possess predictive capability, assessing their worth is problematic; however, simulation studies have been performed to observe how they behave (Maulik and Bandyopadhyay, 2002).

The obvious question is whether validity indices, internal or otherwise, have much to do with validation. One way to address this issue is to evaluate the worth of a validity index by computing Kendall's rank correlation between the error of a cluster operator and the validity index. We can expect the performance of an index to vary according to the cluster operator and the model to which it is applied. If the correlation between the clustering error and the validity index is not near 1, then the index does not say much about the clustering goodness. The overall problem has been addressed by considering the performance of validity indices across numerous clustering algorithms and models (Brun et al., 2006). The results vary widely and as a body raise serious questions about the worth of validity indices.

### 6.3.3 Instability Index

We now discuss an internal validity index based on assessing the stability of a cluster operator (Breckenridge, 1989; Roth et al., 2002; Dudoit and Fridlyand, 2002). It uses cluster-induced classifiers. The notion of a classifier induced with respect to a cluster operator was introduced in the context of conditional-distribution-induced labeled point processes. This was done so that the resulting classifier might perform well. However, suppose we have only a single point set $S$. We can still apply $\zeta$ to partition $S$, randomly assign a label operator $\phi_S$ according to the partition, and apply a classification rule to $S$ to obtain a classifier $\psi_{S,\phi_S}$. We cannot evaluate this classifier because we do not have a labeled point process; nonetheless, a classifier has been defined.

Consider a cluster operator $\zeta$, a random point process $\Xi$, and a realization $S$ of $\Xi$. Randomly pick a subset $U$ including $|S|/2$ points (assuming $|S|$ to be even) and let $V = S - U$. For each label operator $\lambda_U \in F_\zeta$ induced by $\zeta$ as an operator on $U$, obtain the label function $\lambda_U(U)$ and form the induced classifier $\psi_{U,\lambda_U(U)}$. Next, for each label operator $\lambda_V \in F_\zeta$ induced by $\zeta$ as an operator on $V$, obtain the label function $\lambda_V(V)$. A measure of stability for the cluster operator is defined by comparing the label function $\lambda_V(V)$ to $\psi_{U,\lambda_U(U)}$. Define $\gamma_\zeta(\lambda_V(V); \lambda_U(U))$ to be the fraction of points in $V$ misclassified relative to $\lambda_V(V)$ by $\psi_{U,\lambda_U(U)}$, meaning that $\Psi_{U,\lambda_U(U)}(\mathbf{x}) \neq \lambda(V)(\mathbf{x})$. Note that $\gamma_\zeta(\lambda_V(V); \lambda_U(U))$ is not symmetric because the comparison is made between $\lambda(V)$ and $\psi_{U,\lambda_U(U)}$. Here $\gamma_\zeta(\lambda_V(V); \lambda_U(U))$ provides a measure of consistency between the action of $\lambda_V$ and $\lambda_U$, but as usual it needs to be corrected for label permutations to measure the stability of $\zeta$. Mimicking equation (6.28), we define the *instability* of $\zeta$ relative to the partition $\{U, V\}$ of $S$ by

$$\gamma_\zeta(S; U, V) = \min_{\lambda_V \in F_\zeta} \gamma_\zeta(\lambda_V(V); \lambda_U(U)). \tag{6.42}$$

We now define the instability index of $\zeta$ with respect to $\Xi$ by taking the expectation relative to $\Xi$:

$$\gamma_\zeta(\Xi) = E_\Xi[\gamma_\zeta(S; U, V)]. \tag{6.43}$$

Even though $\gamma_\zeta(\lambda_V(V); \lambda_U(U))$ is not symmetric, the random selection of $U$ ensures that the expectation is not dependent on the order of the arguments. Intuitively, if $\zeta$ is able to well cluster the realizations of $\Xi$, then the quantities $\gamma_\zeta(\lambda_V(V); \lambda_U(U))$ should be small because the classifier trained on $U$ should partition the space well and this will be reflected in the labeling of $V$.

The instability can be estimated by taking a sample $S = \{S_1, S_2, \ldots, S_r\}$ of independent realizations of $\Xi$ and obtaining the average $\gamma_\zeta(S)$ of $\gamma_\zeta(S_1; U_1, V_1), \ldots, \gamma_\zeta(S_r; U_r, V_r)$. If we are presented with only a single point set $S$, then resampling can be used to form an average using $\gamma_\zeta(S; U_1, V_1), \ldots, \gamma_\zeta(S; U_r, V_r)$, where the sample pairs are drawn independently from $S$ in a manner reminiscent of cross-validation.

The instability index can be applied to a prominent problem with clustering. With clustering algorithms, the number of clusters must often be specified beforehand. If the specified number is in accordance with the point process, then the algorithm has a better chance of performing as intended. A typical approach to deciding on the number of clusters is to apply a validity index and to choose the number of clusters that maximizes the index upon running the algorithm with different numbers of clusters. An impediment to this approach is that a validity index can be strongly influenced by the number of clusters specified for the algorithm, thereby biasing the decision as to the best number of clusters.

In the case of the instability index, suppose $\zeta$ is set to form $k$ clusters. Then, owing to the random permutations in its computation, $\gamma_\zeta(S; U, V) \leq 1 - 1/k$. This bound is tight (Roth et al., 2002), and therefore it has been recommended that the normalized instability be used,

$$\gamma_\zeta^{\text{nor}}(\Xi) = \frac{E_\Xi[\gamma_\zeta(S; U, V)]}{E_\Xi[\gamma_\xi(S; U, V)]}, \tag{6.44}$$

where $\xi$ is the operator that assigns clusters at random. The expectations are estimated by appropriate resampling averages.

### 6.3.4 Bayes Factor

Comparing performance relative to the number of clusters is a special case of comparing performance based on input parameters to a cluster operator. For model-based clustering, the issue is one of comparing models of different complexities relative to the parameters. This requires a comparison measure. Given models $M_1$ and $M_2$ possessing parameter vectors $\boldsymbol{\theta}^{(1)}$ and $\boldsymbol{\theta}^{(2)}$, respectively, the *integrated likelihood* for the sample data $S$ given the model $M_j$ is

$$P(S|M_j) = \int P(S|\boldsymbol{\theta}^{(j)}, M_j) P(\boldsymbol{\theta}^{(j)}|M_j) \, d\boldsymbol{\theta}^{(j)} \tag{6.45}$$

for $j = 1, 2$. Here $P(S|M_j)$ is the probability of the data given the model. A model comparison measure is given by the *Bayes factor*,

$$B_{12} = \frac{P(S|M_1)}{P(S|M_2)} \tag{6.46}$$

(Kass and Raftery, 1995). If $B_{12} > 1$, then model $M_1$ is chosen over $M_2$. Use of the integrated likelihood easily extends to more than two models.

Because the integration for the integrated likelihood is problematic, an approximation can be used. For applying model-based clustering to gene expression data, the *Bayesian information criterion* (BIC) has been recommended (Yeung et al., 2001). The BIC is given by

$$\text{BIC}_j = 2 \log P(S|\hat{\boldsymbol{\theta}}^{(j)}, M_j) - m_j \log n, \tag{6.47}$$

where $m_j$ is the number of parameters to be estimated in the model $M_j$ and $\hat{\boldsymbol{\theta}}^{(j)}$ is the maximum-likelihood estimate of $\boldsymbol{\theta}^{(j)}$ (Schwarz, 1978). The approximation is given by

$$2 \log P(S|M_j) \approx \text{BIC}_j. \tag{6.48}$$

Owing to this approximation, models can be compared based on their BIC scores.

## 6.4  LEARNING CLUSTER OPERATORS

In recent years there has been an effort to bring learning into clustering in the sense that cluster operators are learned from data. The key point is that a cluster operator is learned using labeled data so that it can be applied on new unlabeled data. In this section we discuss learning in the context of the random-set-based probabilistic theory, but before doing so we briefly mention some proposed learning paradigms.

A parametric technique applied to spectral clustering, which relies on minimizing a cost assigned to the partition of the data and the similarity matrix, proposes to learn the similarity matrix from training examples (Bach and Jordan, 2004). Given a family of labeled sets, their associated similarity matrices are the ones that minimize the difference between the partition obtained by spectral clustering from these matrices and the original partitioning. The pairs of labeled sets and matrices are used to learn a parametric mapping that assigns matrices to unlabeled data sets. Another parametric approach learns the parameters of a Mahalanobis distance by optimizing the clustering results on the training data (Xing et al., 2003). A learning procedure for learning clustering applied to image segmentation relies on the specification of attributes for the points (like their positions), pairs of points (like their similarity or distance), and partitions (like the average, maximum, or minimum number of objects in the clusters) (Kamishima, 2003). Having defined the attributes, a maximum a posteriori approach is used to estimate the partition of a new set of points, assigning the partition that maximizes the joint probability of the attributes. This technique allows a systematic selection of attributes and relies completely on the training examples and the definition of the attributes. A drawback is that the number of examples needed for training grows exponentially with the number of objects.

In the context of random labeled point processes, learning cluster operators is analogous to learning classifiers: from sample data, design a cluster operator to estimate the partitions created by a labeled point process $(\Xi, \Lambda)$. A training set of size $n$ is a family $D_n = \{(S_1, \phi_1), (S_2, \phi_2), \ldots, (S_n, \phi_n)\}$ of realizations of the labeled point process. The set $S_i$ is a realization of $\Xi$, and the label function $\phi_i$ is a realization of the random label function $\Phi_{S_i}$. A *clustering rule* (learning algorithm) assigns a label operator $\lambda_n$ to each training set of size $n$, and $\lambda_n$ induces a cluster operator $\zeta_n = \zeta_{\lambda_n}$. Like classification rules, $\lambda_n$ and $\zeta_n$ are random because they depend on the training set.

### 6.4.1  Empirical Error Cluster Operator

A basic learning rule is to select the best cluster operator based on the empirical error on the training data, $D_n$ (Brun and Dougherty, 2005). Referring to equation (6.28), if $\{\zeta^1, \zeta^2, \ldots, \zeta^r\}$ is a family of cluster operators, then the *empirical error cluster operator* is the one minimizing the error

$$\varepsilon(\zeta^i) = \min_{\lambda \in F_{\zeta^i}} \hat{\varepsilon}(\lambda), \tag{6.49}$$

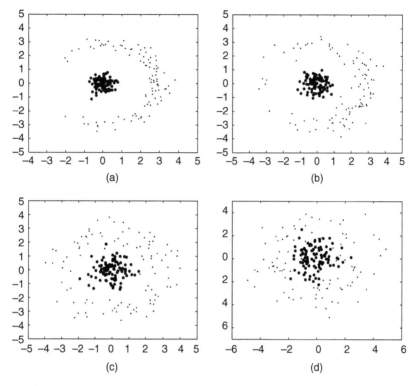

Figure 6.1. Labeled point process. (a) $\sigma_i = 1$; (b) $\sigma_i = 2$; (c) $\sigma_i = 4$; (d) $\sigma_i = 8$.

where

$$\hat{\varepsilon}(\lambda) = \frac{1}{n} \sum_{j=1}^{n} \sum_{\mathbf{x} \in S_j} \frac{1}{|S_j|} I_{\phi_j(\mathbf{x}) \neq \lambda(\mathbf{x})}, \tag{6.50}$$

where $I$ is the indicator function. In the event that the underlying random labeled point process is unknown but samples are available to compute the empirical error, one can use the empirical error rule to select a best cluster operator from among a family of operators.

To illustrate the empirical error cluster rule, we consider a model defined as a mixture of two distributions: a two-dimensional Gaussian distribution with zero mean vector and covariance matrix $\mathbf{K}_i = \sigma_i^2 \mathbf{I}$, and a circular distribution whose radius is normally distributed with mean 3 and variance $\sigma_i^2$ and whose angle is normally distributed with mean 0 and variance 1. We consider four cases, $\sigma_i = 1, 2,$ 4, and 8. Figure 6.1 shows examples of the four processes. Figure 6.2 shows the errors averaged over 100 training labeled sets for $k$-means, fuzzy $k$-means, Euclidean distance hierarchical clustering with complete, single, and average linkage, self-organizing map (so), and random labeling. Fuzzy $k$-means outperforms all the other algorithms for the lower-variance models, but for larger variances, where the data are more mixed, hierarchical clustering with complete linkage performs best.

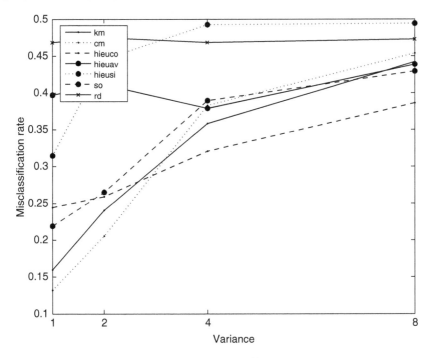

Figure 6.2. Empirical error rates for different cluster operators.

### 6.4.2 Nearest-Neighbor Clustering Rule

A more sophisticated approach is to apply a variant of $k$-nearest-neighbor classification appropriate to random point sets (Dougherty and Brun, 2004). In the context of clustering, a "point" is a set, and we need to interpret the rule accordingly. We let $k = \log_b(n)$, with $b$ an integer, rounding to the closest odd integer (no smaller than 3).

To compute the label function $\phi_{S,n}$ for an observed set $S$, we must find the $k$ nearest sets in the training family $D_n$, say, $S_{i,1}, S_{i,2}, \ldots, S_{i,k}$, and then define the label function for $S$ based on the label functions $\phi_{i,1}, \phi_{i,2}, \ldots, \phi_{i,k}$. In line with the theory of random sets, we use the Hausdorff metric to measure the distance between sets. The *Hausdorff distance* between $S$ and $S_j$ is given by

$$d(S, S_j) = \max \left( \max_{\mathbf{x} \in S} d(\mathbf{x}, S_j), \max_{\mathbf{x} \in S_j} d(\mathbf{x}, S) \right), \tag{6.51}$$

where

$$d(\mathbf{x}, S) = \min_{\mathbf{y} \in S} \; ||\mathbf{x} - \mathbf{y}||. \tag{6.52}$$

To illustrate the method we consider $k = 1$, the nearest-neighbor rule. This means that for a new unlabeled set $S$ we need to (1) find the labeled set $S_{i,1}$ closest to $S$ and (2) use the label function $\phi_{i,1}$ to design the label function $\phi_{S,n}$ for $S$.

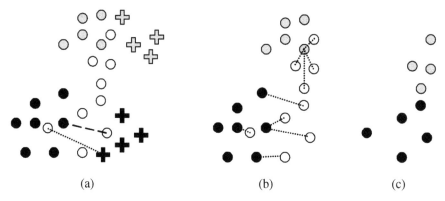

Figure 6.3. Nearest-neighbor $(k = 1)$ class assignment. (a) Example of Hausdorff metric for neighbor selection; (b) example of assignment of the new labels; (c) final assignment of classes.

Step 1 is accomplished by computing the Hausdorff distance between $S$ and each labeled set $S_i \in D$. Figure 6.3a illustrates selection of the nearest neighbor based on the Hausdorff distance. Two training labeled sets are represented by the circles, set $S_1$, and the crosses, set $S_2$, with black for class 0 and gray for class 1. A new unlabeled set $S$ is shown in white circles. The dotted line shows the longest distance between a point of $S$ and the set $S_1$. This defines the distance between $S$ and $S_1$ according to equation (6.51) because all distances from points in $S_1$ to $S$ are less. The dashed line shows the longest distance between a point of $S$ and the set $S_2$. This defines the distance between $S$ and $S_2$ because all distances from points in $S_2$ to $S$ are less. Thus, $S$ is closer to $S_1$, so that $S_1$ is selected as the nearest neighbor.

Having identified $S_1$ as the closest set to $S$, step 2 is to define the label function for $S$ based on $S_1$. For each point $\mathbf{x} \in S$, we find the closest point to it in $S_1$ and assign its label to $\mathbf{x}$. Figure 6.3b shows the nearest neighbor in $S_1$ to each point in $S$. Each point (white circle) in $S$ gets the label of the nearest point in $S_1$. Figure 6.3c shows the labels for $S$.

In the general case of $k > 1$, having determined $S_{i,1}, S_{i,2}, \ldots, S_{i,k}$, the label function $\phi_{S,n}$ for $S$ is defined by voting among the label functions $\phi_{i,1}, \phi_{i,2}, \ldots, \phi_{i,k}$. For each example $(S_{i,h}, \phi_{i,h})$, $h = 1, 2, \ldots, k$, define a label function $\phi_{S,n}^h$ on $S$ by

$$\phi_{S,n}^h(\mathbf{x}) = \phi_{i,h}(\mathbf{y}), \tag{6.53}$$

where $\mathbf{y}$ is the closest point in $S_{i,h}$ to $\mathbf{x}$.

Owing to the label-switching problem, to define a label operator we apply permutations to the $\phi_{S,n}^h$ in an incremental fashion. Let $\psi_{S,n}^1 = \phi_{S,n}^1$. The label functions $\psi_{S,n}^h$, $h = 2, 3, \ldots, k$, are defined in the following way: for all permutations of the labels, we select the permutation $\pi$ such that $\pi \phi_{S,n}^h$ is closest to $\psi_{S,n}^{h-1}$ relative to the number of points $\mathbf{x} \in S$ for which $\pi \phi_{S,n}^h(\mathbf{x}) \neq \psi_{S,n}^{h-1}(\mathbf{x})$, and we define

$$\psi_{S,n}^h = \pi \phi_{S,n}^h. \tag{6.54}$$

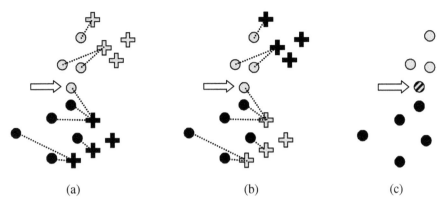

(a)                          (b)                          (c)

Figure 6.4. Possible permutations of the labels for nearest-neighbor ($k=2$) class assignment. (a) First permutation of the label for set $S_2$; (b) second permutation of the
labels for set $S_2$; (c) final voting using two training sets $S_1$ and $S_2$.

$\phi_{S,n}$ is then defined by majority vote: $\phi_{S,n}(\mathbf{x}) = 1$ if

$$|\{\psi^h_{S,n}(\mathbf{x}) = 1\}| \geq k/2, \qquad (6.55)$$

and $\phi_{S,n}(\mathbf{x}) = 0$ otherwise.

To illustrate the case with $k=2$, we can assume that the two closest sets to $S$ are
$S_1$ and $S_2$, in that order. This means that the labels of $S_2$ may need to be switched
to fit closer to the new labels for the set $S$ computed from $S_1$.

Figure 6.4 shows the two possible permutations of labels for set $S_2$. In figure
6.4a only the point indicated by the arrow would be given a different class label
from set $S_2$; in figure 6.4b eight of the nine points would be given a wrong label
(relative to the labels assigned previously by set $S_1$). Therefore, the best choice
of labels for set $S_2$ is the first one. The final labels for the points are obtained by
adding the votes from sets $S_1$ and $S_2$. In this case, the votes agree for all points
except for the one indicated by the arrow, which gets a vote for class 1 from set $S_1$
and a vote for class 0 from set $S_2$. Vote ties are solved by precedence: the closest
set (or the first one on the list) defines the winner in case of ties. The final result is
shown in figure 6.4c.

We now illustrate the Hausdorff distance–based $k$NN clustering rule for several
models. For each model, we independently generate a training set

$$D = \{(S^D_1, \phi^D_1), (S^D_2, \phi^D_2), \ldots, (S^D_t, \phi^D_t)\} \qquad (6.56)$$

and a testing set

$$T = \{(S^T_1, \phi^T_1), (S^T_2, \phi^T_2), \ldots, (S^T_r, \phi^T_r)\}. \qquad (6.57)$$

For $n = 1, 2, \ldots, t$, the label operator $\lambda_n$ is estimated from the training set

$$D_n = \{(S^D_1, \phi^D_1), (S^D_2, \phi^D_2), \ldots, (S^D_n, \phi^D_n)\} \qquad (6.58)$$

consisting of the first $n$ elements of $D$. Here $\lambda_n$ is applied to each pair $(S_i^T, \phi_i^T) \in T$, and its error is the number of points $\mathbf{x} \in S_i^T$ for which $\phi_{S_i^T, n}(\mathbf{x}) \neq \phi_{S_i^T}(\mathbf{x})$. The average error rate $\hat{\varepsilon}(\lambda_n)$ is an estimator of the error $\varepsilon(\lambda_n)$ of the designed label operator. This process results in a graph of error rates as functions of $n$. We compare the errors of the label operator $\lambda_n$ with some standard clustering algorithms: $k$-means, fuzzy $k$-means, and Euclidean distance hierarchical clustering with both complete and single linkage. Each algorithm is applied to each example in $T$, and the errors are averaged over $T$. To compare cluster operators and classifiers, we train a mean-square-error linear classifier and a $k$NN classifier.

Consider a distribution $\Omega$ on $\mathfrak{R}^2 \times \{0, 1\}$, where the conditional distributions $\mathbf{X}|0$ and $\mathbf{X}|1$ are Gaussian with $\mathbf{X}|0 \sim N(\boldsymbol{\mu}_0, \boldsymbol{\Sigma}_0)$ and $\mathbf{X}|1 \sim N(\boldsymbol{\mu}_1, \boldsymbol{\Sigma}_1)$, where $\boldsymbol{\mu}_0 = (0, 3)$, $\boldsymbol{\mu}_1 = (3, 0)$, and $\boldsymbol{\Sigma}_0 = \boldsymbol{\Sigma}_1 = \mathbf{I}$, the identity matrix. Figure 6.5a shows a set of points drawn from $\Omega$. A random set $\Xi$ is determined by random samples from $\Omega$. Here we assume that each realization contains three points from $\mathbf{X}|0$ and seven points from $\mathbf{X}|1$. We use $k = \log_2 n$, $r = 1000$ test sets, and a maximum of $t = 50$ training sets. As seen in figure 6.5b, where C-means corresponds to fuzzy $k$-means, the $k$NN-trained cluster operator outperforms the classical cluster operators with hardly any training, even though the classical operators are well suited to the mixture model. Because the means of the label distributions are fixed, clustering can be accomplished by just designing a classifier and the clustering based on the decision boundary of the classifier. As seen in figure 6.5b, performance of the $k$NN cluster operator is essentially the same as the performances of the MSE linear classifier and the $k$NN classifier.

Consider the distribution $\Omega$ in which $\mathbf{X}|0 \sim N((0, 0), 0.2\mathbf{I})$ and $\mathbf{X}|1$ is a circular distribution in which the points are expressed in polar form by radius $R$ and angle $\Theta$, with $R \sim N(3, 0.2)$ and $\Theta \sim N(0, 1)$. Figure 6.6a shows a set of points drawn from $\Omega$. Here $\Xi$ is determined by randomly drawing five points from each distribution. We use $k = \log_2 n$, $r = 1000$ test sets, and a maximum of $t = 50$ training sets. As seen in figure 6.6b, the performance of the classical clustering algorithms is poor, except for hierarchical clustering with single linkage, and it is severely affected by the proximity of the two classes. The $k$NN-trained cluster operator outperforms the classical cluster operators with hardly any training. As in the preceding example, the means of the label distributions are fixed, but here the linear classifier does not perform well.

In the two preceding examples, clustering via a classifier has worked well. This has resulted from the spatial stability of the label distributions. In general, a cluster operator takes the sample data into account when labeling a point, whereas a classifier takes only the point itself into account. The advantage of clustering is that a cluster operator can perform well on spatially random label distributions.

Consider an example favorable to the classical algorithms. Let $\Xi$ be the point process of the first, two-Gaussian experiment, but consider $\Xi_\tau$ where $\tau$ is a random vector uniformly distributed in $[0,5]^2$. We use $k = \log_3 n$, $r = 1000$ test sets, and a maximum of $t = 250$ training sets. Figure 6.7a shows a set of points drawn from $\Omega$, and figure 6.7b shows the performance curves. Both classifier-based cluster operators perform poorly. The performances of the classical operators are the same as in the first model because they are not affected by translations of the whole set.

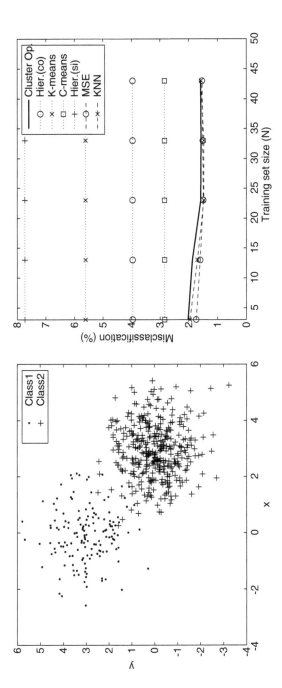

Figure 6.5. Performance for fixed Gaussian distributions. (a) Point sets sampled from the distributions; (b) estimated error rates as a function of training set size.

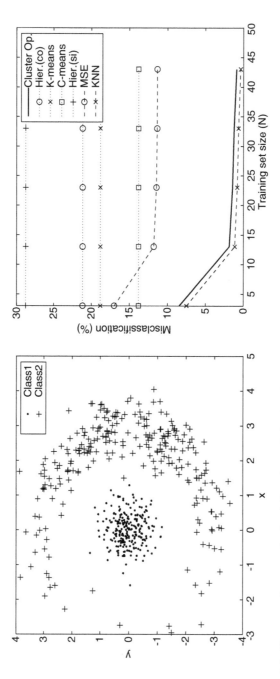

Figure 6.6. Cluster performance for fixed Gaussian and circular distributions. (a) Point sets sampled from the distributions; (b) estimated error rates as a function of training set size.

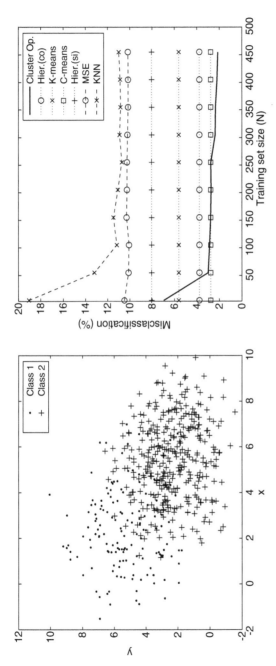

Figure 6.7.  Performance for fixed distributions plus a random translation. (a) Point sets sampled from the distributions; (b) estimated errors as a function of training set size.

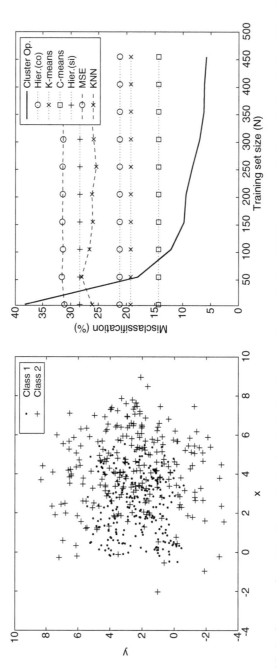

Figure 6.8. Performance for a Gaussian and a circular distribution plus a random translation. (a) Point sets sampled from the distributions; (b) estimated errors as a function of training set size.

With a modest amount of training, the trained cluster operator performs as well as fuzzy $k$-means, and better than the others.

Last, let $\Xi$ be the preceding point process involving a Gaussian and a circular distribution, but now $\Xi_\tau$, where $\tau$ is a random vector uniformly distributed in $[0,5]^2$. Figure 6.8a shows a set of points drawn from $\Omega$, and figure 6.8b shows the performance curves. Both classifier-based cluster operators perform poorly. The performances of the classical operators are the same as in the case of $\Xi$, and they do not perform well. The learned cluster operator outperforms the best of the classical algorithms with about 75 training sets. Very little training is required to outperform single-linkage hierarchical clustering.

# Bibliography

[1] Bach FR, Jordan MI. (2004) Learning spectral clustering. *Proc Adv Neural Inform Process* 16.

[2] Banfield JD, Raftery AE. (1993) Model-based Gaussian and non-Gaussian clustering, *Biometrics* 49:803–21.

[3] Ben-Dor A, Shamir R, Yakhini Z. (1999) Clustering gene expression patterns. *Comput Biol* 6(3/4):281–97.

[4] Bezdek JC. (1981) *Pattern Recognition with Fuzzy Objective Function Algorithms* Plenum Press, New York.

[5] Breckenridge JN. (1989) Replicating cluster analysis: method, consistency, and validity. *Multivariate Behav Res* 24(2):147–61.

[6] Brun M, Dougherty ER. (2005) Clustering algorithms do not learn, but they can be learned. *Proc SPIE Math Meth Pattern Image Anal* 5916.

[7] Brun M, Sima C, Hua J, Lowey J, Carroll B, Suh E, Dougherty ER. (2006) Model-based evaluation of clustering validation measures. *Pattern Recognition* 40(3):807–24.

[8] Celeux G, Govaert G. (1993) Comparison of the mixture and the classification maximum likelihood in cluster analysis. *Stat Comput Simulation* 47:127–46.

[9] Cottrell M, Fort JC, Pages G. (1998) Theoretical aspects of the SOM algorithm. *Neurocomputing* 21:119–38.

[10] Cressie N, Lasslett GM. (1987) Random set theory and problems of modeling. *SIAM Rev.* 29(4):557–574.

[11] Dasgupta A, Raftery AE. (1998) Detecting features in spatial point processes with clutter via model-based clustering. *J Am Stat Assoc* 93:294–302.

[12] Davies DL, Bouldin DW. (1979) A Cluster separation measure. *IEEE Trans Pattern Anal Mach Intell* 1:224–7.

[13] Dunn JC. (1973) A fuzzy relative of the ISODATA process and its use in detecting well-separated clusters. *Cybernetics* 3:32–57.

[14] Dougherty ER. (1999) *Random Processes for Image and Signal Processing*, IEEE and SPIE Presses, Bellingham, WA.

[15] Dougherty ER, Barrera J, Brun M, Kim S, Cesar RM, Chen Y, Bittner M, Trent JM. (2002) Inference from clustering: application to gene-expression time series. *Comput Biol* 9(1):105–26.

[16] Dougherty ER, Brun M. (2004) A probabilistic theory of clustering. *Pattern Recognition* 37:917–25.

[17] Dougherty ER, Braga-Neto U. (2006) Epistemology of computational biology: mathematical models and experimental prediction as the basis of their validity. *J Biol Syst* 14(1):65–90.

[18] Dudoit S, Fridlyand J. (2002) A prediction-based resampling method for estimating the number of clusters in a dataset. *Genome Biol* 3(7):1–32.

[19] Feynman R. (1985) *QED: The Strange Theory of Light and Matter*, Princeton University Press, Princeton, NJ.

[20] Flexer A. (2001) On the use of self-organizing maps for clustering and visualization. *Intell Data Anal* 5:373–84.

[21] Fraley C, Raftery AE. (1998) How many clusterts? Which clustering method? Answers via model-based cluster analysis. *Comput J*, 41(8):578–88.

[22] Jain AK, Murty NM, Flynn PJ. (1999) Data clustering: a review. *ACM Comput Surv* 31(3):264–323.

[23] Kamishima T, Motoyoshi F. (2003) Learning from cluster examples, *Mach Learning* 53(3):199–233.

[24] Kass RE, Raftery AE. (1995) Bayes factors. *J Am Stat Assoc* 90:773–95.

[25] Kerr MK, Churchill GA. (2001) Bootstrapping cluster analysis: assessing the reliability of conclusions from microarray experiments. *Proc Nat Acad Sci, USA* 98(16):8961–6.

[26] Kohonen T. (1982) Self-organized formation of topologically correct feature maps. *Biol Cybern* 43:59–69.

[27] Kohonen T. (1995) *Self-Organizing Maps*, Springer, New York.

[28] Linder T, Lugosi G, Zeger, K. (1994) Rates of convergence in the source coding theorem, empirical quantizer design, and universal lossy source coding. *IEEE Trans Inform Theory* 40:1728–40.

[29] Maulik U, Bandyopadhyay S. (2002) Performance evaluation of some clustering algorithms and validity indices. *IEEE Trans Pattern Anal Mach Intell* 24(12):1650–4.

[30] Matheron G. (1975) *Random Sets and Integral Geometry*, John Wiley & Sons, New York.

[31] Milligan GW, Cooper MC. (1986) A study of the comparability of external criteria for hierarchical cluster analysis. *Multivariate Behav Res* 21:441–58.

[32] Rand WM. (1971) Objective criteria for the evaluation of clustering methods. *J Am Stat Assoc* 66:846–50.

[33] Roth V, Lange T, Braun M, Buhmann J. (2002) A resampling approach to cluster validation, *COMPSTAT 2002*, Berlin.

[34] Schwarz G. (1978) Estimating the dimension of a model. *Ann Stat* 6:461–4.

[35] Stoyan D, Kendall WS, Mecke J. (1987) *Stochastic Geometry and Its Application* John Wiley & Sons, Chichester, UK.

[36] Xing EP, Ng AY, Jordan MI, Stuart R. (2003) Distance-metric learning, with application to clustering with side information. *Adv Neural Inform Process Syst* 15.

[37] Yeung KY, Fraley C, Murua A, Raftery AE, Ruzzo WL. (2001) "Model-based clustering and data transformations for gene expression data. *Bioinformatics* 17(10):977–87.

# Index